ITERATIVE METHODS IN SCIENTIFIC COMPUTING

Springer
Singapore
Berlin
Heidelberg
New York
Barcelona
Budapest
Hong Kong
London
Milan
Paris
Santa Clara
Tokyo

ITERATIVE METHODS IN SCIENTIFIC COMPUTING

Raymond H. Chan, Tony F. Chan
and Gene H. Golub (Eds.)

Raymond H. Chan
Department of Mathematics
The Chinese University of Hong Kong
Shatin
Hong Kong

Tony F. Chan
Department of Mathematics
University of California, Los Angeles
Los Angeles, CA 90024-1555
USA

Gene H. Golub
Department of Computer Science
Stanford University
Stanford, CA 94305
USA

Library of Congress Cataloging-in-Publication Data

Iterative methods in scientific computing / [edited by] Raymond H. Chan, Tony F. Chan, Gene H. Golub.
p. cm.
Includes bibliographical references and index.
ISBN 9813083085
1. Iterative methods (Mathematics)--Data processing--Congresses.
I. Chan, Raymond H., 1958- . II. Chan, Tony F. III. Golub, Gene H. (Gene Howard), 1932- .
QA297.8.I835 1997
511'.4--dc21 96-45461
CIP

ISBN 981-3083-08-5

Printed in Singapore

Typesetting: Camera-ready by Authors using Springer LaTeX style files
SPIN 10560214 5 4 3 2 1 0

Foreword

The *"Winter School on Iterative Methods in Scientific Computing and Their Applications"*, organized by the Institute of Mathematical Sciences at the Chinese University of Hong Kong, took place in Hong Kong on December 14–20, 1995. It was the first major school on scientific computing held in Hong Kong and perhaps even in the Asian region.

There has been a tremendous growth in the area of scientific computing in the Far East in the past few years, as evidenced by the recent establishment of national supercomputing and scientific computing centers in China, Singapore, South Korea and Taiwan and the large turn-outs in regional scientific computing conferences such as the First Asian Computational Fluid Dynamics Conference (Hong Kong, 1995), the First Asian High Performing Computing Conference (Taiwan, 1995) and the Sixth International Conference on Hyperbolic Problems (Hong Kong, 1996).

The Winter School was intended to provide an intensive one week course on the state-of-the-art in iterative methods in scientific computing for graduate students, researchers and practitioners in industry in this part of the world. It is hoped that after the course, the attendees will be familiar with the basic theories and algorithms, as well as potential application areas and the vast and fast developing literature. The School drew a truly international audience with 15 invited speakers and 76 participants from Europe, USA, mainland China, Taiwan, Japan, S. Korea, Macau, Malaysia, Australia, Brazil and Hong Kong.

This book contains nine polished version of the lecture notes from the School and is organized in three major parts: Basic Iterative Methods (Chapters 1–4), Preconditioning (Chapters 5–6) and Applications (Chapter 7–9). Topics include Linear Systems and Iterations, Wavelets, Signal and Image Processing, Multigrid and Domain Decomposition Methods, Software and Tools and Iterative Solvers for Convection-Diffusion and Navier-Stokes Equations. These represent major areas of applications of iterative methods, as well as different forefronts of scientific computing. In addition to surveying the state-of-the-art applications and forefronts in the area, some of the chapters are also written as a general introduction to the field for beginners. More comprehensive introduction of the field can be found in the references listed at the end of this foreword.

The editors of the book are Raymond Chan (The Chinese University of Hong Kong), Tony F. Chan (University of California, Los Angeles) and Gene Golub (Stanford University). On behalf of the organizers of the School, we would like to thank the following sponsors for their generous support: the Institute of Mathematical Sciences at the Chinese University of Hong Kong, the S.H. Ho Foundation, the British Council, the Hong

Kong Pei Hua Education Foundation Ltd., and the Hong Kong Research Grants Council. Finally, we would like to thank the students, C.P. Cheung, H.W. Sun and H.M. Zhou, for their help and dedication in preparing and typesetting the manuscripts.

General References on Iterative Methods and Applications:

1. On Iterative Methods:

    - O. Axelsson, *Iterative Solution Methods*, Cambridge University Press, 1994.
    - R. Barrett, M. Berry, T. Chan, J. Demmel, J. Donato, J. Dongarra, V. Eijkhout, R. Pozo, C. Romine, and H. van der Vorst, *Templates for the Solution of Linear Systems: Building Blocks for Iterative Methods*, SIAM, 1994.
    - G. Golub, and C. van Loan, *Matrix Computations*, 2nd Ed., John Hopkins University Press, 1989.
    - W. Hackbusch, *Iterative Solution of Large Linear Systems of Equations*, Springer Verlag, 1994.
    - Y. Saad, *Iterative Methods for Sparse Linear Systems*, PWS Publ., 1996.

2. On Wavelets:

    - C. Chui, *An Introduction to Wavelets*, Academic Press, 1992.
    - G. Strang, and T. Nguyen, *Wavelets and Filter Banks*, Wellesley-Cambridge Press, 1996.

3. On Multigrid:

    - W. Briggs, *A Multigrid Tutorial*, SIAM, 1987.
    - W. Hashbusch, *Multigrid Methods and Applications*, Springer-Verlag, 1985.
    - P. Wesseling, *An Introduction to Multigrid Methods*, J.W. & Sons, 1992.

4. On Domain Decomposition:

    - B. Smith, P. Bjorstad, and W. Gropp, *Domain Decomposition: Parallel Multilevel Methods for Elliptic Partial Differential Equations*, Cambridge University Press, 1996.

5. On Parallel Numerical Algorithms:

- *Parallel Numerical Algorithms: An Introduction*, ed., D. Keyes, A. Sameh, and V. Venkatakrishnan, from "Parallel Numerical Algorithms" (Hampton, VA, May 1994), Kluwer, 1995.

6. On Image Processing:

    - R. Gonzalez, and R. Wood, *Digital Image Processing*, Addison-Wesley, 1993.

Contents

1

Iterative Methods for Linear Systems and Implementation on Parallel Computers

Henk A. Van der Vorst*

In this chapter we will discuss a number of related iterative methods for the solution of linear systems of equations. These methods are so-called Krylov projection type methods and they include popular methods as Conjugate Gradients, Bi-Conjugate Gradients, CGS, Bi-CGSTAB, QMR, LSQR, and GMRES. We will show how these methods can be derived from simple basic iteration formulas. We will not give convergence proofs, but we will refer for these, as far as available, to litterature.
For the application of the iterative schemes one usually thinks of linear sparse systems, e.g., like those arising in the finite element or finite difference approximations of (systems of) partial differential equations. However, the structure of the operators plays no explicit role in any of these schemes, and these schemes might also successfully be used to solve certain large dense linear systems. Depending on the situation that might be attractive in terms of numbers of floating point operations.

Iterative methods form an alternative for direct methods if the matrix of the system is large and sparse, since the amount of memory space for iterative schemes is limited, and also the amount of computational work can be significantly less. This is supported by the following arguments.
Usually large sparse matrices are related to some grid or network. In a 3D situation this leads typically to a bandwidth $\sim n^{\frac{2}{3}}$ ($= m^2$ and $m^3 = n$, where $1/m$ is a measure for the meshwidth). Standard Gaussian elimination then typically leads to fill-in, and the total amount of flops is then $\mathcal{O}(nm^4) \sim n^{2\frac{1}{3}}$ [37, 27]. For symmetric positive definite systems the error reduction per iteration step of CG is $\sim \frac{\sqrt{\kappa}-1}{\sqrt{\kappa}+1}$, with $\kappa = \|A\|_2\|A^{-1}\|_2$ [15, 2, 37], and for discretized second order pde's, over grids with meshwidth $\sim \frac{1}{m}$ we have that $\kappa \sim m^2$. Hence, for these problems we have that $\kappa \sim n^{\frac{2}{3}}$.

*Mathematical Institute, University of Utrecht, PO Box 80.010, NL-3508 TA Utrecht, the Netherlands vorst@math.ruu.nl

For an error reduction by ϵ we must have that

$$\left(\frac{1-\frac{1}{\sqrt{\kappa}}}{1+\frac{1}{\sqrt{\kappa}}}\right)^j \approx (1-\frac{2}{\sqrt{\kappa}})^j \approx e^{-\frac{2j}{\sqrt{\kappa}}} < \epsilon,$$

from which it follows that

$$j \sim -\frac{\log \epsilon}{2}\sqrt{\kappa} \approx -\frac{\log \epsilon}{2} n^{\frac{1}{3}}.$$

If we assume the number of flops per iteration to be $\sim fn$ (f stands for the number of nonzeros per row of the matrix and the overhead per unknown introduced by the iterative scheme), then the number of flops for a reduction by a factor ϵ is proportional to $-fn^{\frac{4}{3}}\log \epsilon$. We conclude that for these kind of problems the work involved in an iterative scheme may be less by $\mathcal{O}(n)$ asymptotically.

It turns out that all described iterative methods are parallelizable in a straight forward manner. However, especially for computers with a memory hierarchy (i.e., cache or vector registers), and for distributed memory computers, the performance can often be improved significantly through rescheduling of the operations.
Iterative methods are often used in combination with so-called preconditioning operators (approximations for the inverses of the operator of the system to be solved). These preconditioners help to reduce the computational work, but unfortunately they are often not very well parallelizable. We will give an overview of some approaches to improve parallelism for the popular incomplete decomposition (ILU) preconditioners.

1.1 Basic iteration method

A very basic idea, that leads to many effective iterative solvers, is to to split the matrix A of a given linear system in the sum of two matrices, one of which a matrix that would have led to a system that can easily be solved. The most simple splitting we can think of is $A = I - (I - A)$. Given the linear system $Ax = b$, this splitting leads to the well-known Richardson iteration:

$$x_i = b + (I - A)x_{i-1} = x_{i-1} + r_{i-1}.$$

Multiplication by $-A$ and adding b gives

$$b - Ax_i = b - Ax_{i-1} - Ar_{i-1}$$

or

$$r_i = (I - A)r_{i-1} = (I - A)^i r_0 = P_i(A)r_0.$$

Similarly, the error $x - x_i$ can be expressed as

$$x - x_i = P_i(A)(x - x_0).$$

In these expressions P_i is a (special) polynomial of degree i. The Richardson iteration leads to the polynomial $P_i(t) = (1-t)^i$, and as we will see other methods lead to other polynomials. These polynomials are almost never constructed in explicit form, but it is used in formulas to express the special relation between r_i and r_0.

In the derivation of an iterative scheme we have used a very simple splitting. In many cases better approximations K to the given matrix A are available, which define the splitting $A = K - R$, with $R = K - A$. The corresponding iterative method then reads as

$$Kx_i = b + Rx_{i-1} = Kx_{i-1} + b - Ax_{i-1}.$$

Solving for x_i leads formally to

$$x_i = x_{i-1} + K^{-1}(b - Ax_{i-1}).$$

If we define $B \equiv K^{-1}A$ and $c = K^{-1}b$, then the $K-$splitting is equivalent with the standard splitting for $Bx = c$. Therefore, it is no loss of generality if we further restrict ourselves to the standard splitting. Any other splitting can be incorporated easily in the different schemes by replacing the quantities A and b, by B and c, respectively.

From now on we will assume that $x_0 = 0$. This also does not mean a loss of generality, for the situation $x_0 \neq 0$ can, through a simple linear transformation $z = x - x_0$, be transformed to the system

$$Az = b - Ax_0 = \tilde{b},$$

for which obviously $z_0 = 0$.

For the Richardson iteration it follows that

$$x_{i+1} = r_0 + r_1 + r_2 + \cdots + r_i = \sum_{j=0}^{i}(I - A)^j r_0$$

$$\in \{r_0, Ar_0, \ldots, A^i r_0\} = K_{i+1}(A; r_0).$$

The subspace $K_{i+1}(A; r_0)$ is called the *Krylov subspace* of dimension $i+1$, generated by A *and* r_0. Apparently, the Richardson iteration delivers elements of Krylov subspaces of increasing dimension. Including local iteration parameters in the iteration leads to other elements of the same Krylov subspaces, and hence to other polynomial expressions for the error and the residual.

1.1.1 *Working with the Krylov subspace*

We have seen that the basic iteration method constructs an approximate solution that belongs to Krylov subspaces of increasing dimension, and the question arises whether it would be possible to find better approximations without considerably more computational effort. In order to identify better solutions in the Krylov subspace we need a suitable basis for this subspace, one that can be extended in a meaningful way for subspaces of increasing dimension. The obvious basis r_0, Ar_0, ..., $A^{i-1}r_0$, for $K^i(A;r_0)$, is not very attractive from a numerical point of view, since the vectors $A^j r_0$ point more and more in the direction of the dominant eigenvector for increasing j (the power method !), and hence the basis vectors become dependent in finite precision arithmetic.

Instead of the standard basis one usually prefers an orthonormal basis, and Arnoldi [1] suggested to compute this basis as follows. Start with $v_1 \equiv r_0/\|r_0\|_2$. Assume that we have already an orthonormal basis v_1, ..., v_j for $K^j(A;r_0$, then this basis is expanded by computing $t = Av_j$, and by orthonormalizing this vector t with respect to v_1, ..., v_j. In principle the orthonormalization process can be carried out in different ways, bu the most commonly used approach is to do this by a modified Gram-Schmidt procedure [37]. This leads to the following algorithm for the creation of an orthonormal basis for $K^m(A;r_0)$:

$$
\begin{array}{l}
v_1 = r_0/\|r_0\|_2; \\
\text{for } j = 1,..,m-1 \\
\quad t = Av_j; \\
\quad \text{for } i = 1,...,j \\
\qquad h_{i,j} = t^* v_i; \\
\qquad t = t - h_{i,j} v_i; \\
\quad \text{end;} \\
\quad h_{j+1,j} = \|t\|_2; \\
\quad v_{j+1} = t/hj+1,j; \\
\text{end}
\end{array}
$$

It is easily verified that v_1, ..., v_m form an orthonormal basis for $K^m(A;r_0)$ (that is, if the construction does not terminate at a vector $t = 0$). The orthogonalization leads to relations between the v_j, that can be formulated in a compact algebraic form. Let V_j denote the matrix with columns v_1 up to v_j, then it follows that

$$AV_{m-1} = V_m H_{m,m-1}.$$

The m by $m-1$ matrix $H_{m,m-1}$ is upper Hessenberg, and its elements $h_{i,j}$ are defined by the Arnoldi algorithm.

From a computational point of view, this construction is composed from three basic elements: a matrix vector product with A, innerproducts, and updates. We see that this orthogonalization becomes increasingly expensive for increasing dimension of the subspace, since the computation of each $h_{i,j}$ requires an inner product and a vector update.

Note that if A is symmetric, then so is $H_{m-1,m-1} = V_{m-1}^* A V_{m-1}$, so that in this situation $H_{m-1,m-1}$ is tridiagonal. This means that in the orthogonalization process, each new vector has to be orthogonalized with respect to the previous two vectors only, since all other innerproducts vanish. The resulting three term recurrence relation for the basis vectors of $K_m(A; r_0)$ is known as the *Lanczos method* and some very elegant methods are derived from it. In the symmetric case the orthogonalization process involves constant arithmetical costs per iteration step: one matrix vector product, two innerproducts, and two vector updates.

1.1.2 Optimal Krylov subspace methods

Since we are looking for a solution x_k in $K^k(A; r_0)$, that vector can be written as a combination of the basis vectors of the Krylov subspace, and hence

$$x_k = V_k y.$$

(Note that y has k components)
There are three successful approaches to find a suitable approximation:
1. The Ritz-Galerkin approach: require that $b - Ax_k \perp K^k(A; r_0)$
2. The minimum residual approach: require $\|b - Ax_k\|_2$ to be minimal over $K^k(A; r_0)$.
3. The Petrov-Galerkin approach: require that $b - Ax_k$ is orthogonal to some other suitable k-dimensional subspace.

The Ritz-Galerkin approach leads to such popular and well-known methods as Conjugate Gradients, the Lanczos method, FOM, and GENCG. The minimum residual approach leads to methods like GMRES, MINRES, and ORTHODIR. If we select the k-dimensional subspace in the third approach as $K^k(A^T; s_0)$, then we obtain the Bi-CG, and QMR methods. More recently, hybrids of the three approaches have been proposed, like CGS, Bi-CGSTAB, BiCGSTAB(ℓ), TFQMR, FGMRES, and GMRESR.

1.2 The Ritz-Galerkin approach: FOM and CG

Let us assume, for simplicity again, that $x_0 = 0$, and hence $r_0 = b$. The Ritz-Galerkin conditions imply that $r_k \perp K^k(A; r_0)$, and this is equivalent with

$$V_k^*(b - Ax_k) = 0.$$

Since $b = r_0 = \|r_0\|_2 v_1$, it follows that $V_k^* b = \|r_0\|_2 e_1$ with e_1 the first canonical unit vector in $\mathbb{R}^k$. With $x_k = V_k y$ we obtain

$$V_k^* A V_k y = \|r_0\|_2 e_1.$$

This system can be interpreted as the system $Ax = b$ projected onto $K^k(A; r_0)$.

Obviously we have to construct the $k \times k$ matrix $V_k^* A V_k$, but this is, as we have seen readily available from the orthogonalization process:

$$V_k^* A V_k = H_{k,k},$$

so that the x_k for which $r_k \perp K^k(A; r_0)$ can be easily computed by first solving $H_{k,k} y = \|r_0\|_2 e_1$, and forming $x_k = V_k y$. This algorithm is known as FOM or GENCG.

When A is symmetric, then $H_{k,k}$ reduces to a tridiagonal matrix $T_{k,k}$, and the resulting method is known as the *Lanczos* method [44]. In clever implementations it is avoided to store all the vectors v_j.
When A is in addition positive definite then we obtain, at least formally, the *Conjugate Gradient* method. In commonly used implementations of this method one forms directly a LU factorization for $T_{k,k}$ and this leads to very elegant short recurrencies for the x_j and the corresponding r_j, see Section 1.2.1. The positive definiteness is necessary to guarantee the existence of the LU factorization.

Note that for some $j \leq n-1$ the construction of the orthogonal basis must terminate. In that case we have that $AV_{j+1} = V_{j+1} H_{j+1,j+1}$. Let y be the solution of the reduced system $H_{j+1,j+1} y = \|r_0\| e_1$, and $x_{j+1} = V_{j+1} y$. Then it follows that $x_{j+1} = x$, i.e., we have arrived at the exact solution, since $Ax_{j+1} - b = AV_{j+1} y - b = V_{j+1} H_{j+1,j+1} y - b = \|r_0\| V_{j+1} e_1 - b = 0$ (we have assumed that $x_0 = 0$).

1.2.1 The CG-method

The Conjugate Gradients (**CG**) method [40], for symmetric positive definite matrices A, is merely a variant on the Ritz-Galerkin approach, that saves storage and computational effort. Since the residuals are orthogonal by construction, we use them as the basis vectors for the Krylov subspace, rather than normalizing them.
This means that the three-term recurrence relation for the basis vectors is slightly different:

$$\gamma_j r_{j+1} = A r_j - \alpha_j r_j - \beta_j r_{j-1}, \tag{1.1}$$

with γ_j such that r_{j+1} has the proper length.
The constants α_j and β_j follow from the orthogonality requirement. We will now explain how the constant γ_j can be obtained. First note that $x_j \in K^j(A; r_0)$, and hence $x_j = Q_{j-1}(A) r_0$, with Q_{j-1} a $j-1$-degree polynomial. Hence $r_j = b - Ax_j = (I - AQ_{j-1}(A)) r_0 = P_j(A) r_0$, with P_j a j-degree polynomial with constant coefficient 1. If we insert these polynomial expressions in equation (1.1), and compare the coefficients for the constant term, then we obtain the relation

$$\gamma_j = -(\alpha_j + \beta_j).$$

Let R_i be the matrix with columns $r_0, \ldots, r_{i-1}$, then the recurrence relation (1.1) expresses the fact that A applied to a column of R_i results in the combination of three successive columns, or

$$AR_i = R_i \begin{pmatrix} \ddots & \ddots & & & \\ & \ddots & \ddots & \beta_j & & \\ & & \ddots & \alpha_j & \ddots & \\ & & & \gamma_j & \ddots & \ddots \\ & & & & \ddots & \ddots \end{pmatrix} + \gamma_i \begin{pmatrix} 0, 0, \ldots, r_i \end{pmatrix}$$

or

$$AR_i = R_i T_i + \gamma_i r_i e_i^T, \tag{1.2}$$

in which T_i is an i by i tridiagonal matrix and e_i is the ith canonical vector in $I\!\!R^i$.

When solving the projected equations in the sketched way, we see that we have to save all columns of R_i throughout the process in order to recover the current iteration vectors x_i. This can be avoided. If we assume that the matrix A is in addition positive definite then, because of the relation

$$R_i^T A R_i = R_i^T R_i T_i,$$

we conclude that T_i can be transformed by a rowscaling matrix $R_i^T R_i$ into a positive definite symmetric tridiagonal matrix (note that $R_i^T A R_i$ is positive definite for $y \in I\!\!R^i$). This implies that T_i can be decomposed without any pivoting:

$$T_i = L_i U_i,$$

with L_i lower unit bidiagonal and U_i upper bidiagonal. Hence

$$x_i = R_i y = R_i T_i^{-1} e_1 = (R_i U_i^{-1})(L_i^{-1} e_1).$$

We concentrate on the factors, placed between parenthesis, separately.

1.

$$L_i = \begin{pmatrix} 1 & & & & \\ f_1 & 1 & & & \\ & f_2 & \ddots & & \\ & & \ddots & \ddots & \\ & & & f_{i-1} & 1 \end{pmatrix}$$

With $q = L_i^{-1} e_1$ we have that q can be solved recursively from $L_i q = e_1 \Rightarrow f_{i-1} q_{i-2} + q_{i-1} = 0 \Rightarrow q_{i-1}$.

2. Write $B_i \equiv R_i U_i^{-1}$, then we have that

$$R_i = B_i U_i = \begin{pmatrix} & \vdots \\ & \vdots \\ & \vdots \\ & \vdots \end{pmatrix} \begin{pmatrix} d_0 & g_1 & & \\ & d_1 & \ddots & \\ & & \ddots & \ddots \\ & & & g_{i-1} \\ & & & d_{i-1} \end{pmatrix}$$

$$\Rightarrow \quad r_{i-1} = g_{i-1}(B_i)_{i-2} + d_{i-1}(B_i)_{i-1}$$

$$\Rightarrow \quad (B_i)_{i-1}.$$

Glueing these two recurrences together we obtain

$$x_i = \begin{pmatrix} & \vdots \\ & (B_i)_{i-1} \\ & \vdots \end{pmatrix} \begin{pmatrix} \vdots \\ \vdots \\ q_{i-1} \end{pmatrix} = x_{i-1} + q_{i-1}(B_i)_{i-1},$$

and this is in fact the well-known conjugate gradients method. The name stems from the property that the update vectors $(B_i)_{i-1}$, usually notated as p_{i-1}, are A-orthogonal.
Note that the positive definiteness of A is only exploited to guarantee the flawless decomposition of the implictly generated tridiagonal matrix T_i. This suggests that the conjugate gradients method may also work for certain non positive definite systems, but then at our own risk [53].

Another possibility, for symmetric, but not positive definite, systems, is to solve the system $T_i y = \|r_0\|_2 e_1$ by some other method than Choleski decomposition (since A is not positive definite). An alternative is to make an LQ decomposition of T_i. This again leads to simple recurrences; the resulting method is known as SYMMLQ [54].

Computational notes

The standard (unpreconditioned) Conjugate Gradient algorithm for the solution of $Ax = b$ may be represented by the following scheme:

x_0= initial guess; $r_0 = b - Ax_0$;
$p_{-1} = 0; \widehat{\beta}_{-1} = 0$;
$\rho_0 = (r_0, r_0)$;
for $i = 0, 1, 2, ...$
  $p_i = r_i + \widehat{\beta}_{i-1} p_{i-1}$;
  $q_i = Ap_i$;
  $\widehat{\alpha}_i = \frac{\rho_i}{(p_i, q_i)}$
  $x_{i+1} = x_i + \widehat{\alpha}_i p_i$;

$r_{i+1} = r_i - \widehat{\alpha}_i q_i;$
if x_{i+1} accurate enough then quit;
$\rho_{i+1} = (r_{i+1}, r_{i+1});$
$\widehat{\beta}_i = \frac{\rho_{i+1}}{\rho_i};$
end;

CG is most often used in combination with a suitable splitting $A = K - R$, and then K^{-1} is called the preconditioner. We will assume that K is also positive definite.
Note first that the CG method can be derived for any choice of the innerproduct. In our derivation we have used the standard innerproduct $(x, y) = \sum x_i y_i$, but we have not used any specific property of that innerproduct. Now we make a different choice:

$$[x, y] \equiv (x, Ky).$$

It is easy to verify that $K^{-1}A$ is symmetric positive definite with respect to [,]:

$$\begin{aligned} [K^{-1}Ax, y] &= (K^{-1}Ax, Ky) = (Ax, y) \\ &= (x, Ay) = [x, K^{-1}Ay]. \end{aligned} \tag{1.3}$$

Hence, we can follow our CG procedure for solving the preconditioned system $K^{-1}Ax = K^{-1}b$, using the new [,]-innerproduct.
Apparently, we are now minimizing

$$[x_i - x, K^{-1}A(x_i - x)] = (x_i - x, A(x_i - x)),$$

which leads to the remarkable (and known) result that for this preconditioned system we still minimize the error in A-norm, although over a Krylov subspace generated by $K^{-1}r_0$ and $K^{-1}A$.

In the following computational scheme for preconditioned CG, for the solution of $Ax = b$ with preconditioner K^{-1}, we have replaced the [,]-innerproduct again by the familiar standard innerproduct. E.g., note that with $\tilde{r}_{i+1} = K^{-1}Ax_{i+1} - K^{-1}b$ we have that

$$\begin{aligned} \rho_{i+1} &= [\tilde{r}_{i+1}, \tilde{r}_{i+1}] \\ &= [K^{-1}r_{i+1}, K^{-1}r_{i+1}] = [r_{i+1}, K^{-2}r_{i+1}] \\ &= (r_{i+1}, K^{-1}r_{i+1}), \end{aligned}$$

and $K^{-1}r_{i+1}$ is the residual corresponding to the preconditioned system $K^{-1}Ax = K^{-1}b$.

x_0= initial guess; $r_0 = b - Ax_0;$
$p_{-1} = 0; \widehat{\beta}_{-1} = 0;$

Solve w_0 from $Kw_0 = r_0$;
$\rho_0 = (r_0, w_0)$;
for $i = 0, 1, 2, ...$
  $p_i = w_i + \widehat{\beta}_{i-1} p_{i-1}$;
  $q_i = A p_i$;
  $\widehat{\alpha}_i = \frac{\rho_i}{(p_i, q_i)}$
  $x_{i+1} = x_i + \widehat{\alpha}_i p_i$;
  $r_{i+1} = r_i - \widehat{\alpha}_i q_i$;
  if x_{i+1} accurate enough then quit;
  Solve w_{i+1} from $K w_{i+1} = r_{i+1}$;
  $\rho_{i+1} = (r_{i+1}, w_{i+1})$;
  $\widehat{\beta}_i = \frac{\rho_{i+1}}{\rho_i}$;
end;

Note that this formulation, which is quite popular, has the advantage that the preconditioner needs not to be splitt, and it is also avoided to backtransform solutions and residuals, as is necessary when one applies CG to $L^{-1} A L^{-1^T} y = L^{-1} b$.

As we have seen the conjugate gradients algorithm is just an efficient implementation of the Lanczos algorithm. The eigenvalues of the implicitly generated tridiagonal matrix T_i [1] are the Ritz values of A with respect to the current Krylov subspace. It is known from Lanczos theory that these Ritz values converge towards the eigenvalues of A and that in general the extremal eigenvalues of A are first well approximated [43, 52, 57]. Furthermore, the speed of convergence depends on how well these eigenvalues are separated from the others [57]. This helps us to understand the so-called superlinear convergence behaviour of the conjugate gradient method (as well as other Krylov subspace methods). It can be shown that as soon as one of the extremal eigenvalues is modestly well approximated by a Ritz value, the pocedure converges from then on as a process in which this eigenvalue is absent, i.e., a process with a reduced condition number. Note that superlinear convergence behaviour in this connection is used to indicate linear convergence with a factor that is gradually decreased during the process as more and more of the extremal eigenvalues are sufficiently well approximated (for details on this see [73]).

Normal Equations

The simplicity of the CG-method suggest to make a given system explicitly symmetric positive definite, by forming the normal equations:

$$A^T A x = A^T b,$$

[1] see, e.g., [7] for the reconstruction of T_i from the iteration coefficients $\widehat{\alpha}_j$ and $\widehat{\beta}_j$

and to apply CG for this system.
However, this approach has severe disadvantages because of the squaring of the condition number. This has as effect that the solution is more susceptible to errors in the right-hand side and that the rate of convergence of the CG procedure is much slower as for a comparable symmetric system with a matrix with the same condition number as A. Moreover, the amount of work per iteration step, necessary for the matrix vector product, is doubled.
Nevertheless, there are situations for which this approach is attractive, for instance overdetermined systems and systems with almost orthogonal A. Several suggestions have been made to improve the numerical stability of this rather robust approach. The most well-known is by Paige and Saunders [55] and is based upon applying the Lanczos method to the auxiliary system

$$\begin{pmatrix} I & A \\ A^T & 0 \end{pmatrix} \begin{pmatrix} r \\ x \end{pmatrix} = \begin{pmatrix} b \\ 0 \end{pmatrix}.$$

Clever execution of this delivers in fact the factors L and U of the LU-decomposition of the tridiagonal matrix that would have been delivered when carrying out the Lanczos procedure with A^TA.

An interesting variant of LSQR is the so-called Craig's method [55]. The easiest way to think of this method is to apply Conjugate Gradients for the system $A^TAx = A^Tb$, with the following choice of innerproduct

$$[x, y] \equiv (x, (A^TA)^{-1}y),$$

which defines a proper innerproduct if A is of full rank.
First note that the two innerproducts in CG (as in section 1.2.1) can be computed without inverting A^TA:

$$[p_i, A^TAp_i] = (p_i, p_i),$$

and, assuming that $b \in R(A)$ so that $Ax = b$ has a unique solution x (since A has full rank):

$$\begin{aligned} [r_i, r_i] &= [A^T(Ax_i - b), A^T(Ax_i - b] \\ &= [A^TA(x_i - x), A^T(Ax_i - b)] \\ &= (x_i - x, A^T(Ax_i - b)) \end{aligned}$$

$$= (Ax_i - b, Ax_i - b) \tag{1.4}$$

Apparantly, we are minimizing with CG the following innerproduct

$$\begin{aligned} [x_i - x, A^TA(x_i - x)] &= (x_i - x, x_i - x) \\ &= \|x_i - x\|_2^2, \end{aligned} \tag{1.5}$$

that is, in this approach the Euclidean norm of the error is minimized. Note, however, that the rate of convergence of Craig's method is determined by the condition number of $A^T A$, so that this method is only attractive if one has a good preconditioner for $A^T A$.

Further references for CG

A more formal presentation of CG, as well as many theoretical properties, can be found in the textbook by Hackbusch [39]. A shorter presentation is given in [37]. An overview of papers, published in the first 25 years of existence of the method, is given in [36]. Vector processing and parallel computing aspects are discussed in [25] and [51].

Parallelism and data locality in preconditioned CG

Most often, the conjugate gradients method is used in combination with some kind of preconditioning. Usually, K is constructed as an approximation of A, such that systems like $Ky = z$ are much more easy to solve as $Ax = b$. Unfortunately, a popular class of preconditioners, based upon incomplete factorization of A, do not lend themselves very much for parallel implementation. We will discuss some approaches to obtain more parallelism in the preconditioner in section 1.5. For the moment we will assume that the preconditioner is chosen such that the parallelism in solving $Ky = z$ is comparable with the parallelism in computing Ap, for given p.

The scheme for preconditioned CG is given in Section 1.2.1. Note that in that scheme the updating of x and r can only start after the completion of the innerproduct required for α_i. Therefore, this innerproduct is a so-called synchronization point: all further computation has to wait for completion of this operation. One can try to avoid such synchronization points as much as possible, or formulate CG in such a way that synchronization points can be combined.

Since on a distributed memory machine communication is required to assemble the innerproduct, it would be nice if we could proceed with other useful computation while the communication takes place. However, as we see from the CG scheme, there is no possibility to overlap this communication time with useful computation. The same observation can be made for the updating of p, which can only take place after the completion of the innerproduct for β_i. Apart from the computation of Ap, and the computations with K, we need to load 7 vectors for 10 vector floating point operations. This means that for this part of the computation only 10/7 floating point operation can be carried out per memory reference, in average.

Several authors ([12, 46, 47]) have attempted to improve this ratio, and to reduce the number of synchronization points. In our formulation of CG there are two such synchronization points. Meurant [46] (see also [62]) has proposed a variant in which there is only one synchronization point, however at the cost of a possibly reduced numerical stability, and one additional innerproduct. In this scheme the ratio between computations and memory references is about 2.

Chronopoulos and Gear [12] propose to further improve the data locality and parallelism in CG by combining s successive steps. Their algorithm is based upon the following property of CG. The residual vectors $r_0, ..., r_i$ form an orthogonal basis (assuming exact arithmetic) for the Krylov subspace spanned by $r_0, Ar_0, ..., A^{i-1}r_0$. When arrived at r_j, the vectors $r_0, r_1, ..., r_j$, $Ar_j, ..., A^{i-j-1}r_j$ also form a basis for this subspace. Hence, we may combine s successive steps by generating $r_j, Ar_j, ..., A^{s-1}r_j$ first, and then do the orthogonalization and the updating of the current solution with this blockwise extended subspace. This approach leads to a modest increase in flops in comparison with s successive steps of the standard CG.
The main drawback in this approach seems to be the potential numerical instability. Depending on the spectral properties of A, the set $r_j, ..., A^{s-1}r_j$ may tend to converge to a vector in the direction of a dominating eigenvector, or, in other words, may tend to dependence for increasing values of s. The authors claim to have seen successful completion of this approach, with no serious stability problems, for small values of s. Nevertheless, it seems that s-step CG, because of these problems, has a poor reputation (see also [63]). However, a similar approach, suggested by Chronopoulos and Kim [13] for other processes such as GMRES, seems to be more promising. Several authors have pursued this research direction, and we will come back to this in section 1.3.2.

In [19] a variant of CG is proposed, in which there is possibility to overlap all communication time with useful computations. This variant is a rescheduled version of the original CG scheme, and is therefore precisely as stable. The key trick in this approach is to delay the updating of the solution vector by one iteration step. Another advantage is that no additional operations are required.

Parallel performance of CG

From a parallel point of view, CG mimics very well parallel performance properties of a variety of iterative methods such as Bi-CG, CGS, QMR, Bi-CGSTAB, and others.
In this section we study the performance of CG on parallel distributed memory systems and we report on some supporting experiments on actual existing machines. Guided by our experiments we will discuss the suitability of CG for Massively Parallel Processing systems.

All computational intensive elements in preconditioned CG (updates, innerproducts, and matrix vector operations) are trivially parallelizable for shared memory machines [25], except possibly for the preconditioning

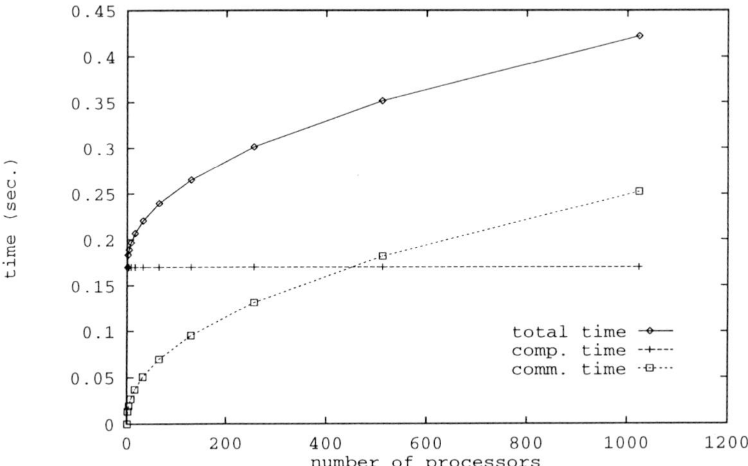

FIGURE 1.1. Modelled timings for 1 iteration with CG

step: *Solve* w_{i+1} *from* $Kw_{i+1} = r_{i+1}$. For the latter operation parallelism depends very much on the choice for K. In this section we restrict ourselves to block Jacobi preconditioning, where the blocks have been chosen so that each processor can handle one block independently of the others. For other preconditioners that allow some degree of parallelism see [25] and Section 1.5.

For a distributed memory machine at least some of the steps require communication between processors: the accumulation of innerproducts and the computation of Ap_i (depending on the non-zero structure of A and the distribution of the non-zero elements over the processors). We consider in some more detail the situation where A is a block-tridiagonal matrix of order N, and we assume that all blocks are of order $\sqrt{N}$:

$$A = \begin{pmatrix} A_1 & D_1 & & & \\ D_1 & A_2 & D_2 & & \\ & D_2 & \ddots & \ddots & \\ & & \ddots & & \\ & & & & \end{pmatrix},$$

in which the D_i are diagonal matrices, and the A_i are tridiagonal matrices. Such systems occur quite frequently in finite difference approximations in 2 space dimensions. Our discussion can easily be adapted to 3 space dimensions.

For simplicity we will assume that the processors are connected as a 2D grid with $p \times p = P$ processors.

The data have been distributed in a straight forward manner over the processor memories and we have not attempted to fully exploit the underlying grid structure for the given type of matrix in order to reduce communication as much as possible. In fact it will turn out that in our case the communication for the matrix vector product plays only a minor role for matrix systems of large size .
Because of symmetry only the 3 non-zero diagonals in the upper triangular part of A need to be stored, and we have chosen to store successive parts of length N/P of each diagonal in consecutive neighbouring processors.

The blocks for block Jacobi are chosen to be the diagonal blocks that are available on each processor, and the various vectors (r_i, p_i, etc.) have been distributed likewise, i.e. each processor holds a section of length N/P of these vectors in its local memory. We will now inspect the ingredients of CG for exploitable parallelism:
matrix vector product It is easily seen for a $2D$ processor grid (as well as for many other configurations, including hypercube and pipeline), that the matrix vector product can be completed with only neighbour-neighbour communication. This means that the communication costs do not increase for increasing values of p. If one follows a domain decomposition way of approach, in which the finite difference discretization grid is subdivided into p by p subgrids (p in x-direction and p in y-direction), then the communication costs are smaller than the computational costs by a factor of $\mathcal{O}(\frac{\sqrt{N}}{p})$.
vector update In our case these operations do not require any communication and we should expect linear speed up when increasing the number of processors P.
inner product For the innerproduct we need global communication for assembly and we need global communication for the distribution of the assembled innerproduct over the processors. For a $p \times p$ processor grid these communication costs are proportional with p. This means that for a constant length of the vectorparts per processor, these communication costs will dominate for values of p large enough. This is quite unlike the situation for the matrix vector product and as we will see it may be a severely limiting factor in achieving high speed-ups in a massively parallel environment.

On a 4-processor MEIKO SP1 we have done some experiments in order to determine the costs of inter processor communication and for computation. Assuming that the costs for communication (for the innerproducts) grow linearly with the length of the path of communication we have modelled the wall-clock time for a similar computer with P processors, for 1 iteration with CG with matrices of order $90000P$, as in Figure 1.1. Note that we have increased the size of the linear system linearly with the number of processors, which seems realistically since with larger computers one aims to solve larger systems. The value 90000 has been chosen since this is more or less the size of the part of the system that could be stored in the local memory of one processor of our MEIKO machine.

From this Figure we learn that for P slightly larger than 400 the communication costs may be expected to dominate, and eventually they will lead to very low speed-ups (even for systems for which the size is as large as the total memory permits). We also see, that if overlap of communication and computation is possible, then potentially the communication can be hidden for values of P less than 400, but this demands for a reformulation of the CG algorithm. Of course, these expectations are based on a model, but we have also carried out similar experiments on the 512 processor Parsytec GCel-3/512 of the University of Amsterdam [16]. In particular we have observed on that machine that the communication time for the innerproduct increases like $\sqrt{P}$, which just explains the behaviour of our model for the MEIKO-type of architecture.

If a given architecture permits the overlap of communication with computation, then we may try to reformulate CG in order to create possibilities for overlap. For the (extrapolated) MEIKO this may help for values of P up to about 400. For larger P we will see communication dominating anyhow, but the adverse effects can be lessened. A stable reformulation of CG which has this effect has been described in [19].

1.3 The minimum residual approach: GMRES and MINRES

The creation of an orthogonal basis for the Krylov subspace, in Section 1.1.1, has led to

$$AV_i = V_{i+1}H_{i+1,i}. \tag{1.6}$$

We look for an $x_i \in K^i(A;r_0)$, that is $x_i = V_i y$, for which $\|b - Ax_i\|_2$ is minimal. This norm can be rewritten as

$$\|b - Ax_i\|_2 = \|b - AV_i y\|_2 = \|b - V_{i+1}H_{i+1,i}y\|_2.$$

Now we exploit the fact that V_{i+1} is an orthonormal transformation with respect to the Krylov subspace $K^{i+1}(A;r_0)$:

$$\|b - Ax_i\|_2 = \|\|r_0\|_2 e_1 - H_{i+1,i}y\|_2,$$

and this final norm can simply be minimized by solving the minimum norm least squares problem:

$$H_{i+1,i}y = \|r_0\|_2 e_1.$$

In GMRES [66] this is done with Givens rotations, that annihilate the subdiagonal elements in the upper Hessenberg matrix $H_{i+1,i}$.
Note that when A is symmetric the upper Hessenberg matrix $H_{i+1,i}$ reduces to a tridiagional system. This simplified structure can be exploited in order

to avoid storage of all the basis vectors for the Krylov subspace, in a way similar as has been pointed out for CG. The resulting method is known as MINRES [54].

In order to avoid excessive storage requirements and computational costs for the orthogonalization, GMRES is usually restarted after each m iteration steps. This algorithm is referred to as GMRES(m); the not-restarted version is often called 'full' GMRES. For a simple algorithmic representation of these methods, see [7].

There is an interesting and simple relation between the Ritz-Galerkin approach (FOM and CG) and the minimum residual approach (GMRES and MINRES). In GMRES the projected system matrix $H_{i+1,i}$ is transformed by Givens rotations to an upper triangular matrix (with last row equal to zero). So, in fact, the major difference between FOM and GMRES is that in FOM the last (($i+1$)-th row is simply discarded, while in GMRES this row is rotated to a zero vector. Let us characterize the Givens rotation, acting on rows i and $i+1$, in order to zero the element in position $(i+1,i)$, by the sine s_i and the cosine c_i. Let us further denote the residuals for FOM with an superscript F and those for GMRES with superscript G. Then the above observations lead to the following results for FOM and GMRES (for details see [66] and [10]).

1. The reduction for successive GMRES residuals is given by

$$\frac{\|r_k^G\|_2}{\|r_{k-1}^G\|_2} = |s_k| \tag{1.7}$$

([66]: p. 862, Proposition 1).

2. If $c_k \neq 0$ then the FOM and the GMRES residuals are related by

$$\|r_k^G\|_2 = |c_k|\ \|r_k^F\|_2 \tag{1.8}$$

([10]: theorem 5.1).

From these relations we see that when GMRES has locally a significant reduction in the norm of the residual (i.e., s_k is small), then FOM gives about the same result as GMRES (since $c_k^2 = 1 - s_k^2$). On the other hand when FOM has a breakdown ($c_k = 0$), then GMRES does not lead to an improvement in the same iteration step.
Because of these relations we can link the convergence behaviour of GMRES with the convergence of Ritz values (the eigenvalues of the "FOM" part of the upper Hessenberg matrix). This has been exploited in [78], for the analysis and explanation of local effects in the convergence behaviour of GMRES.

There are various different implementations of FOM and GMRES. Among those equivalent with GMRES are: Orthomin [81], Orthodir [41], Axelsson's method [3], and GENCR [29]. These methods are often more expensive than GMRES per iteration step. Orthomin seems to be still popular, since this variant can be easily truncated (Orthomin(s)), in contrast to GMRES. The truncated and restarted versions of these algorithms are not necessarily mathematically equivalent.
Methods that are mathematically equivalent with FOM are: Orthores [41] and GENCG [14, 83]. In these methods the approximate solutions are constructed such that they lead to orthogonal residuals (which form a basis for the Krylov subspace; analogously to the CG method). A good overview of all these methods and their relations is given in [65].

1.3.1 GMRESR and GMRES⋆

In [79] it has been shown how the GMRES-method can be combined (or rather preconditioned) with other iterative schemes. The iteration steps of GMRES (or GCR) are called outer iteration steps, while the iteration steps of the preconditioning iterative method are referred to as inner iterations. The combined method is called GMRES⋆, where ⋆ stands for any given iterative scheme; in the case of GMRES as the inner iteration method, the combined scheme is called GMRESR[79].
Similar schemes have been proposed recently. In FGMRES[64] the update directions for the approximate solution are preconditioned, whereas in GMRES⋆ the residuals are preconditioned. The latter approach offers more control over the reduction in the residual, in particular breakdown situations can be easily detected and remedied.
In exact arithmetic GMRES⋆ is very close to the Generalized Conjugate Gradient method[4]; GMRES⋆, however, leads to a more efficient computational scheme.

The GMRES⋆ algorithm can be described by the following computational scheme:

> x_0 is an initial guess; $r_0 = b - Ax_0$;
> for $i = 0, 1, 2, 3, \ldots$
>   Let $z^{(m)}$ be the approximate solution of $Az = r_i$, obtained after m steps of an iterative method.
> $c = Az^{(m)}$ (often available from the iteration method)
>   for $k = 0, \ldots, i-1$
> $\alpha = (c_k, c)$
> $c = c - \alpha c_k$
> $z^{(m)} = z^{(m)} - \alpha u_k$
> $c_i = c/\|c\|_2$; $u_i = z^{(m)}/\|c\|_2$
> $x_{i+1} = x_i + (c_i, r_i)u_i$
> $r_{i+1} = r_i - (c_i, r_i)c_i$

if x_{i+1} is accurate enough then quit
end

A sufficient condition to avoid breakdown in this method ($\|c\|_2 = 0$) is that the norm of the residual at the end of an inner iteration is smaller than the norm of the right-hand side residual: $\|Az^{(m)} - r_i\|_2 < \|r_i\|_2$. This can easily be controlled during the inner iteration process. If stagnation occurs, i.e. no progress at all is made in the inner iteration, then it is suggested in [79] to do one (or more) steps of the LSQR method, which guarantees a reduction (but this reduction is often only small).

The idea behind this combined iteration scheme is that we explore parts of high-dimensional Krylov subspaces, hopefully localizing almost the same approximate solution that full GMRES would find over the entire subspace, but now at much lower computational costs. For the inner iteration we may select any appropriate solver, for instance, one cycle of GMRES(m), since then we have also locally an optimal method, or some other iteration scheme, like for instance Bi-CGSTAB.

In [21] it is proposed to keep the Krylov subspace, that is built in the inner iteration, orthogonal with respect to the Krylov basis vectors generated in the outer iteration.

1.3.2 Parallel performance of GMRES(m)

We will use a simple model for the computation time and the communication cost for the main kernels in GMRES(m); for a more elaborate description and analysis of the performance of Krylov subspace methods we refer to [22].
We use the term *communication cost* to indicate all the wall clock time spent in communication that is not overlapped with useful computation (so that it really contributes to wall-clock time).
The term *communication time* refers to the wall-clock time of the entire communication. In the case of a nonoverlapped communication, the communication time and the communication cost are the same.

Remark: Our quantitative formulas are not meant to give very accurate predictions of the exact execution times, but they will be used to identify the bottlenecks and to evaluate improvements.
Indeed, our experiments show that our models are relatively close to reality, and thus may be used as convenient tools for understanding the negative effects of global communication.

For a processor grid with $P = p^2$ processors, we have that $p_d = \sqrt{P}$. For nearest neighbour communication, let t_s denote the communication start-up time and let the transmission time associated with a single inner product computation be t_w. Then the global accumulation and broadcast time for 1 inner product is taken as $2p_d(t_s + t_w)$, while the global accumulation and broadcast for k simultaneous inner products takes $2p_d(t_s + k\,t_w)$.

For GMRES(m) the communication time for the modified Gram-Schmidt algorithm (with $\frac{1}{2}(m^2 + 3m)$ accumulations and broadcasts) is

$$T^G_{comm} = (m^2 + 3m)p_d(t_s + t_w). \tag{1.9}$$

Note that for other processor configurations one only needs to bring in an appropriate value for p_d in (1.9).

From (1.9) we conclude that the communication cost for GMRES(m) is of the order $\mathcal{O}(m^2\sqrt{P})$ and for large processor grids this may become a bottleneck. Moreover, in the standard implementation we cannot reduce these costs by accumulating multiple inner products together, since the modified Gram-Schmidt orthogonalization of one single vector against a set of vectors and its subsequent normalization is an inherently sequential process.

However, if the modified Gram-Schmidt orthogonalization can be done for a set vectors simultaneously, then we have more possibilities for reduction of communication overhead. We discuss how to generate such a set of vectors in subsection 1.3.2.

Suppose the set of vectors $v_1, \hat{v}_2, \hat{v}_3 \ldots, \hat{v}_{m+1}$ has to be orthogonalized, where $\|v_1\|_2 = 1$. The modified Gram-Schmidt process can be implemented as is sketched in Figure 1.2. This reduces the number of messages to only m

```
for i = 1, 2, ..., m do
    orthogonalize v̂_{i+1}, ..., v̂_{m+1} on v_i;
    v_{i+1} = v̂_{i+1}/||v̂_{i+1}||_2;
end
```

FIGURE 1.2. block-wise modified Gram–Schmidt

instead of $\frac{1}{2}(m^2+3m)$ for the usual implementation of GMRES(m), but the length of the messages has increased. In this way, start-up time is saved by packing small messages, corresponding to one block of orthogonalizations, into one larger message. Note that the computation time for this approach is equal to that for the standard modified Gram–Schmidt algorithm. We will call this variant of GMRES: **parGMRES(m)**.

Creation of the basis before orthogonalization

In order to be able to use this parallel modified Gram–Schmidt algorithm in GMRES(m), a basis for the Krylov subspace has to be generated first. The idea to start with some non-orthogonal basis for the Krylov subspace, and then to orthogonalize this basis afterwards, was already suggested for the CG algorithm, referred to as s-step CG, in [12] for shared (hierarchical)

```
v̂_1 = v_1 = r/||r||_2
for i = 1, 2, ..., m do
    v̂_{i+1} = v̂_i − d_i A v̂_i
end
```

$$\hat{v}_1 = v_1 = r/\|r\|_2$$
$$\text{for } i = 1, 2, \ldots, m \text{ do}$$
$$\quad \hat{v}_{i+1} = \hat{v}_i - d_i A \hat{v}_i$$
$$\text{end}$$

FIGURE 1.3. Generation of a polynomial basis

memory parallel vector processors.

In [12] it is also reported that the s-step CG algorithm may converge slowly due to numerical instability for $s > 5$. In the parGMRES(m) algorithm stability seems to be much less of a problem since each vector is explicitly orthogonalized against all the other vectors, and we first generate a polynomial basis for the Krylov subspace, see [5, 20]. Furthermore, because of the restart, rounding errors made before restart can not have an accumulated effect on iterations carried out after the restart.

The basis vectors for the Krylov subspace $\hat{v}_i$ are generated as indicated in Figure 1.3, where the parameters d_i are chosen to keep the condition number of the matrix $[v_1, \hat{v}_2, \ldots, \hat{v}_{m+1}]$ sufficiently small. In order to achieve this, one might start in the first cycle with a suitable Chebychev recursion, for later cycles one may then take into account the GMRES information of the previous cycle. For more details on this, see [5], and [69]: Section 6.

Experimental observations

We present some experimental observations on the parallel performance of GMRES(m) and parGMRES(m) on a 400-processor Parsytec Supercluster. The T800-20 transputer processors of the Parsytec are connected in a fixed 20×20 mesh, of which arbitrary submeshes can be used. The transputer supports only nearest neighbor synchronous communication; more complicated communication has to be programmed explicitly. The communication rate is fast compared with the flop rate, but to current standards the T800 is a slow processor. For further details on our testing circumstances, see [23].

We will consider the performance of only one (par)GMRES(m) cycle.

We have solved a convection diffusion problem discretized by finite volumes over a 100×100 grid, resulting in the standard five-diagonal matrix with a tridiagonal block-structure, corresponding to the 5-point star. This relatively small problem size was chosen, because for processor grids of increasing size it very well shows the expected degradation of performance for GMRES(m) and the large improvements of parGMRES(m) over GMRES(m). Furthermore, the parallel behavior for this problem size on this slow machine corresponds quite well to much larger problems on more modern machines; see [23].

We give speed-ups and efficiencies in Table 1.1. Since the problem did not fit on a single processor, the speed-up and efficiency values were computed

processor grid	GMRES(30)		parGMRES(30)	
	E (%)	S	E (%)	S
10×10	77.2	77.2	95.9	95.9
14×14	53.9	106.	88.8	174.
17×17	40.5	117.	69.4	201.
20×20	28.6	114.	53.4	214.

TABLE 1.1. Speed-ups for GMRES(30) and parGMRES(30)

against a (fairly accurately) estimated sequential runtime.

We clearly see that the performance of parGMRES(m) is much better than the performance of GMRES(m). For a more elaborate discussion, see [23].

1.4 The Petrov-Galerkin approach: Bi-CG

For unsymmetric systems we can, in general, not reduce the matrix A to a symmetric system in a lower-dimensional subspace, by orthogonal projections. The reason is that we can not create an orthogonal basis for the Krylov subspace by a 3-term recurrence relation [30]. However, we can try to obtain a suitable non-orthogonal basis with a 3-term recurrence, by requiring that this basis is orthogonal to some other basis.

We start by constructing an arbitrary basis for the Krylov subspace:

$$h_{i+1,i}v_{i+1} = Av_i - \sum_{j=1}^{i} h_{j,i}v_j, \tag{1.10}$$

which can be rewritten in matrix notation as $AV_i = V_{i+1}H_{i+1,i}$.

Clearly, we can not use V_i for the projection, but suppose we have a W_i for which $W_i^*V_i = D_i$ (an i by i diagonal matrix with diagonal entries d_i), and for which $W_i^*v_{i+1} = 0$. The coefficients $h_{i+1,i}$ can be chosen, we suggest to choose them such that $\|v_{i+1}\|_2 = 1$.

Then

$$W_i^*AV_i = D_iH_{i,i}, \tag{1.11}$$

and now our goal is to find a W_i for which $H_{i,i}$ is tridiagonal. This means that $V_i^*A^*W_i$ should be tridiagonal too[2]. This last expression has a similar structure as the right-hand side in (1.11), with only W_i and V_i reversed. This suggests to generate the w_i with A^*.

We start with an arbitrary $w_1 \neq 0$, such that $w_1^*v_1 \neq 0$. Then we generate v_2 with (1.10), and orthogonalize it with respect to w_1, which means

[2] A^* denotes the adjoint of A

that $h_{1,1} = w_1^* A v_1 / (w_1^* v_1)$. Since $w_1^* A v_1 = v_1^* A^* w_1$, this implies that w_2, generated with

$$h_{2,1} w_2 = A^* w_1 - h_{1,1} w_1,$$

is also orthogonal to v_1.
This can be continued, and we see that we can create bi-orthogonal basis sets $\{v_j\}$, and $\{w_j\}$, by making the new v_i orthogonal to w_1 up to w_{i-1}, and then by generating w_i with the same recurrence coefficients, but with A^* instead of A.
Now we have that $W_i^* A V_i = D_i H_{i,i}$, and also that $V_i A^* W_i = D_i H_{i,i}$. This implies that $D_i H_{i,i}$ is symmetric, and hence $H_{i,i}$ is a tridiagonal matrix, which gives us the desired 3-term recurrence relation for the v_j's, and the w_j's.

We may proceed in a similar way as in the symmetric case:

$$AV_i = V_{i+1} T_{i+1,i}, \tag{1.12}$$

but here we use the matrix $W_i = [w_1, w_2, ..., w_i]$ for the projection of the system

$$W_i^* (b - Ax_i) = 0,$$

or

$$W_i^* A V_i y - W_i^* b = 0.$$

Using (1.12), we find that y_i satisfies

$$T_{i,i} y = \|r_0\|_2 e_1,$$

and $x_i = V_i y$.
This method is known as the Bi-Lanczos method [44].
We have assumed that $d_i \neq 0$, that is $w_i^* v_i \neq 0$. The generation of the bi-orthogonal basis breaks down if for some i the value of $w_i^* v_i = 0$, this is referred to in litterature as a *serious breakdown*. Likewise, when $w_i^* v_i \approx 0$, we have a near-breakdown. The way to get around this difficulty is the so-called Look-ahead strategy, which comes down to taking a number of successive basis vectors for the Krylov subspace together and to make them blockwise bi-orthogonal. This has been worked out in detail in [56], [33], [34], and [35].
Another way to avoid breakdown is to restart as soon as a diagonal element gets small. Of course, this strategy looks surprisingly simple, but one should realise that at a restart the Krylov subspace, that has been built up so far, is thrown away, which destroys possibilities for faster (i.e., superlinear) convergence.

We can try to construct an LU-decomposition, without pivoting, of $T_{i,i}$. If this decomposition exists, then, similar as for CG, it can be updated from iteration to iteration and this leads to a recursive update of the solution

vector, which avoids to save all intermediate r and w vectors. This variant of Bi-Lanczos is usually called Bi-Conjugate Gradients, or shortly Bi-CG [31]. In Bi-CG, the d_i are chosen such that $v_i = r_{i-1}$, similar as for CG.
Of course one can in general not be certain that an LU decomposition (without pivoting) of the tridiagonal matrix $T_{i,i}$ exists, and this may lead also to breakdown (a breakdown of the *second kind*), of the Bi-CG algorithm. Note that this breakdown can be avoided in the Bi-Lanczos formulation of the iterative solution scheme, e.g., by making an LU-decomposition with 2 by 2 block diagonal elements [6]. It is also avoided in the QMR approach (see Section 1.4.1).
Note that for symmetric matrices Bi-Lanczos generates the same solution as Lanczos, provided that $w_1 = r_0$, and under the same condition Bi-CG delivers the same iterands as CG for positive definite matrices. However, the Bi-orthogonal variants do so at the cost of two matrix vector operations per iteration step.
For a computational scheme for Bi-CG, see [7].

1.4.1 QMR

The QMR method [35] relates to Bi-CG in a similar way as MINRES relates to CG.
We start from the recurrence relations for the v_j:

$$AV_i = V_{i+1}T_{i+1,i}.$$

Similar as for MINRES we would like to construct the x_i, with $x_i \in K^i(A;r_0)$, or $x_i = V_i y$, for which

$$\|b - Ax_i\|_2 = \|b - AV_i y\|_2 = \|b - V_{i+1}T_{i+1,i}y\|_2.$$

Now pretend that the columns of V_{i+1} are orthogonal, then

$$\|b - Ax_i\|_2 = \|V_{i+1}(\|r_0\|_2 e_1 - T_{i+1,i}y)\|_2 = \|(\|r_0\|_2 e_1 - T_{i+1,i}y)\|_2,$$

and in [35] it is suggested to solve the projected miniminum norm least squares problem at the right-hand side. Since, in general the columns of V_{i+1} are not orthogonal, the computed $x_i = V_i y$ does not solve the minimum residual problem, and therefore this approach is referred to as a Quasi-minimum residual approach [35].
This leads to the simplest form of the QMR method. A more general form arises if the least squares problem is replaced by a weighted least squares problem [35]. No strategies are yet known for optimal weights.
In [35] the QMR method is carried out on top of a look-ahead variant of the bi-orthogonal Lanczos method, which makes the method more robust. Experiments suggest that QMR has a much smoother convergence behaviour than Bi-CG, but it is not essentially faster than Bi-CG.

In fact, it is difficult to make a fair comparison between GMRES and QMR. GMRES really minimizes the 2-norm of the residual, but at the cost of increasing work for keeping all residuals orthogonal and increasing demands for memory space. QMR does not minimize this norm, but often it has a comparable fast convergence as GMRES, at the cost of twice the amount of matrix vector products per iteration step. However, the generation of the basis vectors in QMR is relatively cheap and the memory requirements are limited and modest. In view of the relation between GMRES and FOM, it will be no surprise that there is a similar relation between QMR and Bi-CG, for details of this see [35]. This relation expresses that at a significant local error reduction of QMR, Bi-CG and QMR have arrived almost at the same residual vector (similar as for GMRES and FOM). However, QMR has to be preferred in all cases because of its much smoother convergence behaviour.
Several variants of QMR, or rather Bi-CG, have been proposed, which increase the effectiveness of this class of methods in certain circumstances. These variants will be discussed in coming subsections.

1.4.2 CGS

For the bi-conjugate gradient residual vectors it is well-known that they can be written as $r_j = P_j(A)r_0$ and $\hat{r}_j = P_j(A^*)\hat{r}_0$, and because of the bi-orthogonality relation we have that

$$(r_j, \hat{r}_i) = (P_j(A)r_0, P_i(A^*)\hat{r}_0)$$

$$= (P_i(A)P_j(A)r_0, \hat{r}_0) = 0,$$

for $i < j$.
The iteration parameters for bi-conjugate gradients are computed from innerproducts like the above. Sonneveld observed that we can also construct the vectors $\tilde{r}_j = P_j^2(A)r_0$, using only the latter form of the innerproduct for recovering the bi-conjugate gradients parameters (which implicitly define the polynomial P_j). By doing so, it can be avoided that the vectors $\hat{r}_j$ have to be formed, nor is there any multiplication with the matrix A^*.
The resulting CGS [71] method works in general very well for many unsymmetric linear problems. It converges often much faster than BI-CG (about twice as fast in some cases) and does not have the disadvantage of having to store extra vectors like in GMRES. These three methods have been compared in many studies (see, e.g., [60, 11, 58, 49]).
However, CGS usually shows a very irregular convergence behaviour. This behaviour can even lead to cancellation and a spoiled solution [77]. See also section 1.4.3.

The following scheme carries out the CGS process for the solution of $Ax = b$, with a given preconditioner K:

x_0 is an initial guess; $r_0 = b - Ax_0$;
$\tilde{r}_0$ $(= w_1)$ is an arbitrary vector, such that
$(r_0, \tilde{r}_0) \neq 0$,
e.g., $\tilde{r}_0 = r_0; \rho_0 = (r_0, \tilde{r}_0)$;
$\beta_{-1} = \rho_0; p_{-1} = q_0 = 0$;
for $i = 0, 1, 2, ...$
    $u_i = r_i + \beta_{i-1} q_i$;
    $p_i = u_i + \beta_{i-1}(q_i + \beta_{i-1} p_{i-1})$;
    solve $\hat{p}$ from $K\hat{p} = p_i$;
    $\hat{v} = A\hat{p}$;
    $\alpha_i = \frac{\rho_i}{(\tilde{r}_0, \hat{v})}$;
    $q_{i+1} = u_i - \alpha_i \hat{v}$;
    solve $\hat{u}$ from $K\hat{u} = u_i + q_{i+1}$
    $x_{i+1} = x_i + \alpha_i \hat{u}$;
    if x_{i+1} is accurate enough then quit;
    $r_{i+1} = r_i - \alpha_i A\hat{u}$;
    $\rho_{i+1} = (\tilde{r}_0, r_{i+1})$;
    if $\rho_{i+1} = 0$ then method fails to converge !;
    $\beta_i = \frac{\rho_{i+1}}{\rho_i}$;
end

Similar to Bi-CG, we can associate with CGS a small overdetermined linear system that can be solved, ignoring the fact that the basis vectors are non-orthogonal. This leads to TFQMR [32].

1.4.3 Effects of irregular convergence

By very irregular convergence we refer to the situation where successive residual vectors in the iterative process differ in orders of magnitude in norm, and some of these residuals may be even much bigger in norm than the starting residual. We will try to give an impression why this is a point of concern, even if eventually the (updated) residual satisfies a given tolerance. For more details we refer to [67, 69].

We will say that an algorithm is *accurate* for a certain problem if the *updated residual* r_j and the *true residual* $b - Ax_j$ are of comparable size for the j's of interest.

The best we can hope for is that for each j the error in the residual is only the result of applying A to the update w_{j+1} for x_j in finite precision arithmetic:

$$r_{j+1} = r_j - Aw_{j+1} - \Delta_A w_{j+1} \tag{1.13}$$

if

$$x_{j+1} = x_j + w_{j+1}, \tag{1.14}$$

for each j, where Δ_A is an $n \times n$ matrix for which $|\Delta_A| \preceq n_A \, \bar{\xi} \, |A|$: n_A is the maximum number of non-zero matrix entries per row of A, $|B| \equiv (|b_{ij}|)$

if $B = (b_{ij})$, $\overline{\xi}$ is the relative machine precision, the inequality $\preceq$ refers to element-wise $\leq$. In Bi-CG type methods that we consider, we compute explicitly the update Aw_j for the residual r_j from the update w_j for the approximation x_j by matrix multiplication: for this part, (1.13) describes well the local deviations caused by evaluation errors.

In the "ideal" case (i.e. situation (1.13) whenever we update the approximation) we have that

$$\begin{aligned} r_k - (b - Ax_k) &= \sum_{j=1}^{k} \Delta_A w_j \\ &= \sum_{j=1}^{k} \Delta_A (e_{j-1} - e_j), \end{aligned} \tag{1.15}$$

where the perturbation matrix Δ_A may depend on j and e_j is the approximation error in the jth approximation: $e_j \equiv x - x_j$. Hence,

$$|\|r_k\| - \|b - Ax_k\|| \leq \tag{1.16}$$

$$2k\, n_A\, \overline{\xi}\ |||A|||\ \sum_{j=0}^{k} \|e_j\| \leq$$

$$2\,\Gamma\,\overline{\xi} \sum_{j=0}^{k} \|r_j\|\,,$$

$$\text{where}\quad \Gamma \equiv n_A\, |||A|||\ \|A^{-1}\|.$$

Except for the factor Γ, the last upper–bound appears to be rather sharp. We see that approximations with large approximation errors may ultimately lead to an inaccurate result. Such large local approximation errors are typical for CGS, and in [77] an example of the resulting numerical inaccuracy is given. If there are a number of approximations with comparable large approximation errors, then their multiplicity may replace the factor k, otherwise it will be only the largest approximation error that makes up virtually the bound for the deviation.

Example. Figure 1.4 illustrates the loss of accuracy, discussed above. The convergence history of the updated residuals (the 'circles': ∘∘) and the true residuals (the solid curve: ——) of CGS is given for the matrix SHERMAN4 from the Harwell-Boeing set of test matrices. The norm of the residuals, on log-scale, is plotted (along the vertical axis) against the number of matrix-vector multiplications (along the horizontal axis). The dotted curve ($\cdots$) represents the estimated inaccuracy: $2\,\overline{\xi}\sum_{j\leq i}\|r_j\|$ (here with $\Gamma = 1$; cf. (1.16)).

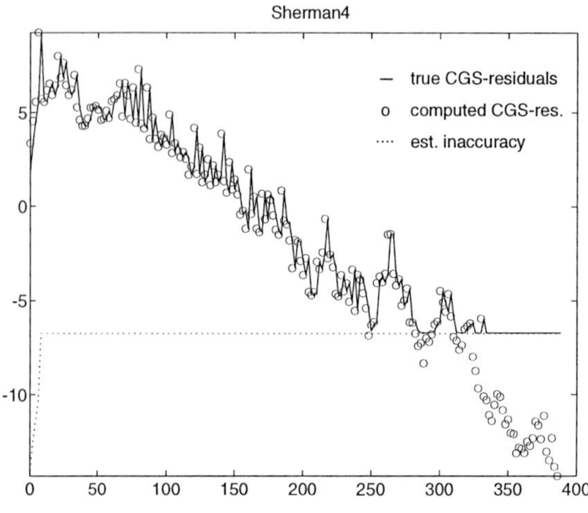

FIGURE 1.4. Convergence plot CGS for the true residuals and the updated residuals

Neumaier [50] has suggested a simple modification for the CGS method, so that the *updated residual* and the *true residual* are of comparable size. In this approach the updates for the approximated solution are collected in an auxiliary vector, and the sum of these updates is added to the approximation only when the norm of the residual has become smaller than the smallest norm of the residual earlier in the process. This means that the large updates, corresponding to large intermediate residuals, may lead to cancelation effects in the auxiliary vector but their collective effect on the best approximation so far (the approximation corresponding to the smallest residual so far) is acceptable.

The implementation of Neumaier's suggestion is given in Fig. 1.5. Note that the norm of the b' (the residuals with respect to the x) strictly decrease: the Neumaier trick also smoothes convergence (without improving its speed!).

Different strategies for updating, for other iterative methods as well, are given in [68]. That publication also contains discussions on the update strategies and it gives examples of the effects. The conclusion it is that it is very advisable to include such a 'reliable' update strategy in an actual implementation.

1.4.4 Bi-CGSTAB

Bi-CGSTAB [77] is based on the observation that the Bi-CG vector r_i is orthogonal to the entire subspace $K^i(A^*, w_1)$. As a result, we can, instead of squaring the Bi-CG polynomial, construct iteration methods, by which x_i are generated so that $r_i = \tilde{P}_i(A)P_i(A)r_0$ with other i^{th} degree polynomials

$$\vdots$$

$x = x_0;\ x' = 0;\ b' = r_0;\ \mu' = \|b'\|;$
for $i = 0, 1, 2, \ldots$

$\vdots$ *Replace all x_i and x by x'.*

Skip the CGS update for r
together with the MV involved
in this update. Compute instead
  $r_{i+1} = b' - Ax';\ \mu = \|r_{i+1}\|;$
  if μ is small enough then quit;
  if $\mu \leq \mu'$
    $x = x + x';\ x' = 0;$
    $b' = r_{i+1};\quad \mu' = \mu;$
  endif
endfor
$x = x + x';$

FIGURE 1.5. Neumaier's strategy for CGS

$\tilde{P}$. An obvious possibility is to take for $\tilde{P}_j$ a polynomial of the form

$$Q_i(x) = (1 - \omega_1 x)(1 - \omega_2 x)\ldots(1 - \omega_i x), \tag{1.17}$$

and to select suitable constants ω_j. This expression leads to an almost trivial recurrence relation for the Q_i.
In Bi-CGSTAB ω_j in the jth iteration step is chosen as to minimize r_j, with respect to ω_j, for residuals that can be written as $r_j = Q_j(A)P_j(A)r_0$.
The preconditioned Bi-CGSTAB algorithm for solving the linear system $Ax = b$, with preconditioning K reads as follows:

x_0 is an initial guess; $r_0 = b - Ax_0$;
$\bar{r}_0$ $(= w_1)$ is an arbitrary vector, such that
  $(\bar{r}_0, r_0) \neq 0$, e.g., $\bar{r}_0 = r_0$;
$\rho_{-1} = \alpha_{-1} = \omega_{-1} = 1$;
$v_{-1} = p_{-1} = 0$;
for $i = 0, 1, 2, \ldots$
  $\rho_i = (\bar{r}_0, r_i); \beta_{i-1} = (\rho_i/\rho_{i-1})(\alpha_{i-1}/\omega_{i-1})$;
  $p_i = r_i + \beta_{i-1}(p_{i-1} - \omega_{i-1}v_{i-1})$;
  Solve $\hat{p}$ from $K\hat{p} = p_i$;
  $v_i = A\hat{p}$;
  $\alpha_i = \rho_i/(\bar{r}_0, v_i)$;
  $s = r_i - \alpha_i v_i$;

if $\|s\|$ small enough then
  $x_{i+1} = x_i + \alpha_i \hat{p}$; quit;
Solve z from $Kz = s$;
$t = Az$;
$\omega_i = (t, s)/(t, t)$;
$x_{i+1} = x_i + \alpha_i \hat{p} + \omega_i z$;
if x_{i+1} is accurate enough then quit;
$r_{i+1} = s - \omega_i t$;
end

The matrix K in this scheme represents the preconditioning matrix and the way of preconditioning [77]. The above scheme carries out the Bi-CGSTAB procedure for the explicitly postconditioned linear system

$$AK^{-1}y = b,$$

but the vectors y_i and the residual have been backtransformed to the vectors x_i and r_i corresponding to the original system $Ax = b$. Compared with CGS, two extra innerproducts need to be calculated.
In exact arithmetic, the α_j and β_j have the same values as those generated by Bi-CG and CGS.
Bi-CGSTAB can be viewed as the product of Bi-CG and GMRES(1). Of course, other product methods can be formulated as well. Gutknecht [38] has proposed BiCGSTAB2, which is constructed as the product of Bi-CG and GMRES(2).

1.4.5 Bi-CGSTAB2 and variants

The residual $r_k = b - Ax_k$ in Bi-CG, when applied to $Ax = b$ with start x_0 can be written formally as $P_k(A)r_0$, where P_k is a k-degree polynomial. These residuals are constructed with one operation with A and one with A^* per iteration step. It was pointed out in [71] that with about the same amount of computational effort one can construct residuals of the form $\tilde{r}_k = P_k^2(A)r_0$, which is the basis for the CGS method. This can be achieved without any operation with A^*. The idea behind the improved efficiency of CGS is that if $P_k(A)$ is viewed as a reduction operator in Bi-CG, then one may hope that the square of this operator will be a twice as powerful reduction operator. Although this is not always observed in practice, one typically has that CGS converges faster than Bi-CG. This, together with the absence of operations with A^*, explains the success of the CGS method. A drawback of CGS is that its convergence behavior can look quite irregular, that is the norms of the residuals converge quite irregularly, and it may easily happen that $\|r_{k+1}\|_2$ is much larger than $\|r_k\|_2$ for certain k (for an explanation of this see [76]).
In [77] it was shown that by a similar approach as for CGS, one can construct methods for which r_k can be interpreted as $r_k = P_k(A)Q_k(A)r_0$, in

which P_k is the polynomial associated with Bi-CG and Q_k can be selected free under the condition that $Q_k(0) = 1$. In [77] it was suggested to construct Q_k as the product of k linear factors $1 - \omega_j A$, where ω_j was taken to minimize locally a residual. This approach leads to the Bi-CGSTAB method. Because of the local minimization, Bi-CGSTAB displays a much smoother convergence behavior than CGS, and more surprisingly, it often also converges (slightly) faster. A weak point in Bi-CGSTAB is that we get breakdown if an ω_j is equal to zero. One may equally expect negative effects when ω_j is small. In fact, BiCGSTAB can be viewed as the combined effect of Bi-CG and GCR(1), or GMRES(1), steps. As soon as the GCR(1) part of the algorithm (nearly) stagnates, then the Bi-CG part in the next iteration step cannot (or only poorly) be constructed. For an analysis, as well as for suggestions to improve the situation, see [67].
Another dubious aspect of Bi-CGSTAB is that the factor Q_k has only real roots by construction. It is well-known that optimal reduction polynomials for matrices with complex eigenvalues may have complex roots as well. If, for instance, the matrix A is real skew-symmetric, then GCR(1) stagnates forever, whereas a method like GCR(2) (or GMRES(2)), in which we minimize over two combined successive search directions, may lead to convergence, and this is mainly due to the fact that then complex eigenvalue components in the error can be effectively reduced.
This point of view was taken in [38] for the construction of the BiCGSTAB2 method. In the odd-numbered iteration steps the Q-polynomial is expanded by a linear factor, as in Bi-CGSTAB, but in the even-numbered steps this linear factor is discarded, and the Q-polynomial from the previous even-numbered step is expanded by a quadratic $1 - \alpha_k A - \beta_k A^2$. For this construction the information from the odd-numbered step is required. It was anticipated that the introduction of quadratic factors in Q might help to improve convergence for systems with complex eigenvalues, and, indeed, some improvement was observed in practical situations (see also [59]).
However, our presentation suggests a possible weakness in the construction of BiCGSTAB2, namely in the odd-numbered steps the same problems may occur as in Bi-CGSTAB. Since the even-numbered steps rely on the results of the odd-numbered steps, this may equally lead to unnecessary break-downs or poor convergence. In [70] another, and even simpler approach was taken to arrive at the desired even-numbered steps, without the necessity of the construction of the intermediate Bi-CGSTAB-type step in the odd-numbered steps. Hence, in this approach the polynomial Q is constructed straight-away as a product of quadratic factors, without ever constructing a linear factor. As a result the new method Bi-CGSTAB(2) leads only to significant residuals in the even-numbered steps and the odd-numbered steps do not lead necessarily to useful approximations.
In fact, it is shown in [70] that the polynomial Q can also be constructed as the product of ℓ-degree factors, without the construction of the intermediate lower degree factors. The main idea is that ℓ successive Bi-CG steps

are carried out, where for the sake of an A^*-free construction the already available part of Q is expanded by simple powers of A. This means that after the Bi-CG part of the algorithm vectors from the Krylov subspace $s, As, A^2s, ..., A^\ell s$, with $s = P_k(A)Q_{k-\ell}(A)r_0$ are available, and it is then relatively easy to minimize the residual over that particular Krylov subspace. There are variants of this approach in which more stable bases for the Krylov subspaces are generated [69], but for low values of ℓ a standard basis satisfies, together with a minimum norm solution obtained through solving the associated normal equations (which requires the solution of an ℓ by ℓ system. In most cases Bi-CGSTAB(2) will already give nice results for problems where Bi-CGSTAB or BiCGSTAB2 may fail. Note, however, that, in exact arithmetic, if no breakdown situation occurs, BiCGSTAB2 would produce exactly the same results as Bi-CGSTAB(2) at the even-numbered steps.

Bi-CGSTAB(2) can be represented by the following algorithm:

x_0 is an initial guess; $r_0 = b - Ax_0$;
$\hat{r}_0$ is an arbitrary vector, such that $(r, \hat{r}_0) \neq 0$,
e.g., $\hat{r}_0 = r$;
$\rho_0 = 1; u = 0; \alpha = 0; \omega_2 = 1$;
for $i = 0, 2, 4, 6, ...$
$\rho_0 = -\omega_2\rho_0$
even BiCG step: $\rho_1 = (\hat{r}_0, r_i); \beta = \alpha\rho_1/\rho_0; \rho_0 = \rho_1$
$u = r_i - \beta u$;
$v = Au$
$\gamma = (v, \hat{r}_0); \alpha = \rho_0/\gamma$;
$r = r_i - \alpha v$;
$s = Ar$
$x = x_i + \alpha u$;
odd BiCG step: $\rho_1 = (\hat{r}_0, s); \beta = \alpha\rho_1/\rho_0; \rho_0 = \rho_1$
$v = s - \beta v$;
$w = Av$
$\gamma = (w, \hat{r}_0); \alpha = \rho_0/\gamma$;
$u = r - \beta u$
$r = r - \alpha v$
$s = s - \alpha w$
$t = As$
GCR(2)-part: $\omega_1 = (r, s); \mu = (s, s); \nu = (s, t); \tau = (t, t)$;
$\omega_2 = (r, t); \tau = \tau - \nu^2/\mu; \omega_2 = (\omega_2 - \nu\omega_1/\mu)/\tau$;
$\omega_1 = (\omega_1 - \nu\omega_2)/\mu$
$x_{i+2} = x + \omega_1 r + \omega_2 s + \alpha u$
$r_{i+2} = r - \omega_1 s - \omega_2 t$
if x_{i+2} accurate enough then quit
$u = u - \omega_1 v - \omega_2 w$

end

For more general Bi-CGSTAB(ℓ) schemes see [70, 69].
Another advantage of Bi-CGSTAB(2) over BiCGSTAB2 is in its efficiency. The Bi-CGSTAB(2) algorithm requires 14 vector updates, 9 innerproducts and 4 matrix vector products per full cycle. This has to be compared with a combined odd-numbered and even-numbered step in BiCGSTAB2, which requires 22 vector updates, 11 innerproducts, and 4 matrix vector products, and with two steps of Bi-CGSTAB which require 4 matrix vector products, 8 innerproducts and 12 vector updates. The numbers for BiCGSTAB2 are based on an implementation described in [59].
Also with respect to memory requirements, Bi-CGSTAB(2) takes an intermediate position: it requires 2 n-vectors more than Bi-CGSTAB and 2 n-vectors less than BiCGSTAB2.

For distributed memory machines the innerproducts may cause communication overhead problems (see, e.g., [17]). We note that the Bi-CG steps are very similar to conjugate gradient iteration steps, so that we may consider all kind of tricks that have been suggested to reduce the number of synchronization points caused by the 4 innerproducts in the Bi-CG parts. For an overview of these approaches see [7]. If on a specific computer it is possible to overlap communication with communication, then the Bi-CG parts can be rescheduled as to create overlap possibillities: 1. the computation of ρ_1 in the even Bi-CG step may be done just before the update of u at the end of the GCR part.
2. The update of x_{i+2} may be delayed until after the computation of γ in the even Bi-CG step.
3. The computation of ρ_1 for the odd Bi-CG step can be done just before the update for x at the end of the even Bi-CG step.
4. The computation of γ in the odd Bi-CG step has already overlap possibillities with the update for u.
For the GCR(2) part we note that the 5 innerproducts can be taken together, in order to reduce start-up times for their global assembling. This gives the method Bi-CGSTAB(2) a (slight) advantage over Bi-CGSTAB. Furthermore, we note that the updates in the GCR(2) may lead to more efficient code than for BiCGSTAB, since some of them can be combined.

1.5 Parallelism in the preconditioner

In this section we consider a number of possibilities to obtain parallelism in the standard Incomplete Choleski preconditioner [45]. The linear systems are supposed to arise from standard finite difference discretisations of second order pde's over rectangular grids in two or three dimensional space.

All of the discussed iterative methods may be combined with preconditioning. The easiest way to explain this is the following one. Given the linear system $Ax = b$, we construct an operator K that approximates A but that leads to simpler to solve systems. Often K is given in factored form (incomplete decompositions [45]): $K = LU$, and then we apply the iterative scheme for the preconditioned system $K^{-1}Ax = K^{-1}b$. Note that we can avoid the explicit inversion of K, since the operator $K^{-1}A$ is only used in matrix vector operations like

$$g = K^{-1}Ap.$$

The vector q is constructed in two steps:
1. $\widetilde{p} = Ap$
2. Solve q from $Kq = \widetilde{p}$.

Given the LU factorization of K, the second step can be carried out again in two successive operations:
2a. Solve z from $Lz = \widetilde{p}$
2b. Solve q from $Uq = \widetilde{z}$.

The lower triangular systems with L and U lead to back substitutions and this leads to major problems with parallelism (it also leads to a lesser extent to problems on vector computers).

In this section we will briefly overview some possibilities for exploiting or creating parallelism in the preconditioner

In particular, we will discuss parallelization techniques, including re-ordering, series expansion and domain decomposition techniques. Generally, the class of incomplete LU preconditioners does not possess a high degree of parallelism in its original form. Re-ordering and approximations by truncating certain series expansion will increase the parallelism, but usually with a deterioration in convergence rate. Domain decomposition offers a compromise.

There are two general approaches for parallelizing numerical methods:

1. extract maximal parallelism from a method which works well on sequential computer, *without* changing its numerical properties,

2. modify or approximate a good sequential method to increase the parallelism available, thus possibly degrading its numerical properties.

There is a fundamental difficulty when applying the above general principles to incomplete factorization methods. The major parallel bottleneck lies in the backsolves involving the triangular LU factors. The same bottleneck arises in computing the LU factors themselves but this occurs only once in the beginning of the iteration process. Even though these backsolves possess some degree of parallelism which can be exploited, this is often not sufficient to efficiently exploit many parallel architectures, especially massively parallel ones. On the other hand, modifying or approximating the sequential method in order to increase the amount of parallelism invariably leads to slower convergence rates. This should not be too surprising; it is just an instance of the fundamental trade-off between parallelism (which prefers locality) and fast convergence rate (which prefers global dependence) which governs many genuinely globally coupled systems (e.g. elliptic PDEs). The goal is to make the right trade-off for a given architecture - algorithm configuration.

There are three basic methodologies to extract or increase the parallelism in ILU methods: re-ordering, series expansions (including polynomial preconditioners), and domain decomposition. We shall briefly discuss them next.

1.5.1 Re-ordering

The Wavefront Ordering: Assume that the ILU factors have been computed and consider now the task of computing the product $y = L^{-1}v$. The goal is to find an ordering with which the components of y can be computed with maximal parallel efficiency. This can be accomplished by exploiting the dependency graph for the computations. A similar ordering can be used for computing $U^{-1}v$.

It is obvious that if one uses this idea for finite difference operators on a regular d dimensional grid with n grid points in each direction, then we can extract $O(n^{d-1})$ degree of parallelism. Thus, for a fixed number of processors, higher dimensional problems are easier to parallelise. On the other hand, potential degradation in performance can be caused by memory addressing with unequal stride for $d > 2$ and cache problems with the indirect addressing needed to access data on the wavefronts when the grid is stored as a 2D array.

The use of wavefront ordering has been investigated in [61] for the IBM 3090; numerical experiments on the CM2 can be found in [9]. The wavefront (or hyperplane) ordering for regular 3D grids has been described in detail for vector computers in [75]. Recently, in [8] a data mapping is described for the wavefront ordering in 3D useful for distributed memory architectures together with performance results for an 8-processor Cray Y-MP, a 128 processor Intel iPSC/860, and a 32K processor CM-2.

Multi-color Orderings: Since the degree of parallelism for ILU methods are limited in the natural ordering, a popular alternative is to use orderings that are designed to be more parallel. However, it must be emphasized that most of these orderings are *not* equivalent to the natural ordering, in the sense that the ILU factors computed using these are generally different from those generated using the natural ordering. Thus, the goal is to tradeoff the relatively fast and well-understood convergence rate of the natural ordering for orderings with a high degree of parallelism.

An example is the well-known red-black ordering for 5-point stencils in 2D. Because the red points depend only on the black points but not on each other, they can all be updated in parallel. Thus the degree of parallelism is $n^2/2$, a substantial increase from $O(n)$ for the natural ordering. However, since the data dependence are completely local and there is no global sharing of information, the convergence rate is poor [28]. In fact, it can be shown that the condition number of the preconditioned system in the red-black ordering is only about 1/4 that of the *unpreconditioned* system for ILU, MILU and SSOR, with no asymptotic improvement as h tends to zero [42].

One way to strike a better balance between parallelism and fast convergence is to use more colors. In principle, since the different colors are updated sequentially, using more colors decreases the parallelism but increases the global dependence and hence the convergence. The key is to choose the number of colors to match the architecture. For example, In [24] up to 75 colors are used for a 76^2 grid on the NEC SX-3/14 resulting in a 2 Gflops performance, which is much better than that for the wavefront ordering.

DeLong and Ortega [18] exploit the fact that SOR works well in combination with red-black ordering, and they suggest to take a fixed number of SOR iterations as a preconditioner for GMRES or Bi-CGSTAB. This results in a highly parallel method that is quite competitive with ILU preconditioning in terms of matrix vector operations.

Multi-wavefront Orderings: A different approach to increase parallelism is to use several hyperplane wavefronts to sweep through the grid, the idea being that all wavefronts can be updated in parallel. For example, in [75] it was suggested to starte wavefronts from each of the four corners in a 2D rectangular grid, or from the eight corners in a 3D grid. Earlier, a similar idea was proposed in which the grid is divided into equal parts (e.g. halves or quadrants) and each part is ordered in its own natural ordering [48].

1.5.2 Series Expansions

Instead of using an ordering with more parallelism, a quite different approach to increase the parallelism in the naturally ordered ILU method is to replace it by an *approximation* which can be evaluated more efficiently in parallel.

In order to illustrate this, consider the computation of $(I-L)^{-1}v$, which is needed in applying the preconditioner. Here we have assumed without loss of generality that the diagonal entries of the lower triangular factor has been scaled to unity. It can be easily proved that if the spectral radius $\rho(L)$ satisfies $\rho(L) < 1$ and L is n by n strictly lower triangular, then we have the following finite expansion:

Neumann Expansion: $$(I-L)^{-1}v = (I + L + L^2 + ... + L^{n-1})v.$$

Note that $L^n = 0$. Each of the terms on the right-hand-side in the above expansion can be evaluated in parallel efficiently because they only involve

repeated multiplication of v by sparse matrices. The idea is to then truncate the expansion but keeping enough terms so that the convergence rate is not too adversely affected.
In [74] this idea was applied for a truncated Neumann expansion to the diagonal blocks (which correspond to grid lines) in the point ILU factorization in order to increase the degree of vectorization.

Finally, a related method is the class of *polynomial preconditioners* in which A^{-1} is approximated by a low degree polynomial in A, chosen in some optimal manner [26]. In [80] it is shown how GMRES can often be effectively preconditioned by a Chebyshev matrix polynomial, for which the coefficients are obtained from eigenvalue approximations from a limited number of GMRES steps. In particular, the harmonic Ritz values have been employed as useful approximations and a surprisingly simple algorithm is presented in [80] for the computation of the Chebyshev parameters of the Chebyshev polynomial over a piecewise linear contour that encloses the eigenvalue approximations. Using this type of relatively expensive polynomial preconditioners often leads to a significant reduction in GMRES steps and hence the required communication-intensive innerproducts have a lesser degrading effect on the parallel performance of the preconditioned GMRES algorithm on distributed memory machines.

1.5.3 Domain Decomposition

In this general approach, the physical domain or grid is decomposed into a number of overlapping or non-overlapping subdomains on each of which an independent incomplete factorization can be computed and applied in parallel. The main idea is to obtain more parallelism at the subdomain level rather than at the grid point level. Usually, the interfaces or overlapping region between the subdomains must be treated in a special manner. The advantage of this approach is that it is quite general and can be used with different methods used within different subdomains.
Radicati and Robert [60] used an algebraic version of this approach by computing ILU factors within overlapping block diagonals of a given matrix A. When applying the preconditioner to a vector v, the values on the overlapped region is averaged from the two values computed from the two overlapping ILU factors.
The approach of Radicati and Robert has been further perfectioned in [22], who studies the effects of overlap from the point of view of geometric domain decompositioning. He introduces artificial mixed boundary conditions on the internal boundaries of the subdomains. In [22]:Table 5.8 experimental results are shown for a decomposition in 20×20 slightly overlapping subdomains of a 200×400 mesh for a discretized convection-diffusion equation (5-point stencil). When taking ILU preconditioning for each subdomain, it is shown that the complete linear system can be solved by GMRES on a 400-processor distributed memory Parsytec system with an efficiency in the

order of 80% (compared with $\frac{1}{400}$-th of the CPU time of ILU preconditioned GMRES for the unpartitioned system on 1 single processor).

In [72] interface conditions along subdomains are studied and continuity for the solution at the interface is forced up to some degree. It is proposed to include also mixed derivatives in these relations. The involved parameters can be determined locally by mmeans of normal mode analysis, and they are adapted to the discretized problem. It is shown that the resulting domain decomposition method defines a standard iterative method for some splitting $A = M - N$, and the local coupling aimes at minimizing the largest eigenvalues of $I - AM^{-1}$. Of course, this method can be accelerated and impressive results for GMRES acceleration are shown in [72]. Some attention is paid to the case where the solutions for the subdomains are obtained in only modest accuracy per iteration step.

Recently, Washio and Hayami [82] employed a domain decomposition approach for a rectangular grid by which one step of SSOR is done for the interior part of each subdomain. In order to make this domain-decoupled SSOR more resemble the global SSOR, the SSOR iteration matrix for each subdomain is modified by premultiplying it with a matrix $(I - X_L)^{-1}$ and postmultiplying it by $(I - X_U)^{-1}$. The matrices X_L and X_U depend on the couplings between adjacent subdomains. In order to further improve the parallel performance, the inverses are approximated by low-order truncated Neumann series. Experimental results have been shown for a 32-processor NEC-Cenju distributed memory computer.

1.6 References

[1] W. E. Arnoldi. The principle of minimized iteration in the solution of the matrix eigenproblem *Quart. Appl. Math.*, 9:17–29, 1951.

[2] O. Axelsson. Solution of linear systems of equations: iterative methods. In V. A. Barker, editor, *Sparse Matrix Techniques*, Berlin, 1977. Copenhagen 1976, Springer Verlag.

[3] O. Axelsson. Conjugate gradient type methods for unsymmetric and inconsistent systems of equations. *Lin. Alg. and its Appl.*, 29:1–16, 1980.

[4] O. Axelsson and P. S. Vassilevski. A black box generalized conjugate gradient solver with inner iterations and variable-step preconditioning. *SIAM J. Matrix Anal. Appl.*, 12(4):625–644, 1991.

[5] Zhaojun Bai, Dan Hu, and Lothar Reichel. A Newton basis GMRES implementation. Technical Report 91-03, University of Kentucky, 1991.

[6] R. Bank and T. F. Chan. An analysis of the composite step biconjugate gradient method. *Numer. Math.*, 66:295–319, 1993.

[7] R. Barrett, M. Berry, T. Chan, J. Demmel, J. Donato, J. Dongarra, V. Eijkhout, R. Pozo, C. Romine, and H. van der Vorst. *Templates for the Solution of Linear Systems: Building Blocks for Iterative Methods.* SIAM, Philadelphia, PA, 1994.

[8] E. Barszcz, R. Fatoohi, V. Venkatakrishnan, and S. Weeratunga. Triangular systems for CFD applications on parallel architectures. Technical report, NAS Applied Research Branch, NASA Ames Research Center, 1994.

[9] H. Berryman, J. Saltz, W. Gropp, and R. Mirchandaney. Krylov methods preconditioned with incompletely factored matrices on the CM-2. Technical Report 89-54, NASA Langley Research Center, ICASE, Hampton, VA, 1989.

[10] P. N. Brown. A theoretical comparison of the Arnoldi and GMRES algorithms. *SIAM J. Sci. Statist. Comput.*, 12:58–78, 1991.

[11] G. Brussino and V. Sonnad. A comparison of direct and preconditioned iterative techniques for sparse unsymmetric systems of linear equations. *Int. J. for Num. Methods in Eng.*, 28:801–815, 1989.

[12] A. T. Chronopoulos and C. W. Gear. s-Step iterative methods for symmetric linear systems. *J. on Comp. and Appl. Math.*, 25:153–168, 1989.

[13] A. T. Chronopoulos and S. K. Kim. s-Step Orthomin and GMRES implemented on parallel computers. Technical Report 90/43R, UMSI, Minneapolis, 1990.

[14] P. Concus and G. H. Golub. A generalized Conjugate Gradient method for nonsymmetric systems of linear equations. Technical Report STAN-CS-76-535, Stanford University, Stanford, CA, 1976.

[15] P. Concus, G. H. Golub, and D. P. O'Leary. A generalized conjugate gradient method for the numerical solution of elliptic partial differential equations. In J. R. Bunch and D. J. Rose, editors, *Sparse Matrix Computations.* Academic Press, New York, 1976.

[16] G. C. (Lianne) Crone. The conjugate gradient method on the parsytec GCel-3/512. *FGCS*, 11:161–166, 1995.

[17] L. Crone and H. van der Vorst. Communication aspects of the conjugate gradient method on distributed-memory machines. *Supercomputer*, X(6):4–9, 1993.

[18] M. A. DeLong and J. M. Ortega. SOR as a Preconditioner. *Appl. Num. Math.*, 18:431–440, 1995.

[19] J. Demmel, M. Heath, and H. van der Vorst. Parallel numerical linear algebra. In *Acta Numerica 1993*. Cambridge University Press, Cambridge, 1993.

[20] E. de Sturler. A parallel variant of GMRES(m). In R. Miller, editor, *Proc. of the fifth Int.Symp. on Numer. Methods in Eng.*, 1991.

[21] E. de Sturler and D. R. Fokkema. Nested Krylov methods and preserving the orthogonality. In N. Duane Melson, T.A. Manteuffel, and S.F. McCormick, editors, *Sixth Copper Mountain Conference on Multigrid Methods*, volume Part 1 of *NASA Conference Publication 3324*, pages 111–126. NASA, 1993.

[22] E. de Sturler. *Iterative methods on distributed memory computers.* PhD thesis, Delft University of Technology, Delft, the Netherlands, 1994.

[23] E. de Sturler and H.A. van der Vorst. Reducing the effect of global communication in GMRES(m) and CG on parallel distributed memory computers. *J. Appl. Num. Math.*, 1995.

[24] S. Doi and A. Hoshi. Large numbered multicolor MILU preconditioning on SX-3/14. *Int'l J. Computer Math.*, 44:143–152, 1992.

[25] J. J. Dongarra, I. S. Duff, D. C. Sorensen, and H. A. van der Vorst. *Solving Linear Systems on Vector and Shared Memory Computers.* SIAM, Philadelphia, PA, 1991.

[26] P. F. Dubois, A. Greenbaum, and G. H. Rodrigue. Approximating the inverse of a matrix for use in iterative algorithms on vector processors. *Computing*, 22:257–268, 1979.

[27] I. S. Duff, A. M. Erisman, and J.K.Reid. *Direct methods for sparse matrices.* Oxford University Press, London, 1986.

[28] I. S. Duff and G. A. Meurant. The effect of ordering on preconditioned conjugate gradient. *BIT*, 29:635–657, 1989.

[29] H. C. Elman. *Iterative methods for large sparse nonsymmetric systems of linear equations.* PhD thesis, Yale University, New Haven, CT, 1982.

[30] V. Faber and T. Manteuffel. Necessary and sufficient conditions for the existence of a conjugate gradient method. *SIAM J. Numer. Anal.*, 21:315–339, 1984.

[31] R. Fletcher. *Conjugate gradient methods for indefinite systems*, volume 506 of *Lecture Notes Math.*, pages 73–89. Springer-Verlag, Berlin–Heidelberg–New York, 1976.

[32] R. W. Freund. A transpose-free quasi-minimum residual algorithm for non-Hermitian linear systems. *SIAM J. Sci. Comput.*, 14:470–482, 1993.

[33] R. W. Freund, M. H. Gutknecht, and N. M. Nachtigal. An implementation of the look-ahead Lanczos algorithm for non-Hermitian matrices. *SIAM J. Sci. Comput.*, 14:137–158, 1993.

[34] R. W. Freund and N. M. Nachtigal. An implementation of the look-ahead Lanczos algorithm for non-Hermitian matrices, part 2. Technical Report 90.46, RIACS, NASA Ames Research Center, 1990.

[35] R. W. Freund and N. M. Nachtigal. QMR: a quasi-minimal residual method for non-Hermitian linear systems. *Num. Math.*, 60:315–339, 1991.

[36] G. H. Golub and D.P. O'Leary. Some history of the conjugate gradient and lanczos algorithms: 1948-1976. *SIAM Review*, 31:50–102, 1989.

[37] G. H. Golub and C. F. van Loan. *Matrix Computations.* The Johns Hopkins University Press, Baltimore, 1989.

[38] M. H. Gutknecht. Variants of BICGSTAB for matrices with complex spectrum. *SIAM J. Sci. Comput.*, 14:1020–1033, 1993.

[39] W. Hackbusch. *Iterative Lösung großer schwachbesetzter Gleichungssysteme.* Teubner, Stuttgart, 1991.

[40] M. R. Hestenes and E. Stiefel. Methods of conjugate gradients for solving linear systems. *J. Res. Natl. Bur. Stand.*, 49:409–436, 1954.

[41] K. C. Jea and D. M. Young. Generalized conjugate-gradient acceleration of nonsym- metrizable iterative methods. *Lin. Algebra Appl.*, 34:159–194, 1980.

[42] J.C.C. Kuo and T.F. Chan. Two-color fourier analysis of iterative algorithms for elliptic problems with red/black ordering. *SIAM J. Sci. Stat. Comp.*, 11:767–793, 1990.

[43] C. Lanczos. An iteration method for the solution of the eigenvalue problem of linear differential and integral operators. *J. Res. Natl. Bur. Stand*, 45:225–280, 1950.

[44] C. Lanczos. Solution of systems of linear equations by minimized iterations. *J. Res. Natl. Bur. Stand*, 49:33–53, 1952.

[45] J. A. Meijerink and H. A. van der Vorst. An iterative solution method for linear systems of which the coefficient matrix is a symmetric M-matrix. *Math.Comp.*, 31:148–162, 1977.

[46] G. Meurant. The block preconditioned conjugate gradient method on vector computers. *BIT*, 24:623–633, 1984.

[47] G. Meurant. Numerical experiments for the preconditioned conjugate gradient method on the CRAY X-MP/2. Technical Report LBL-18023, University of California, Berkeley, CA, 1984.

[48] G. Meurant. Domain decomposition methods for partial differential equations on parallel computers. *Int. J. Supercomputing Appls.*, 2:5–12, 1988.

[49] N. M. Nachtigal, S. C. Reddy, and L. N. Trefethen. How fast are nonsymmetric matrix iterations? *SIAM J. Matrix Anal. Appl.*, 13:778–795, 1992.

[50] A. Neumaier. Oral presentation at the Oberwolfach meeting: Numerical Linear Algebra, Oberwolfach, 1994.

[51] J. M. Ortega. *Introduction to Parallel and Vector Solution of Linear Systems.* Plenum Press, New York and London, 1988.

[52] C. C. Paige. Computational variants of the Lanczos method for the eigenproblem. *J. Inst. Math. Appl.*, 10:373–381, 1972.

[53] C. C. Paige, B. N. Parlett, and H. A. van der Vorst. Approximate solutions and eigenvalue bounds from Krylov subspaces. *Num. Lin. Alg with Appl.*, 2(2):115–134, 1995.

[54] C. C. Paige and M. A. Saunders. Solution of sparse indefinite systems of linear equations. *SIAM J. Numer. Anal.*, 12:617–629, 1975.

[55] C. C. Paige and M. A. Saunders. LSQR: An algorithm for sparse linear equations and sparse least squares. *ACM Trans. Math. Soft.*, 8:43–71, 1982.

[56] B. N. Parlett, D. R. Taylor, and Z. A. Liu. A look-ahead Lanczos algorithm for unsymmetric matrices. *Math. Comp.*, 44:105–124, 1985.

[57] Beresford N. Parlett. *The Symmetric Eigenvalue Problem.* Prentice-Hall, Englewood Cliffs, N.J., 1980.

[58] C. Pommerell and W. Fichtner. PILS: An iterative linear solver package for ill-conditioned systems. In *Supercomputing '91*, pages 588–599, Los Alamitos, CA., 1991. IEEE Computer Society.

[59] Claude Pommerell. *Solution of large unsymmetric systems of linear equations.* PhD thesis, Swiss Federal Institute of Technology, Zürich, 1992.

[60] G. Radicati di Brozolo and Y. Robert. Parallel conjugate gradient-like algorithms for solving sparse non-symmetric systems on a vector multiprocessor. *Parallel Computing*, 11:223–239, 1989.

[61] G. Radicati di Brozolo and M. Vitaletti. Sparse matrix-vector product and storage representations on the IBM 3090 with Vector Facility. Technical Report 513-4098, IBM-ECSEC, Rome, July 1986.

[62] Y. Saad. Practical use of polynomial preconditionings for the conjugate gradient method. *SIAM J. Sci. Stat. Comput.*, 6:865–881, 1985.

[63] Y. Saad. Krylov subspace methods on supercomputers. Technical report, RIACS, Moffett Field, CA, September 1988.

[64] Y. Saad. A flexible inner-outer preconditioned GMRES algorithm. *SIAM J. Sci. Comput.*, 14:461–469, 1993.

[65] Y. Saad and M. H. Schultz. Conjugate Gradient-like algorithms for solving nonsymmetric linear systems. *Math. of Comp.*, 44:417–424, 1985.

[66] Y. Saad and M. H. Schultz. GMRES: a generalized minimal residual algorithm for solving nonsymmetric linear systems. *SIAM J. Sci. Statist. Comput.*, 7:856–869, 1986.

[67] G. L. G. Sleijpen and H.A. van der Vorst. Maintaining convergence properties of BICGSTAB methods in finite precision arithmetic. *Numerical Algorithms*, 10:203–223, 1995.

[68] G. L. G. Sleijpen and H.A. van der Vorst. Reliable updated residuals in hybrid Bi-CG methods. Preprint 886, Dept. Math., University Utrecht, 1994 (to appear in Computing).

[69] G. L. G. Sleijpen, H.A. Van der Vorst, and D. R. Fokkema. Bi-CGSTAB(ℓ) and other hybrid bi-cg methods. *Numerical Algorithms*, 7:75–109, 1994.

[70] G. L. G. Sleijpen and D. R. Fokkema. BICGSTAB(ℓ) for linear equations involving unsymmetric matrices with complex spectrum. *ETNA*, 1:11–32, 1993.

[71] P. Sonneveld. CGS: a fast Lanczos-type solver for nonsymmetric linear systems. *SIAM J. Sci. Statist. Comput.*, 10:36–52, 1989.

[72] K.H. Tan. *Local coupling in domain decomposition.* PhD thesis, Utrecht University, Utrecht, the Netherlands, 1995.

[73] A. van der Sluis and H. A. van der Vorst. The rate of convergence of conjugate gradients. *Numer. Math.*, 48:543–560, 1986.

[74] H. A. van der Vorst. A vectorizable variant of some ICCG methods. *SIAM J. Sci. Stat. Comput.*, 3:86–92, 1982.

[75] H. A. van der Vorst. High performance preconditioning. *SIAM J. Sci. Statist. Comput.*, 10:1174–1185, 1989.

[76] H. A. van der Vorst. The convergence behaviour of preconditioned CG and CG-S in the presence of rounding errors. In O. Axelsson and L. Yu. Kolotilina, editors, *Preconditioned Conjugate Gradient Methods*, Berlin, 1990. Nijmegen 1989, Springer Verlag. Lecture Notes in Mathematics 1457.

[77] H. A. van der Vorst. Bi-CGSTAB: A fast and smoothly converging variant of Bi-CG for the solution of non-symmetric linear systems. *SIAM J. Sci. Statist. Comput.*, 13:631–644, 1992.

[78] H. A. van der Vorst and C. Vuik. The superlinear convergence behaviour of GMRES. *JCAM*, 48:327–341, 1993.

[79] H. A. van der Vorst and C. Vuik. GMRESR: A family of nested GMRES methods. *Num. Lin. Alg. with Appl.*, 1:369–386, 1994.

[80] M.B. van Gijzen. *Iterative solution methods for linear equations in finite element computations.* PhD thesis, Delft University of Technology, Delft, the Netherlands, 1994.

[81] P. K. W. Vinsome. ORTOMIN: an iterative method for solving sparse sets of simultaneous linear equations. In *Proc.Fourth Symposium on Reservoir Simulation*, pages 149–159. Society of Petroleum Engineers of AIME, 1976.

[82] T. Washio and K. Hayami. Parallel block preconditioning based on SSOR and MILU. *Numer. Lin. Alg. with Applic.*, 1:533–553, 1994.

[83] O. Widlund. A Lanczos method for a class of nonsymmetric systems of linear equations. *SIAM J. Numer. Anal.*, 15:801–812, 1978.

2

Matrices, Moments and Quadrature

Gene H. Golub*

In this chapter we will study methods to obtain bounds or approximations of elements of a matrix $F(A)$, where A is a symmetric positive definite matrix and F is a smooth function. This kind of question can be formulated as a problem of finding an estimate or upper and lower bounds on $u^T F(A)u$ where u is a given real vector. The numerical methods are based on the use of Gauss-type quadrature rules [1, 3] and the Lanczos algorithm [4] for diagonal elements and the block Lanczos for the non-diagonal elements [6]. We will briefly investigate some theoretical results on the behavior of these methods based on results for orthogonal polynomials as well as analytical bounds and numerical experiments on a set of matrices for several functions F. This chapter is a summary of [6].

2.1 Introduction to the Problem

Let A be a real symmetric positive definite matrix of order n. We want to find upper and lower bounds (or approximations, if bounds are not available) for the entries of a function of the matrix. This problem leads us to determine U and L such that

$$L \leq u^T F(A)u \leq U, \tag{2.1}$$

where u is given and F is some smooth (possibly C^∞) function on a given interval of the real line.

This kind of problem and some of the techniques we will introduce here have been considered (without any mathematical justification) in solid state physics, particularly to computing elements of the resolvant of a Hamiltonian modeling the interaction of atoms in a solid, see [8], [10], [11].

Example 1:
Determine upper and lower bounds on $\{A^{-1}\}_{jj}$ or find its approximation.

---

*Computer Science Department, Stanford University, Stanford CA 94305, USA. golub@sccm.stanford.edu

In order to do this, we take $F(\lambda) = \lambda^{-1}$ and $u^T = e_j^T$, here e_j is the jth unit vector.

The analytic bounds for this problem using different techniques have been obtained in [12].

Example 2:
We consider error estimation of a linear system:

$$Ax = b.$$

Suppose ξ is a given approximation of the solution x and e is the error vector, i.e.

$$x = \xi + e.$$

The residual vector r can be expressed as

$$r = b - A\xi = Ae.$$

Therefore, we have the error norm

$$||e||_2^2 = r^T A^{-2} r.$$

In this case, $F(\lambda) = \lambda^{-2}$ and $u = r$.

Example 3:
Consider the following optimization problem:

$$\begin{array}{ll} minimize & x^T Ax - 2b^T x \\ subject \quad to & ||x||_2 = \alpha \end{array}$$

where $\alpha \le ||A^{-1}b||_2$ is a given constant. By using a Lagrangian multiplier this constraint optimization problem can be changed to an unconstrained optimization problem:

$$minimize \quad \phi(x;\mu) = x^T Ax - 2b^T x + \mu(x^T x - \alpha^2)$$

where $\mu \ge 0$ is the Lagrangian multiplier.

The solution satisfies the following equations:

$$\nabla\phi(x;\mu) = 0.$$

This implies that

$$(A + \mu I)x = b,$$

where μ is an undetermined parameter. In order to satisfy the constraint $||x||^2 = \alpha^2$, μ must be selected such that it satisfies:

$$b^T (A + \mu I)^{-2} b = \alpha^2.$$

We note that this μ can be viewed as the regularization parameter when an ill-posed linear system $Ax = b$ is solved. In this case, we take $u = b$ and $F(\lambda) = (\lambda - \mu)^{-2}$.

Since $A = A^T$ and A is positive definite, we can write A as

$$A = Q\Lambda Q^T,$$

where Q is the orthonormal matrix whose columns are the normalized eigenvectors of A and Λ is a diagonal matrix whose diagonal elements are the eigenvalues λ_i which we order as

$$0 \leq a \leq \lambda_1 \leq \lambda_2 \leq \cdots \leq \lambda_n \leq b.$$

Let $\beta = Q^T b$, then we have

$$u^T F(A) u = \beta^T (\Lambda + \mu I)^{-2} \beta = \alpha^2.$$

The second equation shows that $\mu = \mu^*$ can be chosen such that it satisfies

$$F(A; \mu^*) = \sum_{i=1}^{n} \frac{\beta_i^2}{(\lambda_i + \mu^*)^2} = \alpha^2.$$

How can we determine μ^* without computing the eigenvalues and eigenvectors of A? A numerical method will be introduced in Section 3.

2.2 Bounds on Matrix Functions as Integrals

In this section, we convert finding the upper and lower bounds on matrix functions to estimating the bounds of a Riemann-Stieltjes integral. Then we describe some basic quadrature rules to obtain the bounds or approximation.

As in Example 3, since A is positive definite, it can be decomposed into $A = Q\Lambda Q^T$. By definition, we have

$$\begin{aligned} u^T F(A) u &= u^T F(Q\Lambda Q^T) u \\ &= u^T Q F(\Lambda) Q^T u \\ &\equiv w^T F(\Lambda) w, \end{aligned}$$

where $w = Q^T u$. More precisely, we have

$$u^T F(A) u = \sum_{i=1}^{n} F(\lambda_i) w_i^2.$$

This sum can be considered as a Riemann-Stieltjes integral

$$u^T F(A) u = \int F(\lambda) dw(\lambda), \tag{2.2}$$

where the measure $w(\lambda)$ is piecewise constant and is defined by

$$w(\lambda) = \begin{cases} 0 & \text{if } \lambda \le \lambda_1, \\ \sum_{j=1}^{i} w_j^2 & \text{if } \lambda_i \le \lambda \le \lambda_{i+1}, \\ \sum_{j=1}^{n} w_j^2 & \text{if } \lambda_n \le \lambda. \end{cases}$$

2.2.1 *Construction of the Orthogonal Polynomials*

In order to use quadrature rules to compute the integrals, we need to construct a sequence of orthonormal polynomials with respect to the measure $w(\lambda)$. In this section, we consider the problem of computing such sequence of polynomials. A very natural and elegant way to do this is use the classical Lanczos algorithms.

Let $x_{-1} = 0$ and x_0 be given such that $||x_0|| = 1$. The Lanczos algorithm is defined by the following relations,

$$\eta_j x_j = r_j = (A - \xi_j I)x_{j-1} - \eta_{j-1}x_{j-2}, \quad j = 1, \ldots$$

$$\xi_j = x_{j-1}^T A x_{j-1},$$

$$\eta_j = ||r_j||.$$

By induction, it is easy to verify the following facts,

$$x_{j+1}^T x_j = 0,$$

and

$$x_{j+1}^T x_{j-1} = 0.$$

These relations imply that

$$x_{j+1}^T x_l = 0, \quad \text{for } l \le j.$$

It also can be proved that the sequence $\{x_i\}_{i=0}^{j}$ is an orthonormal basis of the Krylov space

$$\text{span } \{x_0, Ax_0, \ldots, A^j x_0\}.$$

Using two sequences $\{\xi_j\}_{j=1}^{k}$ and $\{\eta_j\}_{j=0}^{k}$ generated in Lanczos process, and if $\int_a^b dw(\lambda) = 1$ (otherwise, it can be rescaled to make it true), we construct a series of polynomials with the following three term recurrence relationship:

$$\eta_j p_j(\lambda) = (\lambda - \xi_j)p_{j-1}(\lambda) - \eta_{j-1}p_{j-2}(\lambda), \quad j = 1, 2, \cdots, k, \tag{2.3}$$

with

$$p_{-1}(\lambda) \equiv 0, \quad p_0(\lambda) \equiv 1.$$

Theorem 1 *The vector x_j in Lanczos process is given by*

$$x_j = p_j(A)x_0,$$

where p_j is a polynomial of degree j defined by the relationship (2.3).

Proof: The first order polynomial in A is

$$\eta_1 x_1 = (A - \xi_1 I)x_0.$$

Therefore, by induction, the result is easily obtained.

Theorem 2 *If $x_0 = u$, we have*

$$x_k^T x_l = \int_a^b p_k(\lambda)p_l(\lambda)dw(\lambda).$$

Proof: Let $x_0 = u$, we have

$$\begin{aligned} x_k^T x_l &= (p_k(A)u, p_l(A)u) \\ &= u^T p_k(A)^T p_l(A)u \\ &= u^T Q p_k(\Lambda)Q^T Q p_l(\Lambda)Q^T u \\ &= u^T Q p_k(\Lambda)p_l(\Lambda)Q^T u \\ &= \sum_{j=1}^n p_k(\lambda_j)p_l(\lambda_j)w_j^2 \\ &= \int_a^b p_k(\lambda)p_l(\lambda)dw(\lambda), \end{aligned}$$

where $w = Q^T u$. Therefore, the p_j's are the orthonormal polynomials related to the measure $w(\lambda)$, i.e. p_j's satisfy

$$\int_a^b p_i(\lambda)p_j(\lambda)dw(\lambda) = \begin{cases} 1 & \text{if } i = j, \\ 0 & \text{otherwise.} \end{cases}$$

Obviously, p_k is of exact degree k. Moreover, it can be proved that the roots of p_k are distinct, real and lie in the interval $[a, b]$.

2.2.2 Quadrature Rules

A way to obtain bounds for the Stieltjes integrals is to use Gauss, or Gauss-Radau quadrature formulas. The general formula is

$$\int_a^b F(\lambda)dw(\lambda) = I[F] + R[F], \tag{2.4}$$

where

$$I[F] = \sum_{i=1}^k C_i F(t_i) + \sum_{j=1}^m B_j F(z_j) \tag{2.5}$$

is the quadrature rule and

$$R[F] = \frac{F^{(2k+m)}(\eta)}{(2k+m)!} \int_a^b \prod_{j-1}^{m} (\lambda - z_j) \left[\prod_{i=1}^{k} (\lambda - t_i) \right]^2 dw(\lambda), \quad a \le \eta \le b, \tag{2.6}$$

is the error term.

In this expression the weights $\{C_i\}_{i=1}^k$, $\{B_j\}_{j=1}^m$ and the nodes $\{t_i\}_{i=1}^k$ are unknowns and the nodes $\{z_j\}_{j=1}^m$ are prescribed. If $m = 0$, (2.5) leads to the Gauss rule with no prescribed nodes. If $m = 1$ and $z_1 = a$ or $z_1 = b$, (2.5) is the Gauss-Radau formula.

Let us recall briefly how the nodes and weights are obtained in these quadrature rules.

In the previous section, we construct a set of orthonormal polynomials with respect to the measure $w(\lambda)$ by Lanczos algorithm and the three term recurrence (2.3). The recurrence relationship can be written in matrix form as

$$\lambda p(\lambda) = J_k p(\lambda) + \eta_k p_k(\lambda) e_k,$$

where

$$p(\lambda)^T = [p_0(\lambda) \quad p_1(\lambda) \quad \cdots \quad p_{k-1}(\lambda)],$$
$$e_k^T = (0 \quad 0 \quad \cdots \quad 0 \quad 1),$$

and

$$J_k = \begin{pmatrix} \xi_1 & \eta_1 & & & \\ \eta_1 & \xi_2 & \eta_2 & & \\ & \eta_2 & \ddots & \ddots & \\ & & \ddots & \ddots & \eta_{k-1} \\ & & & \eta_{k-1} & \xi_k \end{pmatrix}. \tag{2.7}$$

The eigenvalues of J_k (which are the zeroes of p_k) are the nodes of the Gauss quadrature rule (i.e. $m = 0$). The weights are the squares of the first elements of the normalized eigenvectors of J_k, i.e. $t_i, i = 1, 2, \ldots, k$ satisfy

$$p_k(t_i) = 0.$$

This is equivalent to

$$J_k v_j = t_j v_j, \quad j = 1, 2, \ldots, k,$$

where $\{v_j\}_1^k$ are the normalized eigenvectors of J_k and

$$C_j = v_{1j}^2, \quad j = 1, 2, \ldots, k$$

Theorem 3 *Suppose $F^{(2n)}(\xi) \ge 0$, for all $n \ge 0$, and $\xi \in [a, b]$. Let*

$$I[F] = \sum_{j=1}^{k} C_j F(t_j).$$

Then, for all k, there exists $\eta \in [a, b]$ *such that*

$$I[F] \le \int_a^b F(\lambda)dw(\lambda),$$

and

$$\int_a^b F(\lambda)dw(\lambda) - I[F] = \frac{F^{(2k)}(\eta)}{(2k)!}.$$

Proof: For a complete proof, see [13]. The main idea of the proof is to use a Hermite interpolatory polynomial of degree $2k-1$ on the k nodes which allows us to express the remainder as an integral of the difference between the function and its interpolatory polynomial and to apply the mean value theorem (as the measure is positive and increasing). As we know the sign of the remainder, we easily obtain the bounds.

To obtain the Gauss-Radau rule ($m = 1$), we should extend the matrix J_k to $\bar{J}_k$ in such a way that it has one prescribed eigenvalue $z_1 = a$ or $z_1 = b$.

Assume $z_1 = a$, we wish to construct p_{k+1} such that $p_{k+1}(a) = 0$. From the recurrence relation, we have

$$0 = \eta_{k+1}p_{k+1}(a) = (a - \xi_{k+1})p_k(a) - \eta_k p_{k-1}(a).$$

This gives

$$\xi_{k+1} = a - \eta_k \frac{p_{k-1}(a)}{p_k(a)}.$$

In matrix form, it is

$$(J_k - aI)p(a) = -\eta_k p_k(a)e_k.$$

Let us denote $\delta(a) = [\delta_1(a), \cdots, \delta_k(a)]^T$ with

$$\delta_l(a) = -\eta_k \frac{p_{l-1}(a)}{p_k(a)} \quad l = 1, \ldots, k.$$

This gives $\xi_{k+1} = a + \delta_k(a)$ and

$$(J_k - aI)\delta(a) = \eta_k^2 e_k. \tag{2.8}$$

From these relations we have the solution of the problem as: 1) we generate η_k by the Lanczos process, 2) we solve the tridiagonal system for $\delta(a)$ and 3) we compute ξ_{k+1}. Then the tridiagonal matrix $\bar{J}_{k+1}$ defined as

$$\bar{J}_{k+1} = \begin{pmatrix} J_k & \eta_k e_k \\ \eta_k e_k^T & \xi_{k+1} \end{pmatrix},$$

will have a as an eigenvalue and gives the weights and the nodes of the corresponding Gauss-Radau quadrature rule. Therefore, the recipe is to compute as for the Gauss quadrature rule and then to modify the last step to obtain the prescribed node.

Theorem 4 *In Gauss-Radau quadrature rule, if $F^{(2k+1)}(\eta) \leq 0$ and $z_1 = a$, $I[F]$ is defined as*

$$IU[F] = \sum_{j=1}^{k} C_j F(t_j) + B_1 F(a);$$

or if $F^{(2k+1)}(\eta) \leq 0$ and $z_1 = b$, define $I[F]$ as

$$IL[F] = \sum_{j=1}^{k} C_j F(t_j) + B_1 F(b).$$

Then

$$IL[F] \leq \int_a^b F(\lambda) dw(\lambda) \leq IU[F].$$

and

$$\int_a^b F(\lambda) dw(\lambda) - IU[F] = \frac{F^{(2k+1)}(\eta)}{(2k+1)!} \int_a^b (\lambda - a) \left[\prod_{j=1}^{k} (\lambda - t_j) \right]^2 dw(\lambda),$$

$$\int_a^b F(\lambda) dw(\lambda) - IL[F] = \frac{F^{(2k+1)}(\eta)}{(2k+1)!} \int_a^b (\lambda - b) \left[\prod_{j=1}^{k} (\lambda - t_j) \right]^2 dw(\lambda).$$

Proof: With our hypothesis the sign of the remainder is easily obtained. It is negative if we choose $z_1 = a$, positive if we choose $z_1 = b$.

We take F as all the monomials $\lambda^r, r = 1, \cdots, 2k + m - 1$. Let μ_r the moment, be defined as

$$\mu_r = \int_a^b \lambda^r dw(\lambda).$$

According to the definition and the quadrature rules like Gauss quadrature being exact for polynomials of degree $2k + m - 1$, we have

$$I[\lambda^r] = \mu_r = \sum_{i=1}^{k} A_i t_i^r + \sum_{j=1}^{m} B_j z_j^r.$$

We can write an arbitrary polynomial p_j with degree j as

$$p_j(\lambda) = \sum_{k=0}^{j} p_k^{(j)} \lambda_k.$$

Then, we have

$$\int_a^b p_j(\lambda)dw(\lambda) = \sum_{k=0}^{j} p_k^{(j)} \int_a^b \lambda^k dw(\lambda) = \sum_{k=0}^{j} p_k^{(j)} \mu_k,$$

and more generally

$$\int_a^b p_j(\lambda)\lambda^q dw(\lambda) = \sum_{k=0}^{j} p_k^{(j)} \mu_k.$$

2.2.3 *Evaluation*

Once the tridiagonal matrix $\bar{J}_k$ is formed, the quadrature rules can be evaluated as follows:

Decompose $\bar{J}_k$ to diagonal form, i.e.:

$$\bar{J}_{k+1} = VTV^T$$

where T is a diagonal matrix, and V is a orthonormal matrix. The first component of V is $V^T e_1$, so we have

$$\begin{aligned} I[F] &= \textstyle\sum_{i=0}^{k} v_{1i}^2 F(t_i) \\ &= e_1^T V F(T) V^T e_1 \\ &= e_1^T F(VTV^T) e_1 \\ &= e_1^T F(\bar{J}_{k+1}) e_1. \end{aligned} \tag{2.9}$$

So computing the Riemann-Stieltjes integrals is reduced to computing the first entry of matrix $F(\bar{J}_{k+1})$.

Example 1: We consider obtaining numerical bounds and approximations for the entries of the inverse of a given matrix. As shown before, they are the same as finding upper and lower bounds on $e_j^T A^{-1} e_j$.

We start Lanczos process with $u = e_j$, and the first step of the Lanczos algorithm gives us

$$\xi_1 = e_j^T A e_j = a_{jj},$$

$$\eta_1 x_1 = r_1 = (A - \xi_1 I)e_j.$$

Let s_j be defined by

$$s_j^2 = \sum_{i \neq j} a_{ij}^2.$$

After one more Lanczos iteration, we note that

$$F^{(2k+1)}(\lambda) = -(2k+1)!\lambda^{-(2k+2)}, \quad 0 \leq a \leq \lambda \leq b.$$

Apply Theorems 3 , 4 and by computing Gauss-Radau rule with $z_1 = a$ and $z_1 = b$, we get the upper bounds and lower bounds on the diagonal elements of the inverse matrix respectively,

$$\frac{a_{jj} - b + \frac{s_j^2}{b}}{a_{jj}^2 - a_{jj}b + s_j^2} \leq (A^{-1})_{jj} \leq \frac{a_{jj} - a + \frac{s_j^2}{a}}{a_{jj}^2 - a_{jj}a + s_j^2}.$$

These results can be generalized for the off-diagonal elements $(A^{-1})_{i,k}$. The details can be found in [6].

In the computations using the Lanczos algorithm for the Gauss and Gauss-Radau rules, we need to compute the (1, 1) element of the inverse of a tridiagonal matrix. This may be done in many different ways, see for instance [9] and [6].

Example 2: Find error bounds of the approximation to the linear system. From Section 1 we know that it is equivalent to find the upper and lower bounds on

$$\mu_{-2} = r^T A^{-2} r.$$

Starting the Lanczos process with initial vector r and like Example 1, use Gauss-Radau rule with $z_1 = a$ and $z_1 = b$ to compute

$$\bar{J}_k = \begin{pmatrix} \xi_1 & \eta_1 & & 0 \\ \eta_1 & \ddots & \ddots & \\ & \ddots & \xi_k & \eta_k \\ 0 & & \eta_k & \bar{\xi}_{k+1} \end{pmatrix}$$

respectively. Then $e_1^T \bar{J}_k^{-2} e_1$ will yield upper and lower bounds on μ_{-2} depending on $\bar{\xi}_{k+1}$. By doing so, we reduce the error estimation to computing the inverse of a tridiagonal matrix.

Example 3: Determine μ^* such that it satisfies

$$b^T (A + \mu^* I)^{-2} b = \alpha^2.$$

Like Example 2, we begin Lanczos process with $u = b$. Then we construct the tridiagonal matrix $\bar{J}_{k+1}$. According to (2.9), we need to solve the following equation:

$$e_1^T (\bar{J}_{k+1} + \mu I)^{-2} e_1 = \alpha^2.$$

Solving this equation can be accomplished in various ways. For instance, one can use Newton's method, or it is better to use the method of undetermined coefficient to approximate $e_1^T (\bar{J}_{k+1} + \mu I)^{-2} e_1$ by $\left(\frac{a}{b+c\mu}\right)^2$ where

a, b, c are undetermined coefficients. Then we only need to find μ^* to satisfy

$$\left(\frac{a}{b + c\mu^*}\right)^2 = \alpha^2.$$

We can also use direction generated by Lanczos process for updating solution [6]. Details of the computation are given in [7].

2.3 Other Extensions

We can consider

$$W^T F(A) W,$$

where W is an $n \times p$ matrix.

The problem is to determine two $p \times p$ matrix L and U, such that the following inequality is satisfied:

$$L \leq W^T F(A) W \leq U$$

Like the 1-dimension case, we can define the Riemann-Stieltjes integral, construct the orthogonal polynomials by using block Lanczos and non-symmetric Lanczos methods and then obtain a numerical method for computing bounds that we need, see [6].

For the more general case, we consider

$$u^T F(A) v,$$

where u and v are given vectors and $u \neq v$. It can be easily converted into the symmetric case by using the identity

$$u^T F(A) v = \frac{1}{2}[u^T F(A) u + v^T F(A) v - (u - v)^T F(A)(u - v)],$$

see [5]. Or the Riemann-Stieltjes integral can be defined like,

$$u^T F(A) v = \sum_{i=1}^{n} F(\lambda_i) \alpha_i \beta_i = \int_a^b F(\lambda) dw(\lambda), \tag{2.10}$$

where $\alpha = Q^T u$, $\beta = Q^T v$ and the measure $w(\lambda)$ is piecewise constant defined by

$$w(\lambda) = \begin{cases} 0 & \text{if } \lambda \leq a = \lambda_1, \\ \sum_{j=1}^{i} \alpha_j \beta_j & \text{if } \lambda_i \leq \lambda \leq \lambda_{i+1}, \\ \sum_{j=1}^{n} \alpha_j \beta_j & \text{if } b = \lambda_n \leq \lambda. \end{cases}$$

Here, in order to make $w(\lambda)$ a measure, we should choose u and v such that $\alpha_i \beta_i \geq 0$. Then by using Lanczos algorithms, the orthogonal polynomials can be constructed and we can obtain the numerical methods for computing bounds or approximations, see[6].

2.4 Numerical Examples

Finally, let us look at a numerical example [6]. When one uses domain decomposition methods for matrices arising from the finite difference approximation of partial differential equations in a rectangle, it is known that

$$Z = \left(A + \frac{1}{4}A^2\right)^{\frac{1}{2}},$$

where A is the matrix of the one dimensional Laplacian,

$$A = \begin{pmatrix} 2 & -1 & & & \\ -1 & 2 & -1 & & \\ & \ddots & \ddots & \ddots & \\ & & -1 & 2 & -1 \\ & & & -1 & 2 \end{pmatrix},$$

is a good preconditioner for the Schur complement matrix. It is interesting to see if we can estimate some elements of the matrix Z to generate a Toeplitz tridiagonal approximation to A. We choose an example of dimension 125. We estimate the $(50, 50)$ element whose exact value is 1.6367 with the Gauss-Radau rule. We obtain the following results.

Nit	2	3	4	5	10	20
LB	1.6014	1.6196	1.6269	1.6305	1.6355	1.6365
UB	1.6569	1.6471	1.6430	1.6409	1.6378	1.6369

where Nit represents number of Lanczos iteration, LB is the lower bound and UB is the upper bound. It shows that very good approximations are obtained in a few iterations.

Now let us consider the problem of estimating one of the elements of the error vector of the classical five point approximation to Poisson's equation over a rectangle with $u = 0$ on the boundary. Thus we are solving $Ax = 0$. We choose as our initial vector $\xi = \frac{\tau}{\|\tau\|_2}$, where τ is a random vector which is $\eta(0, I)$. We ahve a system of $10,000$ equations, and we wish to estimate the element e_j where $j = 5,000$. The following tabular show the numerical results by using our methods.

Lanczos iteration	Gauss	Gauss-Radau upper	Gauss-Radau lower
10	0.00896165	17240.7595	-17456.0484
25	0.00931480	2663.37220	-2642.63124
50	0.00916682	654.197104	-653.958756
100	0.00912767	164.063323	-164.331294
200	0.00915408	0.01048046	0.00781305
300	0.00915408	0.00915410	0.00915406

We see that the estimate of the error converges quite fast but the bounds converge more slowly.

2.5 References

[1] P. Davis and P. Rabinowitz, *Methods of numerical integration*, 2nd Ed. (1984) Academic Press, London.

[2] G. Golub, *Bounds for matrix moments*, Rocky Mnt. J. of Math., 4 (1974) 207–211.

[3] G. Golub and J.H. Welsch, *Calculation of Gauss quadrature rule*, Math. Comp., 23 (1969) 221–230.

[4] G. Golub and C. van Loan, *Matrix Computations*, 2nd Ed., (1989) Johns Hopkins University Press.

[5] G. Golub and Z. Strakos, *Estimates in quadratic formulas*, Technical reports, SCCM-93-08, Computer Science Department, Stanford University.

[6] G. Golub and G. Meurant, *Matrices, moments and quadrature*, in Proceedings of the 15-th Durdee conference, June-July 1993, D.F. Griffiths and G.A. Watson, Eds., Longman Scientific & Technical, 1994.

[7] G. Golub and Urs Von Matt, *Quadratically constrained least squares and quadratic problems*, Numer. Math., 59 (1991) 561–580.

[8] R. Haydock, *Accuracy of the recursion method and basis non-orthogonality*, Computer Physics Communications, 53 (1989) 133–139.

[9] G. Meurant, *A review of the inverse of tridiagonal and block tridiagonal matrices*, SIAM J. Matrix Anal. Appl., 13 (1992) 707–728.

[10] C.M. Nex, *Estimation of integrals with respect to a density of states*, J. Phys. A, 11 (1978) 653–663.

[11] C.M. Nex, *The block Lanczos algorithm and the calculation of matrix resolvents*, Computer Physics Communications, 53 (1989) 141–146.

[12] P.D. Robinson and A. Wathen, *Variational bounds on the entries of the inverse of a matrix*, IMA J. of Numer. Anal., 12 (1992) 463–486

[13] J. Stoer, R. Bulirsch, *Introduction to numerical analysis*, 2nd Ed. (1983) Springer Verlag, New York.

3

Wavelets from Filter Banks

Gilbert Strang*

3.1 Introduction

This subject has two parts. One part is discrete, the other is continuous. In discrete time we develop the idea and applications of **filter banks**. In continuous time we have **scaling functions** $\phi(t)$ and **wavelets** $w(t)$. By a natural limiting process, iteration of the lowpass filter leads to the scaling function. One highpass filter then produces a wavelet. Our goal is to make this connection clear. We find the conditions on the discrete coefficients that lead to good filter banks and good wavelets.

Historically and mathematically, the filters come first. Perfect reconstruction filter banks were developed in the early 1980's. The excitement around wavelets started later (and grew quickly). This excitement was not universal – designers of filter banks naturally asked what was new. Part of the answer is precisely in that process of **iteration**. For a filter to behave well in practice, when it is combined with subsampling and repeated five times, it must have an extra property – not built into earlier designs. This property expresses itself in the frequency domain by a sufficient number of "zeros at π". Then the frequency band can be successfully separated into five octaves.

The underlying problem is to choose a good basis. We want to represent a signal well, by a small number of basic signals. These can be sinusoids and they can be wavelets. On a discrete grid, $\omega = \pi$ is the highest frequency at which a signal can oscillate. Those oscillations $\boldsymbol{x}(n) = e^{i\pi n} = (-1)^n$ are stopped by the lowpass filter with a "zero at π". The highpass filter lets fast oscillations through, and the synthesis filters can reconstruct the exact input. But **compression** may come between analysis and synthesis. Frequencies that are barely represented will be intentionally lost. That mostly means high frequencies but the filter bank is impartial – it keeps the basis functions that are important to the specific signal. We want to show

---

*Department of Mathematics, Massachusetts Institute of Technology, Cambridge, MA 02139, U.S.A. gs@math.mit.edu

when, and why, filter banks and wavelets are effective in reconstruction and signal representation and compression.

Filter Banks

For filter banks, we identify the two conditions for perfect reconstruction (in the absence of lossy compression). One condition removes distortion, the other condition removes aliasing. The anti-distortion condition applies to the products $\boldsymbol{F}_0\boldsymbol{H}_0$ and $\boldsymbol{F}_1\boldsymbol{H}_1$ along the channels of the filter bank. Then the anti-aliasing condition controls how those products can be separated into the four filters.

The design of a perfect reconstruction filter bank is a choice of $\boldsymbol{F}_0\boldsymbol{H}_0$ and then a factorization. To understand the conditions on distortion and aliasing, we apply the techniques of multirate filtering. The algebra is moved into the "z-domain". We are selecting and factoring a polynomial $P_0(z) = F_0(z)H_0(z)$. We go forward now, to illustrate a filter bank that gives perfect reconstruction.

The analysis bank is on the left. It has a lowpass filter $\boldsymbol{H}_0$, and a highpass filter $\boldsymbol{H}_1$, and decimation by $(\downarrow 2)$ – which removes the odd-numbered components after filtering. The analysis bank yields two "half-length" outputs. Then the synthesis bank on the right begins with the upsampling operation $(\uparrow 2)$ – which inserts zeros in those odd components:

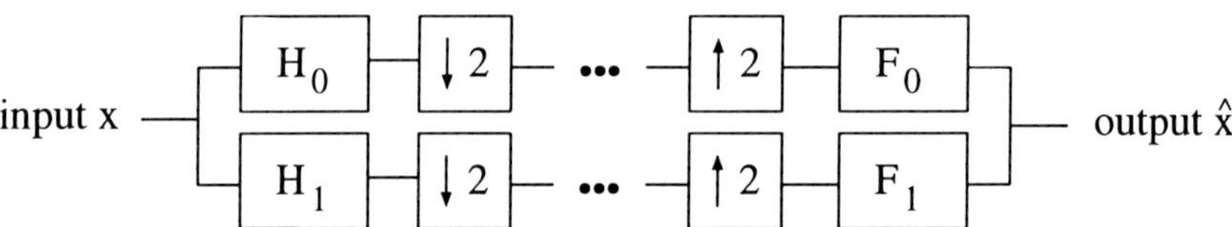

Two-channel filter bank: Separate the input into frequency bands (filter and downsample). Then reassemble (upsample and filter).

The gap in the center indicates where the subband signals are compressed or enhanced. The applications of this structure are extremely widespread. We believe that any reader interested in signal processing (and image processing) will find that filter bank analysis is extremely useful.

The filters $\boldsymbol{H}_0$, $\boldsymbol{H}_1$, $\boldsymbol{F}_0$ and $\boldsymbol{F}_1$ are linear and time-invariant. The operators $(\downarrow 2)$ and $(\uparrow 2)$ are **not** time-invariant. These multirate operations are responsible for *aliasing* and for *imaging* – they create undesirable and extraneous signals that the filters must cancel. The structure of an **orthogonal** bank is very special, and the next figure shows how the filters are related. For length 4 all filters use the four coefficients, a, b, c, d that Daubechies derived:

The form of an orthogonal filter bank with four coefficients.

How did she choose a, b, c, d? Part of the answer will have to wait, but here

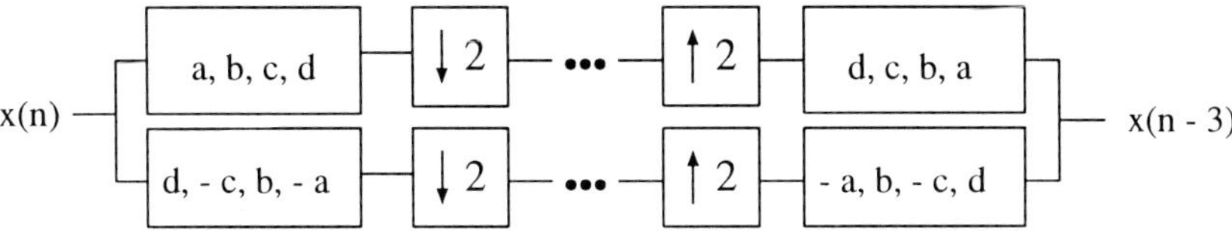

is the essential idea. The product along the top channel gives a particular "halfband filter" as $\boldsymbol{P}_0$:

$$(a,\ b,\ c,\ d)\ *\ (d,\ c,\ b,\ a)\ =\ (-1,\ 0,\ 9,\ 16,\ 9,\ 0,\ -1)/16.$$

This convolution is a multiplication of two polynomials, when a, b, c, d are the coefficients:

$$(a+bz^{-1}+cz^{-2}+dz^{-3})\quad(d+cz^{-1}+bz^{-2}+az^{-3})=$$

$$(-1+9z^{-2}+16z^{-3}+9z^{-4}-z^{-6})/16.$$

The four coefficients are pleasant to calculate. The serious job is to explain what is special about that 6th degree polynomial in which z^{-1} and z^{-5} are missing.

A filter bank also gives perfect reconstruction if it is **biorthogonal**. This design is less restricted. The product $\boldsymbol{F}_0\boldsymbol{H}_0$ must skip the same odd powers of z^{-1}, but $\boldsymbol{F}_0$ does not have to be the transpose (the flip) of $\boldsymbol{H}_0$. Here are specific numbers for the filter coefficients – not the only choice and maybe not the best. They show how the filters $\boldsymbol{F}_0$ and $\boldsymbol{F}_1$ on the synthesis side are related to the analysis filters $\boldsymbol{H}_1$ and $\boldsymbol{H}_0$ (by alternating signs):

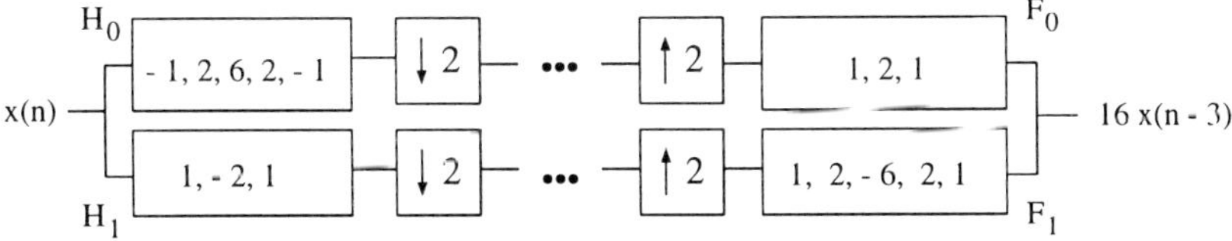

A biorthogonal filter bank: Perfect reconstruction with 3 delays.

For filters, we stop here. This time $F_0(z) = 1 + 2z^{-1} + z^{-2}$. Multiplied by $H_0(z)$ it gives the same important 6th degree polynomial as before. To understand why the zero coefficients are necessary in that polynomial, and why $-\frac{1}{16}$ and $\frac{9}{16}$ are desirable, we refer to the new book [SN].

Our discussion went this far so as to make a basic and encouraging point: *The construction of new filter banks need not be complicated.* This subject is accessible to new ideas and experiments.

Wavelets

Wavelets are localized waves. Instead of oscillating forever, they drop to zero. They come from the **iteration** of filters (with rescaling). The link between discrete-time filters and continuous-time wavelets is in the limit of a logarithmic filter tree:

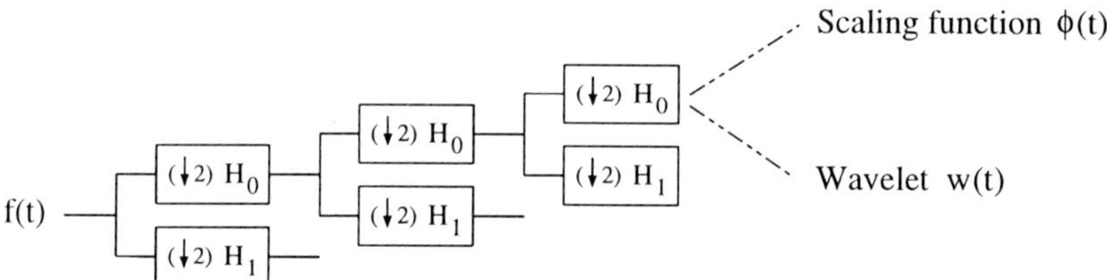

Scaling function and wavelets from iteration of the lowpass filter.

Scaling functions and wavelets have remarkable properties. They inherit orthogonality, or biorthogonality, from the filter bank. Because of the repeated rescaling that produces them, wavelets decompose a signal into details at all scales. The wavelet $w(t)$ and its shifts $w(t-k)$ are at unit scale. The wavelets $w(2^j t)$ and $w(2^j t-k)$ are at scale 2^{-j}. The biorthogonal functions $\tilde{\phi}(t)$ and $\tilde{w}(t)$ come from iterating the synthesis bank.

Wavelets produces a natural "**multiresolution**" of every image, including the all-important edges. Where the low frequency part of the Fourier transform is often a blur, the output from the lowpass channel is a useful compression.

The wavelets created by Ingrid Daubechies are orthogonal, with the advantages and limitations that this property brings. The biorthogonal alternatives come from different factorizations of the same polynomial (as above). This polynomial corresponds to a "maxflat halfband filter", and we hope you will like the connections. This subject is a beautiful combination of mathematical analysis and signal processing applications.

Summary of the Theory

There are four conditions that play a central part in the analysis. Because of their importance we highlight them here. They apply directly to the coefficients in the filter banks – and the consequences are felt (after iteration!) in the scaling functions and wavelets. Here are the four conditions – some might say in decreasing order of importance:

PR Condition *Perfect reconstruction.*
The synthesis bank inverts the analysis bank, with ℓ delays. Biorthogonal banks with no aliasing and no distortion.

Condition O *Orthogonality.*
The analysis bank is inverted by its transpose.
The wavelets are orthogonal to their dilates and translates.

Condition A_p *Accuracy of order p from the scaling functions.*
p vanishing moments in the wavelets.
pth order decay of wavelet coefficients for smooth $f(t)$.

Condition E *Eigenvalue condition on the cascade algorithm.*
Determines convergence to $\phi(t)$ and smoothness of wavelets. Equivalent to stability of the wavelet basis.

The four fundamental conditions will be stated explicitly for a two-channel filter bank. We continue to use the polynomials $H_0(z)$, $H_1(z)$, $F_0(z)$, and $F_1(z)$, whose coefficients come directly from the filters. By convention, these are polynomials in z^{-1}, and the lowpass analysis filter is represented by $H_0(z) = \boldsymbol{h}(0) + \boldsymbol{h}(1)z^{-1} + \ldots + \boldsymbol{h}(N)z^{-N}$. Here are the conditions that give filters and wavelets with good properties:

1. Perfect Reconstruction (**PR condition**)

$$F_0(z)H_0(z)+F_1(z)H_1(z) = 2z^{-\ell} \quad \text{and} \quad F_0(z)H_0(-z)+F_1(z)H_1(-z) = 0.$$

The second equation gives the anti-aliasing choices $F_0(z) = H_1(-z)$ and $F_1(z) = -H_0(-z)$.

2. Orthogonality (**Condition O**)

The filter coefficients are reversed by $F_0(z) = z^{-N}H_0(z^{-1})$ and $F_1(z) = z^{-N}H_1(z^{-1})$. Then perfect reconstruction depends on the "double-shift orthogonality" of the lowpass coefficients $\boldsymbol{h}(k)$:

$$\sum \boldsymbol{h}(k)\, \boldsymbol{h}(k+2n) = \boldsymbol{\delta}(n).$$

In terms of the polynomials this is $H_0(z)H_0(z^{-1}) + H_0(-z)H_0(-z^{-1}) = 2$.

3. Accuracy of order p (**Condition $\boldsymbol{A_p}$**)

The lowpass filter has a zero of order p at $z = -1$:

$$H_0(z) = \left(\frac{1+z^{-1}}{2}\right)^p Q(z).$$

4. Convergence and Stability (**Condition E**)
The transition matrix $\boldsymbol{T}$ has $\lambda = 1$ as simple eigenvalue and all other $|\lambda(\boldsymbol{T})| < 1$.

Final note: The sixth degree polynomial in the examples above has *four zeros* at $z = -1$:

$$-1 + 9z^{-2} + 16z^{-3} + 9z^{-4} - z^{-6} = (1 + z^{-1})^4 \, (-1 + 4z^{-1} - z^{-2}).$$

These zeros give flat responses near $\omega = \pi$ and also $\omega = 0$. The absence of z^{-1} and z^{-5} is the key to perfect reconstruction. Polynomials of higher degree, also with zeros at $z = -1$ and also with only one odd power, factor into $F_0(z)H_0(z)$ give the best filters for iteration. In the limit of the iterations, these filters give the best wavelets.

Note to the reader: The four sections of these notes are adapted from recent papers and also from our new textbook. The lectures could not cover the subject in full, and it may be helpful to indicate the sources from which these notes were drawn. The Guide to the Book and the journal articles are available by e-mail from `gs@math.mit.edu`.

I. Introduction (from the Guide to the Book: *Wavelets and Filter Banks* by Gilbert Strang and Truong Nguyen, Wellesley-Cambridge Press)

II. Filter Banks and Perfect Reconstruction (from Chapters 4 and 5 of *Wavelets and Filter Banks*, [SN])

III. Eigenvalues of $(\downarrow 2)H$ and Convergence of the Cascade Algorithm (from an article to appear in IEEE Transactions on Signal Processing)

IV. Zeros of the Daubechies Polynomials (from a paper with Jianhong Shen to appear in Proceedings of the American Mathematical Society)

The cascade algorithm studied in Section III of these notes connects filter banks to wavelets. May I emphasize that the new textbook [SN] studies wavelets in full !

3.2 Filter Banks and Perfect Reconstruction

We begin with an overview of **filters, filter banks** and **wavelets**. We want to indicate, first in rough outline and then in detail, the connections between these three topics. Our immediate purpose is to open up the problem and the language — starting with the filter coefficients $\boldsymbol{h}(n)$. The choice of those coefficients is the crucial decision. Their properties govern all that follows.

Each step is a natural development from the one before:

(1) A **filter** is a linear time-invariant operator. It acts on input vectors $\boldsymbol{x}$. The output vector $\boldsymbol{y}$ is the convolution of $\boldsymbol{x}$ with a fixed vector $\boldsymbol{h}$. The vector $\boldsymbol{h}$ contains the filter coefficients $\boldsymbol{h}(0), \boldsymbol{h}(1), \boldsymbol{h}(2), \ldots$. Our filters are digital, not analog, so the coefficients $\boldsymbol{h}(n)$ come at discrete times $t = nT$. The sampling period T is assumed to be 1 here. The inputs $\boldsymbol{x}(n)$ and outputs $\boldsymbol{y}(n)$ come at all times $t = 0, \pm 1, \pm 2, \ldots$:

$$\boldsymbol{y}(n) \quad = \quad \sum_k \boldsymbol{h}(k)\,\boldsymbol{x}(n-k) \quad = \quad \text{convolution } \boldsymbol{h} * \boldsymbol{x} \text{ in the time domain.}$$

One input $\boldsymbol{x} = (\ldots, 0, 1, 0, \ldots)$ has special importance — a unit impulse at time zero. The input has $\boldsymbol{x}(n-k) = 0$ except when $n = k$. The sum in the convolution has only one term, and that term is $\boldsymbol{h}(n)$. This output $\boldsymbol{y}(n) = \boldsymbol{h}(n)$ is the response at time n to the unit impulse $\boldsymbol{x}(0) = 1$. It is the *impulse response* $\boldsymbol{h}(0)$, $\boldsymbol{h}(1)$, $,\ldots,$ $\boldsymbol{h}(N)$.

In a moment the same filter will be described in the frequency domain. Convolution with the vector $\boldsymbol{h}$ will become *multiplication* by a function H. It is the simplicity of multiplication that makes this subject a success. The action of a filter in time and frequency is the foundation on which signal processing is built.

(2) A **filter bank** is a set of filters. The analysis bank often has two filters, lowpass and highpass. They separate the input signal into frequency bands. Those subsignals can be compressed much more efficiently than the original signal. Then they can be transmitted or stored. We are describing "subband coding" and its applications. At any time the signals can be recombined (by the *synthesis bank*).

It is not necessary to preserve the full outputs from the analysis filters. Normally they are *downsampled* by the operator $(\downarrow 2)$. **We keep only the even components of the lowpass and highpass filter outputs.** If there are M filters, then keeping every Mth component of each output gives a total of the same length as the input. Critical sampling is the key to subband coding.

This book explains how two or more filters, with downsampling, can jointly achieve properties that are impossible for a single filter. We are particularly interested in "perfect reconstruction FIR filter banks". In this case the reconstructed output $\hat{x}(n)$ from the synthesis bank is identical to the original input x to the analysis bank (with only a time delay). In matrix language, a banded matrix (for the analysis bank) has a banded inverse (the synthesis bank).

In the frequency domain, each filter leads to a multiplication. But downsampling is *not a time-invariant operation.* If we delay all components of $\boldsymbol{y}$ by one time unit, the output from downsampling is totally different. The new samples $\boldsymbol{y}(-1)$, $\boldsymbol{y}(1)$, $\boldsymbol{y}(3)$ are entirely separate and independent from the original samples $\boldsymbol{y}(0)$, $\boldsymbol{y}(2)$, $\boldsymbol{y}(4)$. Those two subsampled signals are two "phases" of $\boldsymbol{y}$, not connected. Therefore downsampling alters the multiplication picture in the frequency domain. In fact it introduces *aliasing*.

The simplicity of multiplication can be rescued by looking at each phase separately. Each phase of $\boldsymbol{y}$ comes from filtering the phases of $\boldsymbol{x}$ (using phases of $\boldsymbol{h}$). These separate pieces are multiplications in the frequency domain. The whole operation together, filtering followed by downsampling, becomes a matrix multiplication — by the *polyphase matrix.*

This is the foundation of filter bank theory (still to be explained in detail!). The analysis polyphase matrix $\boldsymbol{H}_p$ will reveal the correct synthesis bank for perfect reconstruction. That synthesis filter bank uses $\boldsymbol{H}_p^{-1}$.

(3) Wavelets are basis functions $w_{jk}(t)$ in continuous time. A basis is a set of linearly independent functions that can be used to produce all admissible functions $f(t)$:

$$f(t) = \text{ combination of basis functions } = \sum_{j,k} b_{jk}\, w_{jk}(t). \tag{3.1}$$

The special feature of the wavelet basis is that all functions $w_{jk}(t)$ are constructed from a single mother wavelet $w(t)$. This wavelet is a small wave (a pulse). Normally it starts at time $t = 0$ and ends at time $t = N$.

The shifted wavelets w_{0k} start at time $t = k$ and end at time $t = k + N$. The rescaled wavelets w_{j0} start at time $t = 0$ and end at time $t = N/2^j$. Their graphs are compressed by the factor 2^j, where the graphs of w_{0k} are translated (shifted to the right) by k:

$$\textit{compressed:}\quad w_{j0} = w\left(2^j t\right) \qquad\qquad \textit{shifted:}\quad w_{0k}(t) = w(t-k).$$

A typical wavelet w_{jk} is compressed j times and shifted k times. Its formula is

$$w_{jk}(t) = w\left(2^j t - k\right).$$

The remarkable property that is achieved by many wavelets is *orthogonality*. The wavelets are orthogonal when their "inner products" are zero:

$$\int_{-\infty}^{\infty} w_{jk}(t)\, w_{JK}(t)\, dt = \text{ inner product of } w_{jk} \text{ and } w_{JK} = 0. \tag{3.2}$$

In this case the wavelets form an *orthogonal basis* for the space of admissible functions. This basis corresponds to a set of axes that meet at 90° angles — as most good axes do. Orthogonality leads to a simple formula for each coefficient b_{JK} in the expansion for $f(t)$. Multiply the expansion displayed in equation (3.1) by $w_{JK}(t)$ and integrate:

$$\int_{-\infty}^{\infty} f(t)\, w_{JK}(t)\, dt = b_{JK} \int_{-\infty}^{\infty} \left(w_{JK}(t)\right)^2 dt. \tag{3.3}$$

All other terms in the sum disappear because of orthogonality. Equation (3.2) eliminates all integrals of w_{jk} times w_{JK}, except the one term that has $j = J$ and $k = K$. That term produces $(w_{JK}(t))^2$. Then b_{JK} is the ratio of the two integrals in equation (3.3).

As we describe the connection between filter banks and wavelets, you will see that it is the "*highpass filter*" that leads to $w(t)$. The "*lowpass filter*" leads to a scaling function $\phi(t)$. In most constructions the lowpass filter comes first — **the scaling function is obtained before the wavelet.** In fact the scaling function (in continuous time) comes from infinite repetition $\boldsymbol{L}\,\boldsymbol{L}\,\ldots\,\boldsymbol{L}$ of the lowpass filter, with rescaling at each iteration. The wavelet follows from $\phi(t)$ by just *one* application of the highpass filter.

Multiresolution

At a given resolution of a signal or an image, the scaling functions $\phi\left(2^j t - k\right)$ are a basis for the set of signals. The level is set by j, and the time steps at that level are 2^{-j}. The new details at level j are represented by the wavelets $w\left(2^j t - k\right)$. Then the smooth signal plus the details, the ϕ's plus the w's, combine into a **multiresolution** of the signal at the finer level $j + 1$. Averages come from the scaling functions, details come from the wavelets:

$$\begin{array}{rcl} \text{signal at level } j \text{ (local averages)} & \searrow & \\ + & & \text{signal at level } j+1 \\ \text{details at level } j \text{ (local differences)} & \nearrow & \end{array}$$

That is multiresolution for one signal. When we apply it to all signals, we have multiresolution for *spaces* of functions:

$$\begin{array}{rcl} V_j = \text{scaling space at level } j & \searrow & \\ \oplus & & V_{j+1} = \text{scaling space at level } j+1 \\ W_j = \text{wavelet space at level } j & \nearrow & \end{array}$$

This idea of multiresolution is absolutely basic to wavelet analysis. Again, we are only introducing it. We are sending a coarse signal to the reader, not the details. You only have the input at level 1.

Thus the signal is divided into different **scales** of resolution, rather than different frequencies. The "time-scale plane" takes the place for wavelets that the "time-frequency plane" takes for filters. Multiresolution divides the frequencies into *octave bands*, from ω to 2ω, instead of uniform bands from ω to $\omega + \Delta\omega$. The compression of a graph, when $f(t)$ is replaced by $f(2t)$, means expansion of its Fourier transform from $F(\omega)$ to $\frac{1}{2}F\left(\frac{\omega}{2}\right)$. Frequencies shift upward by an octave, when time is rescaled by two. The time-frequency plane is partitioned naturally into *rectangles of constant area*.

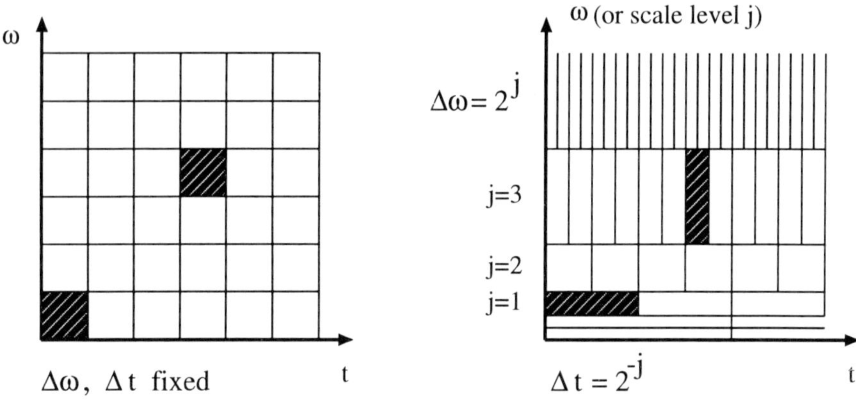

Time-frequency squares for Fourier decompositions become rectangles for wavelets.
Short time intervals are natural for high frequencies.

This matching of long time with low frequency and short time with high frequency occurs in a natural way for wavelets. It is one of the attractions of a wavelet decomposition.

Frequency Domain and Notation

To see a filter as a multiplication, we must take Fourier transforms. This will be the *discrete-time Fourier transform*, since the vectors $\boldsymbol{x}(n)$ and $\boldsymbol{h}(n)$ and $\boldsymbol{y}(n)$ are discrete. The time index n goes from $-\infty$ to ∞. (A vector with zero components at all negative times is called *causal*.) The transform of $\boldsymbol{x}$ has two reasonable notations. They both stand for the same transform, which we denote by X:

$$\begin{aligned} X(e^{j\omega}) &= \sum_{-\infty}^{\infty} \boldsymbol{x}(n)\, e^{-jn\omega} \quad \text{(signal processing notation)} \\ X(\omega) &= \sum_{-\infty}^{\infty} \boldsymbol{x}(n)\, e^{-in\omega} \quad \text{(reduced notation).} \end{aligned}$$

The standard notation allows a direction conversion of the Fourier transform to the *z-transform*. The transform is still X but the variable becomes

z:

$$X(z) = \sum_{-\infty}^{\infty} x(n)\, z^{-n}.$$

We simply replace $e^{j\omega}$ by z, extending the formal definition of X from $e^{j\omega}$ on the unit circle to z in the whole complex plane. (Remember: $e^{j\omega}$ has magnitude 1.) The Fourier transform will dominate the first part of the book, but the z-transform appears more frequently in the end.

Convolution by h in time becomes multiplication by H in frequency:

$$\begin{aligned} Y(e^{j\omega}) &= H(e^{j\omega})\, X(e^{j\omega}) && \text{in signal processing notation} \\ Y(\omega) &= H(\omega)\, X(\omega) && \text{in reduced notation.} \end{aligned}$$

This is the transform of $y(n) = \sum h(k)\, x(n-k)$. It is the "convolution rule." In the z-domain it becomes $Y(z) = H(z)\, X(z)$.

Perfect Reconstruction

A filter bank is a set of filters, linked by sampling operators and sometimes by delays. The downsampling operators are decimators, the upsampling operators are expanders. In a two-channel filter bank, the analysis filters are normally lowpass and highpass. Those are the filters $\boldsymbol{H}_0$ and $\boldsymbol{H}_1$ at the start of the following filter bank:

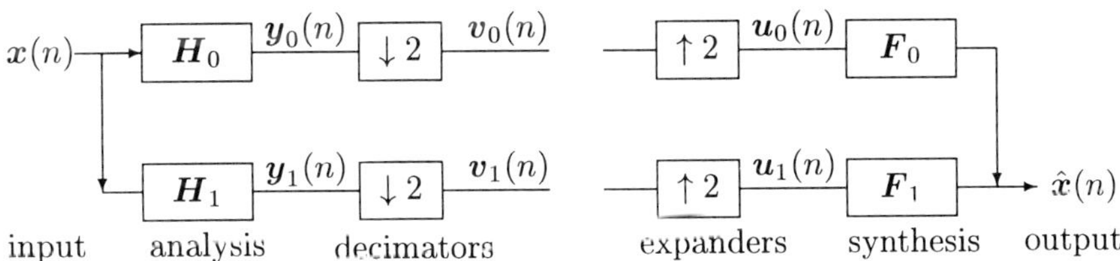

This structure was introduced in the 1980's. It gradually become clear how to choose $\boldsymbol{H}_0$, $\boldsymbol{H}_1$, $\boldsymbol{F}_0$, $\boldsymbol{F}_1$ to get perfect reconstruction: $\hat{x}(n) = x(n-l)$. The gap in the figure indicates where the downsampled signals might be coded for storage or transmission. At that point we may compress the signal and destroy information. Perfect reconstruction assumes no compression, so the gap is closed.

To indicate that $\boldsymbol{H}_0$ is lowpass and $\boldsymbol{H}_1$ is highpass, we often sketch the frequency responses. The drawing shows that they are not ideal brick wall filters. The responses overlap. *There is aliasing in each channel.* There is also amplitude distortion and phase distortion (our drawing does not show the phase). The synthesis filters $\boldsymbol{F}_0$ and $\boldsymbol{F}_1$ must be specially adapted to the analysis filters $\boldsymbol{H}_0$ and $\boldsymbol{H}_1$, in order to cancel the errors in this analysis bank.

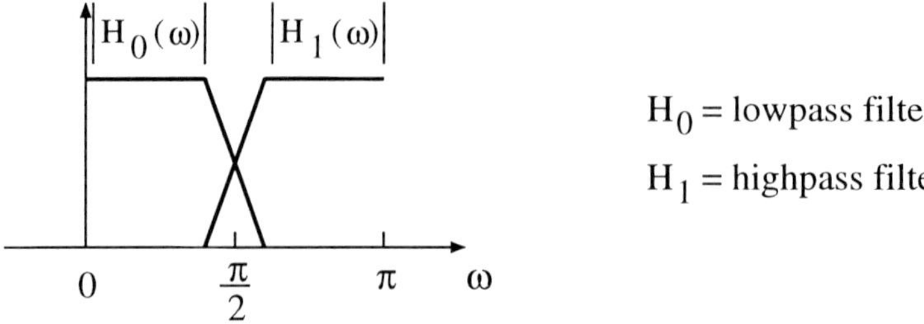

Rough sketch of frequency responses. Do not expect $|H_0(\omega)| + |H_1(\omega)| = 1$.

The goal of this section is to discover the conditions for perfect reconstruction. This means that the filter bank is ***biorthogonal***. The synthesis bank, from $\boldsymbol{F}_0$ and $\boldsymbol{F}_1$ and $\uparrow 2$, is the inverse of the analysis bank. Inverse matrices automatically involve biorthogonality. (The rows of $\boldsymbol{T}$ and the columns of $\boldsymbol{T}^{-1}$ are by definition biorthogonal.) When the analysis bank leads to scaling functions $\phi(t-k)$, and the synthesis bank leads to $\widetilde{\phi}(t-j)$, those are biorthogonal. So are the wavelets.

Perfect reconstruction is a crucial property. If the sampling operators $(\downarrow 2)$ and $(\uparrow 2)$ were not present, a reconstruction without delay would mean that $\boldsymbol{F}_0\boldsymbol{H}_0 + \boldsymbol{F}_1\boldsymbol{H}_1 = \boldsymbol{I}$. A perfect reconstruction with an l-step delay would mean (in the z-domain) that

$$\textit{without } (\downarrow 2) \text{ and } (\uparrow 2): \quad F_0(z)H_0(z) + F_1(z)H_1(z) = z^{-l}. \tag{3.4}$$

We do expect an overall delay z^{-l}, because each individual filter is causal.

Now take account of the sampling operators, which introduce *aliasing*. We recognize aliasing by the appearance of $-z$ as well as z (and $\omega + \pi$ as well as ω). The combination of $(\downarrow 2)$ followed by $(\uparrow 2)$ zeros out the odd-numbered components. In the z-domain it keeps only the *even powers* of $H_0(z)X(z)$:

$$\text{The transform of } (\uparrow 2)(\downarrow 2)\boldsymbol{H}_0\boldsymbol{x} \text{ is } \tfrac{1}{2}(H_0(z)X(z) + H_0(-z)X(-z)).$$

This is an even function, because the odd components are gone. The aliasing term $H_0(-z)X(-z)$ is multiplied by $F_0(z)$ at the synthesis step. This alias has to cancel the alias $F_1(z)H_1(-z)X(-z)$ from the other channel. So there is an alias cancellation condition in addition to a reconstruction condition:

$$\textit{Alias cancellation} \quad F_0(z)H_0(-z) + F_1(z)H_1(-z) = 0. \tag{3.5}$$

Correction The sampling operators also produce a change in equation (3.4). *The right side has an extra factor* 2. You can see this by considering a simple set of filters: $H_0(z) = 1$ and $H_1(z) = z^{-1}$, $F_0(z) = z^{-1}$ and $F_1(z) = 1$. This satisfies (3.5) and cancels aliasing. The left side of equation (3.4) equals $2z^{-1}$ rather than z^{-1}. The overall delay is $l = 1$ for this filter bank, and its perfect reconstruction comes from

$$\textit{No distortion} \quad F_0(z)H_0(z) + F_1(z)H_1(z) = 2z^{-l}. \tag{3.6}$$

The next page establishes these two conditions for perfect reconstruction.

One further point. In a genuine filter bank, the highpass filter has $H_1 = 0$ at $z = 1$ (or $\omega = 0$). Equation (3.6) becomes $F_0(1)H_0(1) = 2$. That equation is more natural if we include an extra factor $\sqrt{2}$ in the filter coefficients. For a similar reason the highpass filters $\boldsymbol{H}_1$ and $\boldsymbol{F}_1$ can be normalized by an extra $\sqrt{2}$.

No Aliasing and No Distortion

Conditions (3.4) and (3.5) for perfect reconstruction come directly from following a signal through the filter bank. The original signal is $\boldsymbol{x}(n)$. The lowpass analysis filter is $\boldsymbol{H}_0$. In the z-domain this produces $H_0(z)X(z)$. Now downsample and upsample:

$$\text{First } (\downarrow 2) \text{ produces} \quad \tfrac{1}{2}[H_0(z^{\frac{1}{2}})X(z^{\frac{1}{2}}) + H_0(-z^{\frac{1}{2}})X(-z^{\frac{1}{2}})]$$

$$\text{Then } (\uparrow 2) \text{ produces} \quad \tfrac{1}{2}[H_0(z)X(z) + H_0(-z)X(-z)].$$

The filtered signal $\boldsymbol{H}_0\boldsymbol{x}$ now has zeros in its odd-numbered components.

Those zeros are produced by averaging $\boldsymbol{H}_0\boldsymbol{x}(n)$ with its alternating alias $(-1)^n\boldsymbol{H}_0\boldsymbol{x}(n)$. In the z-domain this is the average of $H_0(z)X(z)$ with $H_0(-z)X(-z)$. The aliasing term has entered the filter bank.

The final filter multiplies by $F_0(z)$. This yields the output from the lowpass channel. Below it we write the corresponding output from the highpass channel (same formula with subscripts changed to 1):

$$\begin{aligned}\text{lowpass output} &= \tfrac{1}{2}F_0(z)\ \ [H_0(z)X(z) + H_0(-z)\ X(-z)]\\ \text{highpass output} &= \tfrac{1}{2}F_1(z)\ \ [H_1(z)X(z) + H_1(-z)\ X(-z)].\end{aligned}$$

Now add. The filter bank combines the channels to get $\hat{\boldsymbol{x}}(n)$. In the z-domain this is $\hat{X}(z)$. Half the terms involve $X(z)$ and half involve $X(-z)$:

$$\begin{aligned}\hat{X}(z) &= \tfrac{1}{2}[F_0(z)H_0(z) + F_1(z)H_1(z)]X(z)\\ &+ \tfrac{1}{2}[F_0(z)H_0(-z) + F_1(z)H_1(-z)]X(-z).\end{aligned}$$

For perfect reconstruction with l time delays, $\hat{X}(z)$ must be $z^{-l}X(z)$. So the "distortion term" must be z^{-l} and the "alias term" must be zero:

Theorem 3.1 *A 2-channel filter bank gives perfect reconstruction when*

$$F_0(z)H_0(z) + F_1(z)H_1(z) = 2z^{-l} \tag{3.7}$$

$$F_0(z)H_0(-z) + F_1(z)H_1(-z) = 0. \tag{3.8}$$

In vector-matrix form these two conditions involve the "*modulation matrix*" $\boldsymbol{H}_m(z)$:

$$[F_0(z) \quad F_1(z)]\begin{bmatrix} H_0(z) & H_0(-z) \\ H_1(z) & H_1(-z) \end{bmatrix} = [2z^{-l} \quad 0]. \tag{3.9}$$

This matrix $\boldsymbol{H}_m(z)$ will play a very important role. It involves the responses $H_k(z)$ and the alias terms $H_k(-z)$. For an M-channel bank the matrix will be $M \times M$. But the real problem is clearly identified by the separate conditions (3.7) and (3.8) — *how to design filters that meet those conditions?*

Alias Cancellation and the Product Filter $\boldsymbol{P}_0 = \boldsymbol{F}_0\boldsymbol{H}_0$

At this point we have four filters $\boldsymbol{H}_0, \boldsymbol{H}_1, \boldsymbol{F}_0, \boldsymbol{F}_1$ to design. They must satisfy (3.7) and (3.8). It is almost irresistible to determine some of the filters from the others:

For alias cancellation choose $\boxed{F_0(z) = H_1(-z) \text{ and } F_1(z) = -H_0(-z)}$ (3.10)

Important: This choice automatically satisfies $F_0(z)H_0(-z)+F_1(z)H_1(-z) = 0$. Aliasing is removed; it cancels itself! This relation of $\boldsymbol{F}_0$ to $\boldsymbol{H}_1$ and of $\boldsymbol{F}_1$ to $\boldsymbol{H}_0$ gives the *alternating signs* pattern of a 2-channel filter bank:

$\boldsymbol{H}_0$	$a,\ b,\ c$	$p, -q,\ r, -s,\ t$	$F_0(z) = H_1(-z)$
$\boldsymbol{H}_1$	$p,\ q,\ r,\ s,\ t$	$-a,\ b, -c$	$F_1(z) = -H_0(-z)$

Now comes a definition that allows us to rewrite equation (3.7) for no distortion:

Define the "**product filter**" *by* $P_0(z) = F_0(z)H_0(z)$.

This is a lowpass filter. The highpass product filter is $P_1(z) = F_1(z)H_1(z)$. These products P_0 and P_1 are exactly the terms in (3.7). The crucial point is the relation between $P_0(z)$ and $P_1(z)$, when the synthesis filters are determined by $F_0(z) = H_1(-z)$ and $F_1(z) = -H_0(-z)$. We substitute directly to find that $P_1(z) = -P_0(-z)$:

$$P_1(z) = -H_0(-z)H_1(z) = -H_0(-z)F_0(-z) = -P_0(-z). \tag{3.11}$$

The reconstruction equation $F_0(z)H_0(z) + F_1(z)H_1(z) = 2z^{-l}$ simplifies to

$$\boxed{P_0(z) - P_0(-z) = 2z^{-l}.} \tag{3.12}$$

The design of a 2-channel PR filter bank is reduced to two steps:

Step 1. Design a lowpass filter $\boldsymbol{P}_0$ satisfying (3.12).
Step 2. Factor $\boldsymbol{P}_0$ into $\boldsymbol{F}_0\boldsymbol{H}_0$. Then use (3.10) to find $\boldsymbol{F}_1$ and $\boldsymbol{H}_1$.

The length of $\boldsymbol{P}_0$ determines the sum of the lengths of $\boldsymbol{F}_0$ and $\boldsymbol{H}_0$. There are many ways to design $\boldsymbol{P}_0$ in Step 1. And there are many ways to factor it in Step 2. Experiments are going on as this book is written, and undoubtedly they are going on as the book is read, to find the best factors $\boldsymbol{F}_0$ and $\boldsymbol{H}_0$ of the best product filter $\boldsymbol{P}_0$.

Note that (3.12) is a condition on the *odd powers* in $P_0(z)$. Those odd powers must have coefficient zero, except z^{-l} has coefficient one.

A look forward To help the reader find the specific filters that are coming, we point to an outstanding choice for the product filter:

$$P_0(z) = (1 + z^{-1})^{2p} Q(z). \tag{3.13}$$

The polynomial $Q(z)$ of degree $2p - 2$ is chosen so that (3.12) is satisfied. There are $2p - 1$ odd powers in $P_0(z)$, and $2p - 1$ coefficients to choose in $Q(z)$. *Then $Q(z)$ is unique.* This is the Daubechies construction. Since the construction starts with the special factor $(1 + z^{-1})^{2p}$, these filters are called *binomial* or *maxflat*. The binomial factor gives a maximum number p of zeros at $z = -1$, which means that the frequency response is maximally flat at $\omega = \pi$. The binomial by itself, without $Q(z)$, represents a "spline filter". $Q(z)$ is needed to give perfect reconstruction.

Splitting P_0 into $F_0 H_0$ can give linear phase filters (symmetry in F_0 and H_0 separately). It can give orthogonal filters (symmetry between F_0 and H_0). *It cannot give both*, except in the Haar case $p = 1$.

Simplification The equation $P_0(z) - P_0(-z) = 2z^{-l}$ can be made a little more convenient. The left side is an odd function, so l is odd. Normalize $P_0(z)$ by z^l to center it:

$$\textit{The normalized product filter is} \quad P(z) = z^l P_0(z).$$

Then $P(-z) = (-z)^l P_0(-z)$. Since l is odd, this is $-z^l P_0(-z)$. The reconstruction equation $P_0(z) - P_0(-z) = 2z^{-l}$ takes an extremely simple form when we multiply by z^l. The factor z^{-l} disappears and the minus sign becomes plus:

Perfect Reconstruction Condition: $P(z)$ must be a "halfband filter"

$$\boxed{P(z) + P(-z) = 2.} \tag{3.14}$$

This means that *all even powers in $P(z)$ are zero*, except the constant term (which is 1). The odd powers cancel when $P(z)$ combines with $P(-z)$ — so the coefficients of odd powers in $P(z)$ are design variables in 2-channel PR filter banks.

Modulation Matrices

The conditions for perfect reconstruction are expressed in (3.9) by two equations:

$$[F_0(z) \quad F_1(z)]\begin{bmatrix} H_0(z) & H_0(-z) \\ H_1(z) & H_1(-z) \end{bmatrix} = [2z^{-l} \quad 0]. \tag{3.15}$$

This displays the **analysis modulation matrix** $\boldsymbol{H}_m(z)$ — which is central to filter bank theory. With no extra effort we can also produce the synthesis modulation matrix $\boldsymbol{F}_m(z)$. The two matrices should play matching (and even reversible) roles. This balance between $\boldsymbol{F}_m$ and $\boldsymbol{H}_m$ is achieved by expanding (3.15) into a matrix equation:

$$\begin{bmatrix} F_0(z) & F_1(z) \\ F_0(-z) & F_1(-z) \end{bmatrix}\begin{bmatrix} H_0(z) & H_0(-z) \\ H_1(z) & H_1(-z) \end{bmatrix} = \begin{bmatrix} 2z^{-l} & 0 \\ 0 & 2(-z)^{-l} \end{bmatrix}. \tag{3.16}$$

The second row of equations follows from the first, when $-z$ replaces z. Note that the synthesis matrix $\boldsymbol{F}_m$ has $F_1(z)$ in the (1,2) position, while the analysis matrix $\boldsymbol{H}_m$ has $H_1(z)$ in the (2,1) position. This *transpose convention* between analysis and synthesis will appear again for polyphase matrices. The reader sees why it is necessary.

The reconstruction condition (3.16) is now a statement about the matrix product $\boldsymbol{F}_m(z)\boldsymbol{H}_m(z)$. If we "center" the filter coefficients around the zero position, the right side becomes the identity matrix! This is so desirable and memorable that we do it. It is the same normalization that centered $P_0(z)$ into $P(z)$, and it is especially clear when the filters are linear phase (and l is odd):

Theorem 3.2 *If all filters are symmetric (or antisymmetric) around zero, as in $H(z) = H(z^{-1})$ and $\boldsymbol{h}(k) = \boldsymbol{h}(-k)$, then the condition for perfect reconstruction becomes a statement about inverse matrices:*

$$\boxed{\boldsymbol{F}_m(z)\,\boldsymbol{H}_m(z) = 2\boldsymbol{I}.} \tag{3.17}$$

The H's determine the F's. The analysis bank is inverted by the synthesis bank. When we express it that way, equation (3.17) becomes almost obvious.

A Brief History of $\boldsymbol{H}_1$

The reader understands that the filters $\boldsymbol{H}_0$ and $\boldsymbol{H}_1$ are still to be chosen. These choices are connected. Historically, designers chose the lowpass filter coefficients $\boldsymbol{h}(0), \ldots \boldsymbol{h}(N)$ and then constructed $\boldsymbol{H}_1$ from $\boldsymbol{H}_0$. Here are two possibilities that produce *equal length filters.* $\boldsymbol{H}_1$ will be highpass whenever $\boldsymbol{H}_0$ is lowpass:

Alternating signs : $H_1(z) = H_0(-z)$ comes from $(\boldsymbol{h}(0), -\boldsymbol{h}(1), \boldsymbol{h}(2), -\boldsymbol{h}(3), \ldots)$

Alternating flip : $H_1(z) = -z^{-N} H_0(-z^{-1})$ comes from $(\boldsymbol{h}(N), -\boldsymbol{h}(N-1), \ldots)$.

For convenience we are assuming real coefficients. The number N is odd in the alternating flip. The perfect reconstruction condition is *still to be imposed.* When that is satisfied, the overall system delay is $l = N$.

Early choice Croisier-Estaban-Galand (1976) chose alternating signs $H_1(z) = H_0(-z)$. The resulting filter bank was called QMF (Quadrature Mirror Filter). The highpass response $|H_1(e^{j\omega})|$ is a mirror image of the lowpass magnitude $|H_0(e^{j\omega})|$ with respect to the middle frequency $\frac{\pi}{2}$ — the quadrature frequency. Note that IIR filters $\boldsymbol{H}_0$ and $\boldsymbol{H}_1$ are allowed (and needed for PR, except for Haar!). This name QMF has since been extended to a larger class of filter banks, allowing M channels.

Better choice Smith and Barnwell (1984–6) and Mintzer (1985) chose the alternating flip $H_1(z) = -z^{-N} H_0(-z^{-1})$. This leads to orthogonal filter banks, when $\boldsymbol{H}_0$ is correctly chosen. The Daubechies filters will fit this pattern.

General choice The product $F_0(z) H_0(z)$ is a halfband filter. This leads to biorthogonal filter banks, when aliasing is cancelled by the relation of $\boldsymbol{F}_0$ to $\boldsymbol{H}_1$ and $\boldsymbol{F}_1$ to $\boldsymbol{H}_0$.

Actually the synthesis bank has little freedom. Alias cancellation requires $F_0(z)H_0(-z) + F_1(z)H_1(-z) = 0$. Croisier-Estaban-Galand wrote each F_k directly in terms of H_k by

$$F_0(z) = H_0(z) \quad \text{and} \quad F_1(z) = -H_1(z).$$

With alternating signs $H_1(z) = H_0(-z)$ inside the analysis bank, their synthesis construction agrees (as it must) with the anti-aliasing equations

$$F_0(z) = H_1(-z) \quad \text{and} \quad F_1(z) = -H_0(-z). \tag{3.18}$$

Smith-Barnwell also made the anti-aliasing choice (3.18). With the alternating flip in $H_1(z)$, their synthesis filters (remembering that N is odd) are

$$F_0(z) = H_1(-z) = z^{-N} H_0(z^{-1}) \text{ comes from } (\boldsymbol{h}(N), \boldsymbol{h}(N-1), \ldots, \boldsymbol{h}(0)).$$

$$F_1(z) = -H_0(-z) = z^{-N} H_1(z^{-1}) \text{ comes from } (-\boldsymbol{h}(0), \boldsymbol{h}(1), -\boldsymbol{h}(2), \ldots, \boldsymbol{h}(N)).$$

Notice! Each $\boldsymbol{F}_k$ has become the ordinary flip of the corresponding $\boldsymbol{H}_k$. In matrix language the synthesis matrices are the *transposes* of the analysis matrices. A shift by N delays makes them causal. When we flip to get $\boldsymbol{F}_0$ and then alternate signs to get $\boldsymbol{H}_1$, we have the alternating flip from $\boldsymbol{H}_0$ to $\boldsymbol{H}_1$.

The alternating flip automatically gives double-shift orthogonality between highpass and lowpass (to be explained). *Conclusion*: When the design of $\boldsymbol{H}_0$ leads to perfect reconstruction in the alternating flip filter bank, it also leads to orthogonality.

$\boldsymbol{H}_0$	$a,\ b,\ c,\ d$	order flip →	$d,\ c,\ b,\ a$	$F_0(z) = H_1(-z) = z^{-3}H_0(z^{-1})$
	alternating flip ↓	✕	alternating signs	
$\boldsymbol{H}_1$	$d, -c,\ b, -a$		$-a,\ b, -c,\ d$	$F_1(z) = -H_0(-z)$

Relations between the filters allowing orthogonality when $N = 3$.

With aliasing cancelled, we now look at the PR condition

$$F_0(z)H_0(z) + F_1(z)H_1(z) = 2z^{-l}.$$

The early choice was alternating signs $H_1(z) = H_0(-z)$. With $F_0(z) = H_1(-z)$ and $F_1(z) = -H_0(-z)$, PR requires

$$H_0^2(z) - H_1^2(z) = H_0^2(z) - H_0^2(-z) = 2z^{-l}. \tag{3.19}$$

Therefore $H_0^2(z)$ has exactly one odd power z^{-l}. This is not easy for the square of a polynomial. An FIR filter is restricted to *two coefficients* (not good). This forces the filters to be IIR.

The better choice is the alternating flip. Perfect reconstruction is possible. The product filters $\boldsymbol{F}_0\boldsymbol{H}_0$ and $\boldsymbol{F}_1\boldsymbol{H}_1$ become $P_0(z) = z^{-N}H_0(z^{-1})H_0(z)$ and $P_1(z) = -z^{-N}H_0(-z^{-1})H_0(-z)$. Multiply by $z^l = z^N$ to center these filters. The normalized product filter is $P(z)$ and the reconstruction condition is (3.14):

$$P(z) + P(-z) = 2 \quad \text{with} \quad P(z) = H_0(z^{-1})H_0(z). \tag{3.20}$$

This is spectral factorization of a halfband filter! On the unit circle $z = e^{j\omega}$, the product $H_0(e^{-j\omega})H_0(e^{j\omega})$ is a magnitude squared:

$$P(e^{j\omega}) = \sum_{-N}^{N} \boldsymbol{p}(n)e^{-jn\omega} = \left|\sum_{0}^{N} \boldsymbol{h}(n)e^{-jn\omega}\right|^2. \tag{3.21}$$

The halfband coefficients are $\boldsymbol{p}(n) = \boldsymbol{p}(-n)$ for odd n and $\boldsymbol{p}(n) = 0$ for even n (except $\boldsymbol{p}(0) = 1$). *We design $P(z)$ and factor to find $H_0(z)$*. This symmetric factorization coincides with the Smith-Barnwell alternating flip. *It yields orthogonal banks with perfect reconstruction.* The flattest $P(z)$ will lead us to the Daubechies wavelets.

A note on biorthogonality (= PR) with linear phase

Theorem 3.3 *In a biorthogonal linear-phase filter bank with two channels, the filter lengths are all odd or all even. The analysis filters can be*

(a) both symmetric, of odd length

(b) one symmetric and the other antisymmetric, of even length.

Proof: The difference between odd and even lengths comes when we alternate signs:

$$\begin{array}{lccccccl} \text{odd length:} & a & b & c & b & a & \to & a \quad -b \quad c \quad -b \quad a \\ & & & & & & & \text{(remains symmetric)} \\ \text{even length:} & a & b & b & a & & \to & a \quad -b \quad b \quad -a \\ & & & & & & & \text{(becomes antisymmetric).} \end{array}$$

To cancel aliasing, there is sign alternation in $F_0(z) = H_1(-z)$. There is also alternation in $F_1(z) = -H_0(-z)$. (The extra minus sign does not change the symmetry type.) The two successful combinations are

$$\begin{array}{ccccccc} H_0 = \text{symm} & & F_0 = \text{symm} & \qquad & H_0 = \text{symm} & & F_0 = \text{symm} \\ & \times & & & & \times & \\ H_1 = \text{symm} & & F_1 = \text{symm} & & H_1 = \text{anti} & & F_1 = \text{anti} \\ & \textit{odd lengths} & & & & \textit{even lengths} & \end{array}$$

The other possibilities are excluded by the PR condition: $F_0(z)\,H_0(z)$ has to be a **halfband filter**. It must have an odd number of coefficients, and the center coefficient must be 1. For $F_0(z)\,H_0(z)$ to have odd length (which means even degree), the factors $F_0(z)$ and $H_0(z)$ must be both odd length or both even length. If one is symmetric and the other antisymmetric, the product $F_0(z)\,H_0(z)$ will be antisymmetric with zero at the center — not allowed. We conclude that $F_0(z)$ and $H_0(z)$ must match: both odd length or both even length, both symmetric or both antisymmetric.

This leaves the two successful possibilities shown above, and two more: H_0 and F_0 both antisymmetric. But the sum of lowpass coefficients cannot be zero. So antisymmetry of H_0 is ruled out.

Perfect Reconstruction with M Channels

In reality a filter bank can have M channels. Although $M = 2$ is standard in a range of applications, we often see $M > 2$. There are M analysis filters $\boldsymbol{H}_0, \boldsymbol{H}_1, \ldots, \boldsymbol{H}_{M-1}$. The sampling is done at the critical rate by $(\downarrow M)$ and $(\uparrow M)$. There are M filters $\boldsymbol{F}_0, \boldsymbol{F}_1, \ldots, \boldsymbol{F}_{M-1}$ in the synthesis bank. The outputs from all channels are combined into a single output $\hat{\boldsymbol{x}}$. Our standard picture of this implementation is

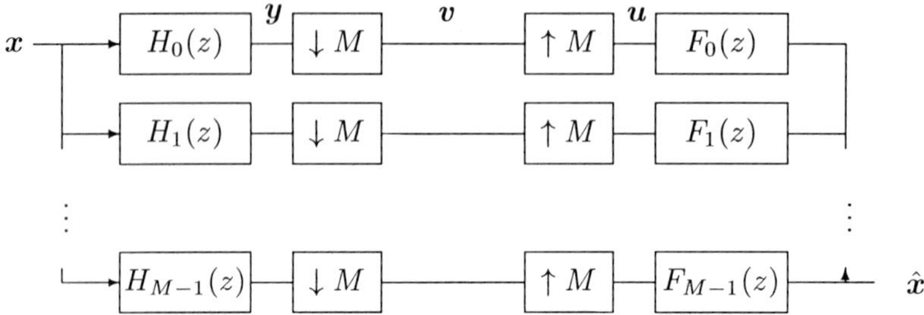

Several perfect reconstruction filter banks deserve special mention. The simplest is the average-difference pair from Haar. This is a useful example but a poor filter. That is the first in a family of "maxflat filters", corresponding to the Daubechies wavelets. The others in the family are orthogonal but not linear phase — since those two properties conflict.

Different factorizations of the product $P_0(z)$ lead to linear phase (not orthogonality). Those filters have become favorites for compression.

For $M > 2$, the design of separate filters $\boldsymbol{H}_0$, $\boldsymbol{H}_1$, $\ldots$, $\boldsymbol{H}_{M-1}$ can become unwieldy. We look for constructions in which these all come from one prototype filter. A particular class is the *cosine-modulated filter banks.* A phase change (= modulation) is the key to their construction. Those are efficient in every way.

The Polyphase Matrix

This section establishes a key idea and a valuable notation. The word "*polyphase*" has gained a certain mystique in the theory of multirate filters. Perhaps we can begin by explaining the meaning of the word, and also the purpose of the idea. Then the notation and applications will come naturally.

Meaning of polyphase: When a vector is downsampled by 2, its even-numbered components are kept. Its odd-numbered components are lost. Those are the two phases, *even* and *odd.* It is natural to follow the two phases of the input vector, $\boldsymbol{x}_{\text{even}}$ and $\boldsymbol{x}_{\text{odd}}$, as they go through the filter bank. They are acted on by the two phases $\boldsymbol{H}_{\text{even}}$ and $\boldsymbol{H}_{\text{odd}}$ of the filter.

For downsampling by M there are M phases. The ideas still apply to this "several-phase" or "polyphase" decomposition. Instead of even and odd inputs we will have M vectors (phases of $\boldsymbol{x}$). Instead of even and odd filters we will have M filters (phases of $\boldsymbol{H}$). The vector of filter coefficients $\boldsymbol{h}(n)$ is separated into phases, exactly as $\boldsymbol{x}(n)$ is separated. Then we watch those phases during downsampling.

The word "phase" is applied because the even filter with coefficients $\boldsymbol{h}(0)$, $\boldsymbol{h}(2)$ has a different delay (phase shift) from the odd phase with coefficients $\boldsymbol{h}(1)$, $\boldsymbol{h}(3)$.

Purpose of polyphase: The operation $(\downarrow 2)\boldsymbol{Hx}$, taken literally, is not efficient. We are computing all components of $\boldsymbol{Hx}$ and then destroying half of them. If we don't compute them, the system is still working at a fast rate (high bandwidth). The output is at half rate, because of downsampling. Each output component needs N additions and $N+1$ multiplications, to apply all the coefficients $\boldsymbol{h}(0), \ldots, \boldsymbol{h}(N)$.

The polyphase implementation works on the different phases separately. The input vector is separated into $\boldsymbol{x}_{\text{even}}$ and $\boldsymbol{x}_{\text{odd}}$. The operator $(\downarrow 2)$ comes *before the filter*! It changes one input at a high rate to two (or M) inputs at a lower rate. Then the separate phases of the filters act simultaneously (*in parallel*) on separate phases of the input.

The notation has to keep track of each phase. Often we find that "even multiplies even" and "odd multiplies odd". The *Noble Identities* justify an interchange of filtering and sampling. For the whole filter this interchange is forbidden, but it is allowed for each phase.

Polyphase in the time domain = block Toeplitz matrix: We can display the infinite matrix for a 2-channel analysis bank. The two filters $\sqrt{2}\boldsymbol{H}_0 = \boldsymbol{C}$ and $\sqrt{2}\boldsymbol{H}_1 = \boldsymbol{D}$ are downsampled by $(\downarrow 2)$. This removes the odd-numbered rows. Then we *interleave the rows* of $\boldsymbol{L} = (\downarrow 2)\boldsymbol{C}$ and $\boldsymbol{B} = (\downarrow 2)\boldsymbol{D}$ to see the analysis bank as a *block-Toeplitz matrix*:

$$\textbf{Block} \quad \boldsymbol{H}_b = \begin{bmatrix} \cdot & \cdot & & & & & & \\ \cdot & \cdot & & & & & & \\ \boldsymbol{c}(3) & \boldsymbol{c}(2) & \boldsymbol{c}(1) & \boldsymbol{c}(0) & & & & \\ \boldsymbol{d}(3) & \boldsymbol{d}(2) & \boldsymbol{d}(1) & \boldsymbol{d}(0) & & & & \\ & & \boldsymbol{c}(3) & \boldsymbol{c}(2) & \boldsymbol{c}(1) & \boldsymbol{c}(0) & & \\ & & \boldsymbol{d}(3) & \boldsymbol{d}(2) & \boldsymbol{d}(1) & \boldsymbol{d}(0) & & \\ & & & & \cdot & \cdot & \cdot & \cdot \\ & & & & \cdot & \cdot & \cdot & \cdot \end{bmatrix}.$$

This takes the input in blocks (two samples at a time). It gives the output in blocks. It is time-invariant in blocks! By block z-transform, multiplication by the infinite matrix $\boldsymbol{H}_b$ (which is block convolution) becomes multiplication by the *polyphase matrix*:

$$\textbf{Polyphase matrix} \quad \boldsymbol{H}_p(z) = \begin{bmatrix} \boldsymbol{c}(0) & \boldsymbol{c}(1) \\ \boldsymbol{d}(0) & \boldsymbol{d}(1) \end{bmatrix} + z^{-1} \begin{bmatrix} \boldsymbol{c}(2) & \boldsymbol{c}(3) \\ \boldsymbol{d}(2) & \boldsymbol{d}(3) \end{bmatrix} \tag{3.22}$$

The polyphase matrix is nothing but *the z-transform of a block of filters*. There are 2^2 or M^2 filters, from M phases of M original filters. Here those filters have four coefficients and their phases have two coefficients.

Notice especially how the block matrix $\boldsymbol{H}_b$ relates to the two separate downsampled filters $(\downarrow 2)\boldsymbol{C}$ and $(\downarrow 2)\boldsymbol{D}$:

> The efficient form downsamples the input *first* (to make blocks for $\boldsymbol{H}_b$)
>
> The inefficient form downsamples *last* (after the filters $\boldsymbol{C}$ and $\boldsymbol{D}$)

The Noble Identities prove the equivalence. It is just a removal of useless odd-numbered rows and an interleaving of the remaining rows. Next we discuss the algebra and the implementation.

Key identity in the z-domain: The even part of $X(z)$ is $\frac{1}{2}(X(z) + X(-z))$. The odd part is $\frac{1}{2}(X(z) - X(-z))$. The first has even powers $1, z^2, z^4$; the second has z, z^3, z^5. The original X is the sum of even plus odd (obviously). The same splitting holds for $C(z)$, and furthermore for $C(z)X(z)$. *The key is to find the even part of* $C(z)X(z)$. It is the even coefficients of $\boldsymbol{Cx}$ that survive downsampling and appear in $(\downarrow 2)\boldsymbol{Cx}$. In most of this section the lowpass filter is denoted by $\boldsymbol{C}$, to avoid the subscripts on $\boldsymbol{H}$.

A simple and important identity shows how the even part of $C(z)X(z)$ comes from even times even plus odd times odd:

$$\begin{aligned}\tfrac{1}{2}[C(z)X(z) + C(-z)X(-z)] &= \tfrac{1}{4}[C(z) + C(-z)][X(z) + X(-z)] \\ &+ \tfrac{1}{4}[C(z) - C(-z)][X(z) - X(-z)].\end{aligned} \tag{3.23}$$

In multiplying numbers, odd times odd is odd. But we are *adding exponents*, as in $(z^3)(z^5) = z^8$. So it is really odd *plus* odd, and even *plus* even, that yield the even part of the z-transform. This is the part that downsampling picks out, when $\boldsymbol{Cx}$ is decimated.

The importance of the key identity is this. The left side involves *all* coefficients of $C(z)$ and $X(z)$. Each product on the right involves only *half* the coefficients. The multiplication in the z-domain, which is $(\downarrow 2)\boldsymbol{Cx}$ in the time domain, becomes computationally efficient. We don't want all of $C(z)X(z)$, only the even half. The right side shows how to do half the work. Better still, it shows how even-even and odd-odd can be executed in parallel at half the rate.

Downsampling an even function effectively replaces z by $z^{1/2}$. It "closes the gaps" in $1, z^2, z^4$ by changing to $1, z, z^2$. For an odd function we will need a delay or an advance. We cannot change z and z^3 to $z^{1/2}$ and $z^{3/2}$. You will see how the coefficients of z^{-1}, z^{-3}, z^{-5} in $C(z)$ become coefficients of $1, z^{-1}, z^{-2}$ in the odd phase $C_{\text{odd}}(z)$. There is a delay for the odd phase and a "delay chain" when there are multiple phases.

This chapter works out the polyphase notation. We concentrate most on $M = 2$; the phases are even and odd. Then the polyphase forms of the analysis and synthesis banks lead quickly to a main goal of the theory. We find the perfect reconstruction condition on the polyphase matrices, when the filters are centered:

$$\boxed{\boldsymbol{F}_p(z)\,\boldsymbol{H}_p(z) = \boldsymbol{I}.}$$

This tells us, clearly and directly, what is required:

1. At a minimum, $\boldsymbol{H}_p(z)$ must be **invertible**. (biorthogonality)

2. Better than that, its inverse $\boldsymbol{F}_p(z)$ should be a **polynomial**. (FIR)

3. Better still, $\boldsymbol{F}_p(z)$ might be the **transpose** of $\boldsymbol{H}_p(z)$. (orthogonality)

In case **3**, the polyphase matrices are "paraunitary". The analysis and synthesis banks are orthogonal. In the more general case **1**, the banks are "biorthogonal". In case **2**, the synthesis bank is biorthogonal and also FIR.

The rows of a matrix are always biorthogonal to the columns of its inverse. When the rows of one are identical to the columns of the other, the matrix is self-orthogonal. Then it is an *orthogonal* matrix if real, a *unitary* matrix if complex, and a *paraunitary* matrix if it is a function of a complex parameter z.

Polyphase for Vectors

Any input vector $\boldsymbol{x}$ and any filter vector $\boldsymbol{c}$ or $\boldsymbol{h}$ can be separated into even and odd:

$$\boldsymbol{x} = (\ldots, \boldsymbol{x}(0), 0, \boldsymbol{x}(2), 0, \ldots) + (\ldots, 0, \boldsymbol{x}(1), 0, \boldsymbol{x}(3), 0, \ldots).$$

The z-transform is separated into even powers and odd powers, as in

$$X(z) = [\boldsymbol{x}(0) + \boldsymbol{x}(2)z^{-2} + \ldots] + z^{-1}[\boldsymbol{x}(1) + \boldsymbol{x}(3)z^{-2} + \ldots]. \tag{3.24}$$

The even part has powers of z^2. So has the odd part, when we factor out z^{-1}. This is the polyphase decomposition of $\boldsymbol{x}$ in the z-domain:

$$\boxed{X(z) = X_{\text{even}}(z^2) + z^{-1}X_{\text{odd}}(z^2)} \tag{3.25}$$

Each phase has its own z-transform:

$$\boldsymbol{x}_{\text{even}} = \begin{bmatrix} \boldsymbol{x}(0) \\ \boldsymbol{x}(2) \\ \cdot \end{bmatrix} \leftrightarrow X_0(z) = \sum \boldsymbol{x}(2k)z^{-k}$$

$$\boldsymbol{x}_{\text{odd}} = \begin{bmatrix} \boldsymbol{x}(1) \\ \boldsymbol{x}(3) \\ \cdot \end{bmatrix} \leftrightarrow X_1(z) = \sum \boldsymbol{x}(2k+1)z^{-k}.$$

Because of the z^2 in the definition, the in-between zeros are gone from $X_0(z)$ and $X_1(z)$. Please verify that the phases of $X(z) = z^{-1} + z^{-2} + z^{-3}$ are $X_{\text{even}} = z^{-1}$ and $X_{\text{odd}} = 1 + z^{-1}$.

Now reverse the process, to recover $\boldsymbol{x}$. Upsampling puts zeros back into $\boldsymbol{x}_{\text{even}}$ and $\boldsymbol{x}_{\text{odd}}$. *Those zeros change z to z^2.* The odd phase is delayed by z^{-1}, to move $\boldsymbol{x}(1)$ from position 0 to position 1. Then addition reconstructs equation (3.25).

Here is the splitting and the reconstruction in block form. Notice that so far the filters are not included, and $(\downarrow 2)\boldsymbol{x}$ is exactly $\boldsymbol{x}_{\text{even}}$:

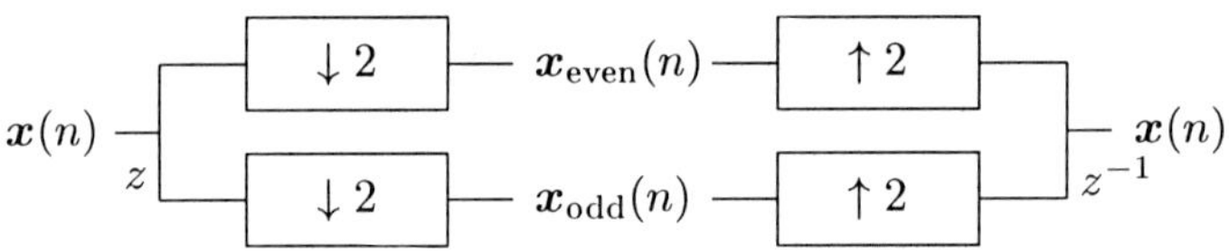

Important! The z at the start of the odd channel is because the odd phase has $\boldsymbol{x}(1)$ in its zeroth position. We have to advance the signal to achieve that. But advances look bad in our flow diagram. So the advance can be replaced by a delay, if we make up for it at the end by delaying the even part too.

Here is the "delay form" that we use in later sections. Please go through that form:

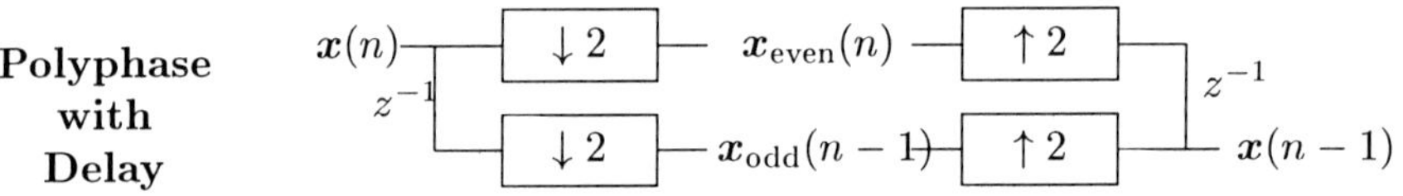

Polyphase Matrices for Filters

The polyphase form of a filter $\boldsymbol{C}$ comes directly from the polyphase form of $\boldsymbol{c}$ (the vector of filter coefficients). That vector separates into $\boldsymbol{c}_{\text{even}}$ and $\boldsymbol{c}_{\text{odd}}$. Its z-transform $C(z)$ separates into phases exactly as $X(z)$ did:

$$C(z) = C_0(z^2) + z^{-1}C_1(z^2). \tag{3.26}$$

The filtering step is $C(z)X(z)$. This is ordinary filtering $\boldsymbol{Cx}$, where even mixes with odd. But when downsampling picks out the even part of the product $C(z)X(z)$, it comes from even times even plus odd times odd. The transform of $(\downarrow 2)\boldsymbol{Cx}$ is

$$\boxed{(C(z)X(z))_{\text{even}} = C_0(z)X_0(z) + z^{-1}C_1(z)X_1(z).} \tag{3.27}$$

The direct multiplication of $C(z)$ times $X(z)$ will have even parts from $C_0(z^2)X_0(z^2)$ and from $z^{-2}C_1(z^2)X_1(z^2)$. Those give the even part of $C(z)X(z)$. Downsampling picks out those terms. It changes z^2 to z in their transform. The result is the z-transform of $(\downarrow 2)\boldsymbol{C}\boldsymbol{x}$.

Example 1. The moving average filter, downsampled.

Chapter 1 introduced the lowpass filter $\frac{1}{2}\boldsymbol{x}(n)+\frac{1}{2}\boldsymbol{x}(n-1)$. The coefficients are $\boldsymbol{h}(0)=\frac{1}{2}$ and $\boldsymbol{h}(1)=\frac{1}{2}$. Thus $\frac{1}{2}$ is the leading and only coefficient in H_{even} and H_{odd}. Both matrices have $\frac{1}{2}$ on *one diagonal*—the main diagonal. Here is the two-phase form of $(\downarrow 2)\boldsymbol{H}\boldsymbol{x}$:

$$\begin{bmatrix} \frac{1}{2} & \frac{1}{2} & & & \\ & & \frac{1}{2} & \frac{1}{2} & \\ & & & & \cdot\ \cdot \end{bmatrix}\begin{bmatrix} \\ \boldsymbol{x} \\ \\ \end{bmatrix} = \begin{bmatrix} \frac{1}{2} & & \\ & \frac{1}{2} & \\ & & \cdot \end{bmatrix}\begin{bmatrix} \boldsymbol{x}(0) \\ \boldsymbol{x}(2) \\ \cdot \end{bmatrix} + \begin{bmatrix} \frac{1}{2} & & \\ & \frac{1}{2} & \\ & & \cdot \end{bmatrix}\begin{bmatrix} \boldsymbol{x}(-1) \\ \boldsymbol{x}(1) \\ \cdot \end{bmatrix}$$

The polyphase components of $H(z)=\frac{1}{2}+\frac{1}{2}z^{-1}$ are constants: $H_{\text{even}}(z)=\frac{1}{2}$ and $H_{\text{odd}}(z)=\frac{1}{2}$.

The same splitting occurs for the highpass filter $\boldsymbol{D}$. Its even and odd phases are represented by $D_0(z)$ and $D_1(z)$. Those go into the 1 by 2 polyphase matrix $\boldsymbol{D}_p(z)$. Then the whole analysis bank comes together when we combine the polyphase matrices for $\boldsymbol{C}=\boldsymbol{H}_0$ and $\boldsymbol{D}=\boldsymbol{H}_1$ into a single polyphase matrix $\boldsymbol{H}_p(z)$:

Polyphase Matrix

$$\boldsymbol{H}_p(z) = \begin{bmatrix} \boldsymbol{C}_p(z) \\ \boldsymbol{D}_p(z) \end{bmatrix} = \begin{bmatrix} C_0(z) & C_1(z) \\ D_0(z) & D_1(z) \end{bmatrix} = \begin{bmatrix} H_{0,\text{even}}(z) & H_{0,\text{odd}}(z) \\ H_{1,\text{even}}(z) & H_{1,\text{odd}}(z) \end{bmatrix}. \tag{3.28}$$

This shows the matrix that we are aiming for. Now we go back for the close look at $(\downarrow 2)\boldsymbol{C}$. This operator is fundamental in the theory of multirate filters and wavelets.

Polyphase in the time domain: When downsampling follows the filter $\boldsymbol{C}$, we get the crucial matrix $\boldsymbol{L}=(\downarrow 2)\boldsymbol{C}$. This has to display the separation of even and odd, and I would like to show how this happens. Most of polyphase theory is developed in the z-domain, and we will do that too.

But first, look at the filter matrix as it produces $\boldsymbol{y} = C\boldsymbol{x}$:

$$\begin{bmatrix} \cdot \\ y(0) \\ y(1) \\ y(2) \\ y(3) \\ \cdot \end{bmatrix} = \begin{bmatrix} \cdot & & & & & \\ \cdot & c(0) & & & & \\ \cdot & c(1) & c(0) & & & \\ \cdot & c(2) & c(1) & c(0) & & \\ \cdot & c(3) & c(2) & c(1) & c(0) & \\ \cdot & \cdot & \cdot & \cdot & \cdot & \cdot \end{bmatrix} \begin{bmatrix} \cdot \\ x(0) \\ x(1) \\ x(2) \\ x(3) \\ \cdot \end{bmatrix}. \tag{3.29}$$

Downsampling leaves only the even-numbered components $\boldsymbol{y}(2n)$. To reach $\boldsymbol{v} = (\downarrow 2)\boldsymbol{y}$, we throw away every other row (the odd-numbered rows). This leaves the matrix $\boldsymbol{L} = (\downarrow 2)\boldsymbol{C}$:

$$\begin{bmatrix} \cdot \\ y(0) \\ y(2) \\ y(4) \\ \cdot \end{bmatrix} = \begin{bmatrix} \cdot & & & & & & \\ \cdot & c(1) & c(0) & & & & \\ \cdot & c(3) & c(2) & c(1) & c(0) & & \\ \cdot & c(5) & c(4) & c(3) & c(2) & c(1) & c(0) \\ \cdot & \cdot & \cdot & \cdot & \cdot & \cdot & \cdot \end{bmatrix} \begin{bmatrix} \cdot \\ x(-1) \\ x(0) \\ x(1) \\ x(2) \\ \cdot \end{bmatrix}. \tag{3.30}$$

For polyphase here is the important point. *Only the even-numbered coefficients* $\boldsymbol{c}(2n)$ *are multiplying the even-numbered coefficients* $\boldsymbol{x}(2n)$. The even and odd $\boldsymbol{c}$'s are in separate columns. The even-numbered $\boldsymbol{x}(0)$ is multiplying the column that starts with $\boldsymbol{c}(0)$. The odd-numbered component $\boldsymbol{x}(1)$ is multiplying the column containing $\boldsymbol{c}(1), \boldsymbol{c}(3), \ldots$. We can separate the matrix multiplication $(\downarrow 2)C\boldsymbol{x}$ into *even times even and odd times odd*:

$$\begin{aligned} \begin{bmatrix} \cdot \\ y(0) \\ y(2) \\ y(4) \\ \cdot \end{bmatrix} &= \begin{bmatrix} \cdot & & & \\ \cdot & c(0) & & \\ \cdot & c(2) & c(0) & \\ \cdot & c(4) & c(2) & c(0) \\ \cdot & \cdot & \cdot & \cdot \end{bmatrix} \begin{bmatrix} \cdot \\ x(0) \\ x(2) \\ \cdot \end{bmatrix} \\ &+ \begin{bmatrix} \cdot & & & \\ \cdot & c(1) & & \\ \cdot & c(3) & c(1) & \\ \cdot & c(5) & c(3) & c(1) \\ \cdot & \cdot & \cdot & \cdot \end{bmatrix} \begin{bmatrix} \cdot \\ x(-1) \\ x(1) \\ \cdot \end{bmatrix} \end{aligned} \tag{3.31}$$

This is a matrix display of equation (3.27):

$$(\downarrow 2)C\boldsymbol{x} = C_{\text{even}}\, \boldsymbol{x}_{\text{even}} + (\text{delay})\, C_{\text{odd}}\, \boldsymbol{x}_{\text{odd}}. \tag{3.32}$$

The two phases $\boldsymbol{x}_{\text{even}}$ and $\boldsymbol{x}_{\text{odd}}$ are filtered by the two polyphase components C_{even} and C_{odd}. We need a delay in the odd phase, because $\boldsymbol{c}(1)\boldsymbol{x}(1)$ contributes to $\boldsymbol{y}(2)$ and not to $\boldsymbol{y}(0)$. Then $(\downarrow 2)C\boldsymbol{x}$ is the sum from the two phases in (3.32):

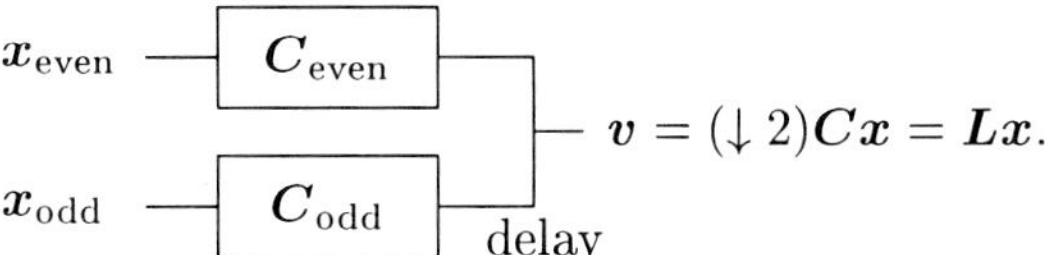

Notice something nice. The two matrices in equation (3.31) have constant diagonals. *The two operators* C_{even} *and* C_{odd} *are time-invariant filters.* They have frequency responses $C_0(z) = C_{\text{even}}(z)$ and $C_1(z) = C_{\text{odd}}(z)$. The delay in the odd channel can go before or after C_1, because it commutes with C_1. (C_1 is time-invariant!) The two filters involve even coefficients $c(2n)$ and odd coefficients $c(2n+1)$, without zeros in between. They can operate in parallel, more efficiently.

Key point The polyphase form puts $(\downarrow 2)$ *before* the filters. This order is more efficient. It was possible to use the Noble Identity on each phase separately, because $C_{\text{even}}(z^2)$ and also $C_{\text{odd}}(z^2)$ appeared in the right place:

$$\text{(direct)} \quad (\downarrow 2)C_{\text{even}}(z) = C_{\text{even}}(z^2)(\downarrow 2) \quad \text{(polyphase)}.$$

Example 2. Four-tap filters yield two taps for each phase. The even phase C_{even} has two coefficients $c(0)$ and $c(2)$. The odd phase has $C_{\text{odd}} = c(1) + c(3)z^{-1}$. The same pattern holds for D. The polyphase matrix for the filter bank is

$$H_p(z) = \begin{bmatrix} c(0) + c(2)z^{-1} & c(1) + c(3)z^{-1} \\ d(0) + d(2)z^{-1} & d(1) + d(3)z^{-1} \end{bmatrix}. \tag{3.33}$$

Even is separated from odd. This reflects what happens in the filter bank, when Cx and Dx are downsampled:

$$\begin{bmatrix} v_0 \\ v_1 \end{bmatrix} = \begin{bmatrix} (\downarrow 2)Cx \\ (\downarrow 2)Dx \end{bmatrix} = \begin{bmatrix} C_{\text{even}} & C_{\text{odd}} \\ D_{\text{even}} & D_{\text{odd}} \end{bmatrix} \begin{bmatrix} 1 & \\ & \text{delay} \end{bmatrix} \begin{bmatrix} x_{\text{even}} \\ x_{\text{odd}} \end{bmatrix}$$

In case you like matrices, I am going to write the time-domain filter bank matrix in three ways. Downsampling is included in all three! First comes the matrix $H_d = H_{\text{direct}}$ that multiplies the input vector x in the direct form:

$$\textbf{Direct} \quad H_d = \begin{bmatrix} \cdot & \cdot & & & & & & \\ c(3) & c(2) & c(1) & c(0) & & & & \\ & & c(3) & c(2) & c(1) & c(0) & & \\ & & & & \cdot & \cdot & \cdot & \cdot \\ \cdot & \cdot & & & & & & \\ d(3) & d(2) & d(1) & d(0) & & & & \\ & & d(3) & d(2) & d(1) & d(0) & & \\ & & & & \cdot & \cdot & \cdot & \cdot \end{bmatrix}. \tag{3.34}$$

Downsampling has removed every other row. That leaves this "square" infinite matrix. Each column is completely odd or completely even.

For the second form I rearrange the *rows* of $\boldsymbol{H}_d$. The highpass outputs are interleaved with the lowpass outputs, both downsampled by 2. This produces the block-diagonal form (or *block-Toeplitz form*) $\boldsymbol{H}_b = \boldsymbol{H}_{\text{block}}$:

$$\textbf{Block} \quad \boldsymbol{H}_b = \begin{bmatrix} \cdot & \cdot & & & & & \\ \cdot & \cdot & & & & & \\ \boldsymbol{c}(3) & \boldsymbol{c}(2) & \boldsymbol{c}(1) & \boldsymbol{c}(0) & & & \\ \boldsymbol{d}(3) & \boldsymbol{d}(2) & \boldsymbol{d}(1) & \boldsymbol{d}(0) & & & \\ & & \boldsymbol{c}(3) & \boldsymbol{c}(2) & \boldsymbol{c}(1) & \boldsymbol{c}(0) & \\ & & \boldsymbol{d}(3) & \boldsymbol{d}(2) & \boldsymbol{d}(1) & \boldsymbol{d}(0) & \\ & & & & \cdot & \cdot & \cdot\ \cdot \\ & & & & \cdot & \cdot & \cdot\ \cdot \end{bmatrix}. \tag{3.35}$$

Your eye will divide that matrix into 2 by 2 blocks. It is like an ordinary time-invariant constant-diagonal matrix, but the entries are blocks instead of scalars. The main diagonal block corresponds to the constants in the polyphase matrix. The subdiagonal block produces the z^{-1} terms. There are only two diagonals because the phases of C and D have *two* coefficients. The original C and D had four coefficients.

The third form is the polyphase form $\boldsymbol{H}_p$. We are still in the time domain. For this third form I rearrange the *columns* of the direct form:

$$\textbf{Polyphase} \quad \boldsymbol{H}_p = \begin{bmatrix} \cdot & & & & \cdot & & & \\ \boldsymbol{c}(2) & \boldsymbol{c}(0) & & & \boldsymbol{c}(3) & \boldsymbol{c}(1) & & \\ & \boldsymbol{c}(2) & \boldsymbol{c}(0) & & & \boldsymbol{c}(3) & \boldsymbol{c}(1) & \\ & & \cdot & \cdot & & & \cdot & \cdot \\ \cdot & & & & \cdot & & & \\ \boldsymbol{d}(2) & \boldsymbol{d}(0) & & & \boldsymbol{d}(3) & \boldsymbol{d}(1) & & \\ & \boldsymbol{d}(2) & \boldsymbol{d}(0) & & & \boldsymbol{d}(3) & \boldsymbol{d}(1) & \\ & & \cdot & \cdot & & & \cdot & \cdot \end{bmatrix}$$

$$= \begin{bmatrix} C_0 & C_1 \\ D_0 & D_1 \end{bmatrix}. \tag{3.36}$$

When the columns are rearranged, the vector $\boldsymbol{x}$ must be rearranged. Here $\boldsymbol{x}_{\text{even}}$ comes above $\boldsymbol{x}_{\text{odd}}$(delayed). The transform of the time-domain matrix $\boldsymbol{H}_p$ is the z-domain polyphase matrix $\boldsymbol{H}_p(z)$. *This* 2 *by* 2 *matrix of filters becomes a* 2 *by* 2 *matrix of functions.*

The block form $\boldsymbol{H}_b$ is an infinite matrix of 2 by 2 blocks. The polyphase form $\boldsymbol{H}_p$ is a 2 by 2 matrix of infinite blocks. Each block is a time-invariant filter. Either of those matrices leads directly by z-transform, to the 2×2 polyphase matrix $\boldsymbol{h}_p(0) + z^{-1}\boldsymbol{h}_p(1)$:

$$\textbf{Polyphase matrix} \quad \boldsymbol{H}_p(z) = \begin{bmatrix} \boldsymbol{c}(0) & \boldsymbol{c}(1) \\ \boldsymbol{d}(0) & \boldsymbol{d}(1) \end{bmatrix} + z^{-1} \begin{bmatrix} \boldsymbol{c}(2) & \boldsymbol{c}(3) \\ \boldsymbol{d}(2) & \boldsymbol{d}(3) \end{bmatrix} \tag{3.37}$$

Paraunitary Matrices

Definition 3.1 *The matrix $\boldsymbol{H}(z)$ is paraunitary if it is unitary for all $|z| = 1$:*

$$\boxed{\boldsymbol{H}^T(e^{-j\omega})\boldsymbol{H}(e^{j\omega}) = \boldsymbol{I} \quad \textit{for all } \omega.} \tag{3.38}$$

This extends to all $z \neq 0$ by $\widetilde{\boldsymbol{H}}(z) = \boldsymbol{H}^T(z^{-1})$. Then a paraunitary matrix has

$$\boldsymbol{H}^T(z^{-1})\boldsymbol{H}(z) = \widetilde{\boldsymbol{H}}(z)\ \boldsymbol{H}(z) = \boldsymbol{I} \quad \textit{for all } z. \tag{3.39}$$

When the coefficients $\boldsymbol{h}(k)$ are complex, they are conjugated in $\widetilde{\boldsymbol{H}}(z)$.

The matrix $\boldsymbol{H}$ need not be 2 by 2. If it is 1 by 1, then $|\boldsymbol{H}(e^{j\omega})| = 1$. The corresponding filter is *allpass.* The best allpass examples are ratios of polynomials coming from IIR filters — since only trivial polynomials z^{-l} can have $|\boldsymbol{H}(e^{j\omega})| = 1$.

If $\boldsymbol{H}(z)$ is $M \times M$, it could come from an M-channel filter bank. It might be the polyphase matrix $\boldsymbol{H}_p(z)$ or the modulation matrix $\boldsymbol{H}_m(z)$ (divided by $\sqrt{2}$). We will show that the filter bank is orthogonal if these matrices are paraunitary. That is the important connection for this book.

Equation (3.38) gives the inverse matrix by transposing and conjugating the original. The synthesis bank comes by "reversing" the analysis bank. Note that for a square matrix, $\boldsymbol{H}(z)$ is paraunitary when $\boldsymbol{H}^{-1}(z)$ and $\boldsymbol{H}^T(z)$ and $\widetilde{\boldsymbol{H}}(z)$ are paraunitary. And notice especially what equation (3.39) says about the *determinants* of these matrices:

$$\left(\det \widetilde{\boldsymbol{H}}(z)\right)(\det \boldsymbol{H}(z)) = 1. \tag{3.40}$$

The determinants are 1 by 1 allpass!

Theorem 3.4 *If a square paraunitary matrix $\boldsymbol{H}(z)$ is FIR (= polynomial), then its determinant must be a delay:*

$$\det \boldsymbol{H}(z) = \pm z^{-l}. \tag{3.41}$$

The determinant of $\boldsymbol{H}_p(z)$ is also a delay for any *bi*orthogonal filter bank. Orthogonality requires more; the polyphase matrix $\boldsymbol{H}_p(z)$ must be paraunitary.

Orthonormal Filter Banks

This section brings together the requirements for an orthonormal filter bank. We will see those requirements in the *time domain* and the *polyphase domain* and the *modulation domain*. These requirements are conditions on the filter coefficients $\boldsymbol{c}(k)$ and $\boldsymbol{d}(k)$. Then equation (3.54) indicates a simple choice of the $\boldsymbol{d}$'s coming from the $\boldsymbol{c}$'s. If the lowpass filter meets the orthogonality requirements, it is easy to construct a highpass filter to go with it.

The discussion is in terms of a 2-channel FIR filter bank, $M = 2$. But the conditions extend immediately to any M. *The polyphase matrix and the modulation matrix must be paraunitary.* In the M-channel case, the lowpass filter does not immediately determine the $M-1$ remaining filters.

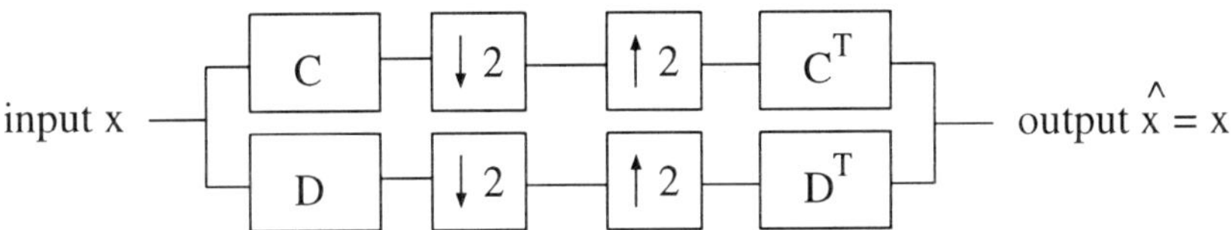

An orthogonal filter bank has synthesis bank = transpose of analysis bank.

The figure shows the structure of an orthogonal filter bank. We intend to achieve $\hat{\boldsymbol{x}} = \boldsymbol{x}$, with synthesis filters $\boldsymbol{C}^T$ and $\boldsymbol{D}^T$ that are time-reversals of the analysis filters:

$$\widetilde{\boldsymbol{C}} = \boldsymbol{C}^T \quad \text{and} \quad \tilde{\boldsymbol{c}}(n) = \boldsymbol{c}(-n) \tag{3.42}$$

$$\widetilde{\boldsymbol{D}} = \boldsymbol{D}^T \quad \text{and} \quad \tilde{\boldsymbol{d}}(n) = \boldsymbol{d}(-n) \tag{3.43}$$

As it stands, $\widetilde{\boldsymbol{C}}$ and $\widetilde{\boldsymbol{D}}$ are anticausal. At the end we make them causal by N delays. The output $\hat{\boldsymbol{x}}(n)$ is equally delayed; it is $\boldsymbol{x}(n-N)$. But the algebra is easiest for $\boldsymbol{C}^T$ and $\boldsymbol{D}^T$ with no delays.

This special structure imposes special conditions on the $\boldsymbol{c}$'s and $\boldsymbol{d}$'s for perfect reconstruction. We will call the requirements **Condition O** (for orthogonality). This section finds four equivalent forms: Condition O on the infinite matrix, on the lowpass coefficients, and on the polyphase and modulation matrices $\boldsymbol{H}_p$ and $\boldsymbol{H}_m$. We look first at the infinite matrices in the time domain.

Time Domain: Condition O and the Alternating Flip

The key matrix in the time domain is $\boldsymbol{H}_t$. It represents the direct form of the analysis bank, with downsampling. The lowpass part $\boldsymbol{L} = (\downarrow 2)\boldsymbol{C}$ comes above the highpass part $\boldsymbol{B} = (\downarrow 2)\boldsymbol{D}$. We display this infinite matrix for filters of length four:

$$\boldsymbol{H}_t = \begin{bmatrix} \boldsymbol{L} \\ \boldsymbol{B} \end{bmatrix} = \begin{bmatrix} \boldsymbol{c}(3) & \boldsymbol{c}(2) & \boldsymbol{c}(1) & \boldsymbol{c}(0) & & \\ & & \boldsymbol{c}(3) & \boldsymbol{c}(2) & \boldsymbol{c}(1) & \boldsymbol{c}(0) \\ & & & & \cdot & \cdot \\ \boldsymbol{d}(3) & \boldsymbol{d}(2) & \boldsymbol{d}(1) & \boldsymbol{d}(0) & & \\ & & \boldsymbol{d}(3) & \boldsymbol{d}(2) & \boldsymbol{d}(1) & \boldsymbol{d}(0) \\ & & & & \cdot & \cdot \end{bmatrix}. \tag{3.44}$$

The shifts by 2 were created by downsampling, which removed the odd-numbered rows.

With orthogonality, the synthesis filters are to be time-reversals of the analysis filters. The infinite synthesis matrix contains the transposes of $(\downarrow 2)\boldsymbol{C}$ and $(\downarrow 2)\boldsymbol{D}$. Those transposes are $\boldsymbol{C}^T(\uparrow 2)$ and $\boldsymbol{D}^T(\uparrow 2)$, since upsampling is the transpose of downsampling:

$$\boldsymbol{H}_t^T = \begin{bmatrix} \boldsymbol{L}^T & \boldsymbol{B}^T \end{bmatrix} = \begin{bmatrix} \boldsymbol{c}(3) & & & \boldsymbol{d}(3) & & \\ \boldsymbol{c}(2) & & & \boldsymbol{d}(2) & & \\ \boldsymbol{c}(1) & \boldsymbol{c}(3) & & \boldsymbol{d}(1) & \boldsymbol{d}(3) & \\ \boldsymbol{c}(0) & \boldsymbol{c}(2) & & \boldsymbol{d}(0) & \boldsymbol{d}(2) & \\ & \boldsymbol{c}(1) & \cdot & & \boldsymbol{d}(1) & \cdot \\ & \boldsymbol{c}(0) & \cdot & & \boldsymbol{d}(0) & \cdot \end{bmatrix}. \tag{3.45}$$

The shifts by 2 in the columns are created by upsampling, which removes every other column. We require $\hat{\boldsymbol{x}} = \boldsymbol{x}$. This means that $\boldsymbol{H}_t^T\boldsymbol{H}_t = \boldsymbol{I}$. The matrix $\boldsymbol{H}_t$ is required to be an *orthogonal matrix*. Its columns are orthonormal and so are its rows: $\boldsymbol{H}_t\boldsymbol{H}_t^T = \boldsymbol{I}$. We can express this Condition O in matrix form, and in block form, and in coefficient form.

Condition O An orthogonal filter bank comes from an orthogonal matrix:

$$\boldsymbol{H}_t^T\boldsymbol{H}_t = \boldsymbol{I} \quad \text{and} \quad \boldsymbol{H}_t\boldsymbol{H}_t^T = \boldsymbol{I}. \tag{3.46}$$

In block form this means that

$$\begin{bmatrix} \boldsymbol{L}^T & \boldsymbol{B}^T \end{bmatrix} \begin{bmatrix} \boldsymbol{L} \\ \boldsymbol{B} \end{bmatrix} = \boldsymbol{L}^T\boldsymbol{L} + \boldsymbol{B}^T\boldsymbol{B} = \boldsymbol{I} \tag{3.47}$$

and

$$\begin{bmatrix} \boldsymbol{L} \\ \boldsymbol{B} \end{bmatrix} \begin{bmatrix} \boldsymbol{L}^T & \boldsymbol{B}^T \end{bmatrix} = \begin{bmatrix} \boldsymbol{L}\boldsymbol{L}^T & \boldsymbol{L}\boldsymbol{B}^T \\ \boldsymbol{B}\boldsymbol{L}^T & \boldsymbol{B}\boldsymbol{B}^T \end{bmatrix} = \begin{bmatrix} \boldsymbol{I} & \boldsymbol{0} \\ \boldsymbol{0} & \boldsymbol{I} \end{bmatrix}. \tag{3.48}$$

For the coefficients $\boldsymbol{c}(k)$ and $\boldsymbol{d}(k)$, equation (3.48) becomes *orthogonality to double shifts*:

$$\boldsymbol{L}\boldsymbol{L}^T = \boldsymbol{I}: \quad \sum \boldsymbol{c}(n)\boldsymbol{c}(n-2k) = \boldsymbol{\delta}(k) \tag{3.49}$$

$$\boldsymbol{L}\boldsymbol{B}^T = \boldsymbol{0}: \quad \sum \boldsymbol{c}(n)\boldsymbol{d}(n-2k) = 0 \tag{3.50}$$

$$\boldsymbol{B}\boldsymbol{B}^T = \boldsymbol{I}: \quad \sum \boldsymbol{d}(n)\boldsymbol{d}(n-2k) = \boldsymbol{\delta}(k). \tag{3.51}$$

Because of (3.49)—(3.51), we refer to Condition O as **"double-shift orthogonality"**. Its equivalent in the frequency domain is presented below. This double-shift orthogonality immediately rules out odd length filters! If the length is $N+1=5$, a shift by 4 gives an inner product that cannot be zero:

$$(\boldsymbol{c}(0), \boldsymbol{c}(1), \boldsymbol{c}(2), \boldsymbol{c}(3), \boldsymbol{c}(4)) \cdot (0,0,0,0,\boldsymbol{c}(0)) = \boldsymbol{c}(0)\boldsymbol{c}(4) \neq 0.$$

N cannot be even (the filter length cannot be odd) because $\frac{N}{2}$ double shifts would give a shift by N — and the inner product $\boldsymbol{c}(0)\boldsymbol{c}(N)$ is not zero. So N is odd. The degree $2N$ of the halfband product filter $\boldsymbol{P}$ is $2, 6, 10, \ldots$ The clearest examples have $N = 3$ and $2N = 6$.

Example Condition O in (3.49) imposes two constraints on four coefficients:

$$\boldsymbol{c}(0)^2 + \boldsymbol{c}(1)^2 + \boldsymbol{c}(2)^2 + \boldsymbol{c}(3)^2 = 1 \quad \text{and} \quad \boldsymbol{c}(0)\boldsymbol{c}(2) + \boldsymbol{c}(1)\boldsymbol{c}(3) = 0. \tag{3.52}$$

Equation (3.51) is an identical condition on the $\boldsymbol{d}$'s, from $\boldsymbol{B}\boldsymbol{B}^T = \boldsymbol{I}$. Equation (3.50) is the orthogonality of the rows of $\boldsymbol{L}$ to the rows of $\boldsymbol{B}$. Those are the fundamental design constraints on an orthogonal filter bank.

The conditions on the $\boldsymbol{c}$'s and $\boldsymbol{d}$'s are independent, but something good happens. *If the* $\boldsymbol{c}$*'s satisfy equation* (3.49), *it is easy to choose the* $\boldsymbol{d}$*'s.* We display a choice of $\boldsymbol{d}$'s that automatically gives orthogonality:

$$\begin{bmatrix} \boldsymbol{L} \\ \boldsymbol{B} \end{bmatrix} = \begin{bmatrix} \boldsymbol{c}(3) & \boldsymbol{c}(2) & \boldsymbol{c}(1) & \boldsymbol{c}(0) & & \\ & & \boldsymbol{c}(3) & \boldsymbol{c}(2) & \boldsymbol{c}(1) & \boldsymbol{c}(0) \\ & & & & \cdot & \cdot \\ -\boldsymbol{c}(0) & \boldsymbol{c}(1) & -\boldsymbol{c}(2) & \boldsymbol{c}(3) & & \\ & & -\boldsymbol{c}(0) & \boldsymbol{c}(1) & -\boldsymbol{c}(2) & \boldsymbol{c}(3) \\ & & & & \cdot & \cdot \end{bmatrix}. \tag{3.53}$$

This is the **alternating flip**. The $\boldsymbol{c}$'s are reversed in order and alternated in sign, to produce the $\boldsymbol{d}$'s. We start with lowpass coefficients $\boldsymbol{c}(0), \ldots, \boldsymbol{c}(N)$, where N is odd. The highpass coefficients are

$$\boxed{\boldsymbol{d}(k) = (-1)^k \boldsymbol{c}(N-k).} \tag{3.54}$$

The essential point can be checked by eye, in the infinite matrix (3.53). *If the* **top** *rows are orthogonal to each other, then* **all** *rows are orthogonal.* The zero dot products in $\boldsymbol{L}\boldsymbol{B}^T$ are:

$$\begin{aligned} -\boldsymbol{c}(3)\boldsymbol{c}(0) + \boldsymbol{c}(2)\boldsymbol{c}(1) - \boldsymbol{c}(1)\boldsymbol{c}(2) + \boldsymbol{c}(0)\boldsymbol{c}(3) &= 0 \\ -\boldsymbol{c}(1)\boldsymbol{c}(0) + \boldsymbol{c}(0)\boldsymbol{c}(1) &= 0. \end{aligned} \tag{3.55}$$

Furthermore, the $\boldsymbol{d}$'s are orthonormal within themselves ($\boldsymbol{B}\boldsymbol{B}^T = \boldsymbol{I}$) because the $\boldsymbol{c}$'s are. Equations (3.52) hold for the $\boldsymbol{d}$'s, when they are constructed by flipping the $\boldsymbol{c}$'s. Minus signs cancel in $\boldsymbol{d}(3)\boldsymbol{d}(1)$ which is $(-\boldsymbol{c}(0))$ $(-\boldsymbol{c}(2))$.

Our four-tap example has $N = 3$. The alternating flip gives $\boldsymbol{L}\boldsymbol{B}^T = 0$ for every odd N. The top rows of $\boldsymbol{H}_t$ in (5.18) are *always* orthogonal to the bottom rows. Also (3.51) for the $\boldsymbol{d}$'s follows from (3.49) for the $\boldsymbol{c}$'s. Thus the alternating flip reduces orthogonality to (3.49):

$$\textbf{Condition O} \text{ on the coefficients:} \quad \sum \boldsymbol{c}(n)\boldsymbol{c}(n-2k) = \boldsymbol{\delta}(k).$$

Polyphase Domain: Condition O and the Alternating Flip

The polyphase form separates $(\downarrow 2)\boldsymbol{C}$ and $(\downarrow 2)\boldsymbol{D}$ into even phase and odd phase. In the time domain, we are rearranging the columns of the matrix $\boldsymbol{H}_t$. All even columns come before the odd columns. The matrix goes into this 2 by 2 block form with time-invariant filters as the blocks:

$$\boldsymbol{H}_{\text{block}} = \begin{bmatrix} \boldsymbol{C}_{\text{even}} & \boldsymbol{C}_{\text{odd}} \\ \boldsymbol{D}_{\text{even}} & \boldsymbol{D}_{\text{odd}} \end{bmatrix}.$$

In the z-domain, we are rearranging the response functions $C(z)$ and $D(z)$:

$$\sum \boldsymbol{c}(k)z^{-k} = C_{\text{even}}(z^{-2}) + z^{-1}C_{\text{odd}}(z^{-2}). \tag{3.56}$$

Those phase responses are written $H_{00}(z)$ and $H_{01}(z)$ when $C(z)$ is $H_0(z)$. The highpass response $D(z) = H_1(z)$ decomposes in the same way. *The polyphase matrix is*

$$\boldsymbol{H}_p(z) = \begin{bmatrix} C_{\text{even}}(z) & C_{\text{odd}}(z) \\ D_{\text{even}}(z) & D_{\text{odd}}(z) \end{bmatrix} = \begin{bmatrix} H_{00}(z) & H_{01}(z) \\ H_{10}(z) & H_{11}(z) \end{bmatrix}. \tag{3.57}$$

Now Condition O translates directly into a requirement on $\boldsymbol{H}_p(z)$. A filter bank is **orthogonal** when its polyphase matrix is **paraunitary**:

$$\boldsymbol{H}_p^T(e^{-j\omega})\boldsymbol{H}_p(e^{j\omega}) = \boldsymbol{I} \text{ for all } \omega \text{ and } \widetilde{\boldsymbol{H}}_p(z)\boldsymbol{H}_p(z) = \boldsymbol{I} \text{ for all } z. \tag{3.58}$$

The inverse of $\boldsymbol{H}_p(z)$ is the synthesis polyphase matrix. The matrix and its inverse can be multiplied in either order — here is analysis times synthesis:

$$\begin{bmatrix} C_{\text{even}}(z) & C_{\text{odd}}(z) \\ D_{\text{even}}(z) & D_{\text{odd}}(z) \end{bmatrix} \begin{bmatrix} C_{\text{even}}(z^{-1}) & D_{\text{even}}(z^{-1}) \\ C_{\text{odd}}(z^{-1}) & D_{\text{odd}}(z^{-1}) \end{bmatrix} = \begin{bmatrix} 1 & 0 \\ 0 & 1 \end{bmatrix}. \tag{3.59}$$

On the unit circle, where z^{-1} is $\bar{z}$, row 1 times column 1 becomes

$$\boxed{|C_{\text{even}}(z)|^2 + |C_{\text{odd}}(z)|^2 = 1 \text{ when } |z| = 1.} \tag{3.60}$$

This is the essence of Condition O in the polyphase domain.

For the example with four coefficients, it is helpful to multiply out equation (3.60):

$$\begin{aligned} &\left(\boldsymbol{c}(0) + \boldsymbol{c}(2)z^{-1}\right)\left(\boldsymbol{c}(0) + \boldsymbol{c}(2)z\right) + \left(\boldsymbol{c}(1) + \boldsymbol{c}(3)z^{-1}\right)\left(\boldsymbol{c}(1) + \boldsymbol{c}(3)z\right) = \\ &\boldsymbol{c}(0)^2 + \boldsymbol{c}(2)^2 + \boldsymbol{c}(1)^2 + \boldsymbol{c}(3)^2 + [\boldsymbol{c}(0)\boldsymbol{c}(2) + \boldsymbol{c}(1)\boldsymbol{c}(3)]\left(z^{-1} + z\right) = 1. \end{aligned}$$

Thus (3.60) is equivalent to the explicit statement (3.52). The sum of squares is 1 and the dot product $\boldsymbol{c}(0)\boldsymbol{c}(2) + \boldsymbol{c}(1)\boldsymbol{c}(3)$ is zero.

The other multiplications in (3.59) give answers 1 or 0 in the same way. All these requirements on the $\boldsymbol{d}$'s are automatically satisfied by the flip construction! We write that choice $\boldsymbol{d}(k) = (-1)^k \boldsymbol{c}(N-k)$ in the z-domain:

$$\sum \boldsymbol{d}(k) z^{-k} = \sum \boldsymbol{c}(N-k)(-z)^{-k} = \sum \boldsymbol{c}(n)(-z)^{n-N}.$$

This relation between highpass and lowpass is an *alternating flip*:

$$D(z) = (-z)^{-N} C(-z^{-1}). \tag{3.61}$$

The number N is odd. Because of $(-z)^{-N}$, the even and odd phases in $\boldsymbol{C}$ are reversed to odd and even phases in $\boldsymbol{D}$. We take $N = 3$ as typical. An alternating flip of $\boldsymbol{c}(0), \boldsymbol{c}(1), \boldsymbol{c}(2), \boldsymbol{c}(3)$ yields

$$\begin{aligned} D(z) &= \boldsymbol{c}(3) - \boldsymbol{c}(2)z^{-1} + \boldsymbol{c}(1)z^{-2} - \boldsymbol{c}(0)z^{-3} \\ &= \left(\boldsymbol{c}(3) + \boldsymbol{c}(1)z^{-2}\right) - z^{-1}\left(\boldsymbol{c}(2) + \boldsymbol{c}(0)z^{-2}\right). \end{aligned} \tag{3.62}$$

With this flip in row 2, the multiplication $\boldsymbol{H}_p(z)\boldsymbol{H}_p^T(z^{-1})$ becomes

$$\begin{bmatrix} \boldsymbol{c}(0) + \boldsymbol{c}(2)z^{-1} & \boldsymbol{c}(1) + \boldsymbol{c}(3)z^{-1} \\ \boldsymbol{c}(3) + \boldsymbol{c}(1)z^{-1} & -\boldsymbol{c}(2) - \boldsymbol{c}(0)z^{-1} \end{bmatrix} \begin{bmatrix} \boldsymbol{c}(0) + \boldsymbol{c}(2)z & \boldsymbol{c}(3) + \boldsymbol{c}(1)z \\ \boldsymbol{c}(1) + \boldsymbol{c}(3)z & -\boldsymbol{c}(2) - \boldsymbol{c}(0)z \end{bmatrix} = \boldsymbol{I}.$$

The off-diagonal entries of the product are *automatically* zero. *The alternating flip achieves* $\boldsymbol{L}\boldsymbol{B}^T = \boldsymbol{0}$ *with or without orthogonality*. The 2, 2 entry of the product is the same as the 1, 1 entry, when $|z| = 1$. The orthogonality requirement (3.60) makes the 1, 1 entry equal to 1.

Modulation Domain: Condition O and the Alternating Flip

The function that arises from modulating $C(z)$ is $C(-z)$. The frequency in $z = e^{j\omega}$ changes by π to produce $-z = e^{j(\omega+\pi)}$. This modulation takes $C(\omega)$ to $C(\omega+\pi)$. The frequency response graph is shifted by π.

Our goal is to relate $C(z)$ to $C(-z)$ and $D(z)$ to $D(-z)$ for an orthogonal filter bank. We already know Condition O on the coefficients:

$$\boldsymbol{c}(0)^2 + \boldsymbol{c}(1)^2 + \boldsymbol{c}(2)^2 + \boldsymbol{c}(3)^2 = 1 \quad \text{and} \quad \boldsymbol{c}(0)\boldsymbol{c}(2) + \boldsymbol{c}(1)\boldsymbol{c}(3) = 0.$$

Watch how 1 and 0 appear in $|C(z)|^2$. Stay on the unit circle where $\bar{z}^{-1} = z$:

$$\begin{aligned} &\left(\boldsymbol{c}(0) + \boldsymbol{c}(1)z^{-1} + \boldsymbol{c}(2)z^{-2} + \boldsymbol{c}(3)z^{-3}\right)\left(\boldsymbol{c}(0) + \boldsymbol{c}(1)z + \boldsymbol{c}(2)z^2 + \boldsymbol{c}(3)z^3\right) = \\ &1 + [\boldsymbol{c}(0)\boldsymbol{c}(1) + \boldsymbol{c}(1)\boldsymbol{c}(2) + \boldsymbol{c}(2)\boldsymbol{c}(3)]\left(z^{-1} + z\right) + \boldsymbol{c}(0)\boldsymbol{c}(3)\left(z^{-3} + z^3\right). \end{aligned}$$

Now change z to $-z$. The odd powers z and z^3 change sign. When we add, those odd powers cancel:

$$\boxed{|C(z)|^2 + |C(-z)|^2 = |C(\omega)|^2 + |C(\omega+\pi)|^2 = 2 \quad \text{for all } z = e^{j\omega}.} \tag{3.63}$$

This is the *halfband condition*, also called the *Nyquist condition*: $|C(z)|^2$ is a halfband filter. Those filters we can design! In other words, the lowpass analysis filter $C(z)$ is a spectral factor of a halfband filter.

Condition O on $\boldsymbol{H}_m(z)$ *The modulation matrix of an orthogonal filter bank is a paraunitary matrix times* $\sqrt{2}$:

$$\boldsymbol{H}_m(z)\widetilde{\boldsymbol{H}}_m(z) = 2\boldsymbol{I} \quad \text{for all } z. \tag{3.64}$$

On the circle $z = e^{j\omega}$, the modulation matrix is a unitary matrix times $\sqrt{2}$:

$$\begin{bmatrix} C(\omega) & C(\omega+\pi) \\ D(\omega) & D(\omega+\pi) \end{bmatrix} \begin{bmatrix} \overline{C(\omega)} & \overline{D(\omega)} \\ \overline{C(\omega+\pi)} & \overline{D(\omega+\pi)} \end{bmatrix} = \begin{bmatrix} 2 & 0 \\ 0 & 2 \end{bmatrix}. \tag{3.65}$$

The 1, 1 entry of this matrix product is $|C(\omega)|^2 + |C(\omega+\pi)|^2 = 2$ by (3.63). The other entries, when we multiply them out, follow immediately from (3.50) and (3.51). Thus Condition O on the coefficients is equivalent to Condition O on the modulation matrix $\boldsymbol{H}_m$. It is also equivalent to Condition O on $\boldsymbol{H}_p$. It is the statement that the analysis bank followed by its transpose gives perfect reconstruction. We summarize:

Theorem 3.5 *For an orthogonal filter bank the lowpass coefficients must satisfy* Condition O *(we give four equivalent forms).*

Matrix form	$\boldsymbol{L}\boldsymbol{L}^T = (\downarrow 2)\boldsymbol{C}\boldsymbol{C}^T(\uparrow 2) = \boldsymbol{I}$
Coefficient form	$\sum \boldsymbol{c}(n)\boldsymbol{c}(n-2k) = \boldsymbol{\delta}(k)$
Polyphase form	$\lvert C_{\text{even}}(e^{j\omega})\rvert^2 + \lvert C_{\text{odd}}(e^{j\omega})\rvert^2 = 1$
Modulation form	$\lvert C(\omega)\rvert^2 + \lvert C(\omega+\pi)\rvert^2 = 2.$

By an alternating flip, $\boldsymbol{L}\boldsymbol{B}^T = \boldsymbol{0}$ and $\boldsymbol{B}\boldsymbol{B}^T = \boldsymbol{I}$ follow immediately from the lowpass part $\boldsymbol{L}\boldsymbol{L}^T = \boldsymbol{I}$. The real problem is the design of the lowpass filter.

Symmetry Prevents Orthogonality

It is natural to want two good properties at once. Symmetry is good for the eye, and orthogonality is good for the algorithm. But the only filters with both properties are averaging filters (Haar filters) with two coefficients. We are forced to use IIR filters, or M channels, or multifilters with matrix coefficients. Extra computation is unavoidable, because the next theorem rules out a perfect filter.

Theorem 3.6 *A symmetric orthogonal FIR filter can only have two nonzero coefficients.*

Proof N is odd for orthogonality. The filter length must be even. With $N = 5$ a symmetric filter of length 6 has the form $(\boldsymbol{c}(0), \boldsymbol{c}(1), \boldsymbol{c}(2), \boldsymbol{c}(2), \boldsymbol{c}(1), \boldsymbol{c}(0))$. This vector must be orthogonal to all its double shifts. The inner product with its shift by *four* must be $2\,\boldsymbol{c}(0)\boldsymbol{c}(1) = 0$. Therefore $\boldsymbol{c}(1) = 0$. Then the inner product with its shift by *two* gives $2\,\boldsymbol{c}(0)\boldsymbol{c}(2) = 0$. The only nonzero coefficient is $\boldsymbol{c}(0)$ at both ends of the filter. This completes the proof.

By convention $\boldsymbol{c}(0)$ is the first nonzero coefficient. Shift the filter if necessary to achieve this. The only symmetric orthogonal possibilities are $\boldsymbol{c} = (1,1)/\sqrt{2}$ and $(1,0,0,1)/\sqrt{2}$ and $(1,0,\ldots,0,1)/\sqrt{2}$. Only the Haar coefficients $(1,1)/\sqrt{2}$ will lead to orthogonal wavelets. Symmetry really conflicts with orthogonality.

A second proof observes that the odd phase is the flip of the even phase:

$$(\boldsymbol{c}(0), \boldsymbol{c}(4), \boldsymbol{c}(2), \boldsymbol{c}(2), \boldsymbol{c}(4), \boldsymbol{c}(0)) \text{ has } |C_{\text{even}}(z)|^2 = |C_{\text{odd}}(z)|^2.$$

Condition O is $|C_{\text{even}}(z)|^2 + |C_{\text{odd}}(z)|^2 = 2$. With symmetry this separates into $|C_{\text{even}}(z)|^2 = 1$ and $|C_{\text{odd}}(z)|^2 = 1$. *The even phase is an allpass filter* ! So is the odd phase. But FIR allpass filters can only have one nonzero coefficient, which completes the second proof.

Spectral Factorization

Out of all these equations we would like to emphasize one. It came at the end. It applied first of all to the lowpass filter $C(z)$, and then by the alternating flip also to $D(z)$. It was equation (3.63), that the frequency response $C(\omega) = \sum \boldsymbol{c}(k)e^{-jk\omega}$ satisfies

$$|C(\omega)|^2 + |C(\omega+\pi)|^2 = 2. \tag{3.66}$$

The key question is, *what does equation* (3.66) *say about* $|C(\omega)|^2$ *itself?*

We assign the symbol $P(\omega)$ to this important quantity $|C(\omega)|^2$. It is the **power spectral response**. Because $C(\omega)$ multiplies $\overline{C(\omega)}$, the filter with this response $P(\omega)$ is *symmetric*. It is the "autocorrelation filter". When $\sum \boldsymbol{c}(k)e^{-jk\omega}$ multiplies its conjugate $\sum \boldsymbol{c}(l)e^{jl\omega}$, we watch for $n = k - l$ which is $l = k - n$. The coefficient $\boldsymbol{p}(n)$ is the sum of $\boldsymbol{c}(k)$ times $\boldsymbol{c}(k-n)$:

$$\boldsymbol{p}(n) = \sum \boldsymbol{c}(k)\boldsymbol{c}(k-n) = \text{autocorrelation of the sequence } \boldsymbol{c}(k). \tag{3.67}$$

Autocorrelation is $\boldsymbol{p} = \boldsymbol{c} * \boldsymbol{c}^T$. This is the convolution of $\boldsymbol{c} = (\boldsymbol{c}(0), \boldsymbol{c}(1), \boldsymbol{c}(2), \ldots)$ with its time reversal $\boldsymbol{c}^T = (\ldots, \boldsymbol{c}(2), \boldsymbol{c}(1), \boldsymbol{c}(0))$. Replacing $-n$ by n in (3.67) brings no change in $\boldsymbol{p}$.

The reason for the surprising notation $\boldsymbol{c}^T$ is that multiplying $C(\omega)$ by $\overline{C(\omega)}$ corresponds exactly to multiplying the infinite filter matrix $\boldsymbol{C}$ by

C^T. $\boldsymbol{P} = \boldsymbol{C}\boldsymbol{C}^T$ is a symmetric positive definite (or semidefinite) Toeplitz matrix.

What does equation (3.66) say about the other coefficients $\boldsymbol{p}(n)$? In a word, it says *nothing* about the odd coefficients and it assigns *zero* to the even coefficients. $\boldsymbol{C}$ is the start of an orthogonal filter bank if and only if the autocorrelation filter $\boldsymbol{P}$ is a **halfband filter**:

$$P(z) + P(-z) = 2. \tag{3.68}$$

The *even coefficients* with $n = 2m$ must be $\boldsymbol{\delta}(m)$:

$$\textbf{Halfband filter} \quad \boldsymbol{p}(2m) = \sum \boldsymbol{c}(k)\boldsymbol{c}(k-2m) = \begin{cases} 1 & \text{if } m = 0 \\ 0 & \text{if } m \neq 0. \end{cases} \tag{3.69}$$

In an orthonormal filter bank, $C(z) = \sqrt{2}\,H(z)$ is a *spectral factor* of a symmetric halfband filter $P(z)$. The factorization is $P(z) = C(z^{-1})C(z)$ and the halfband property is $P(z) + P(-z) = 2$. In frequency, $P(\omega) = |C(\omega)|^2$ achieves the orthogonality condition $|C(\omega)|^2 + |C(\omega+\pi)|^2 = 2$. In the reverse direction, $P(z)$ is the *autocorrelation* of $C(z)$. This intimate relation of spectral factor $C(z)$ and its autocorrelation $P(z)$ is fundamental throughout signal processing.

Two questions arise immediately:

1. (Theory) Can every polynomial with $P(\omega) \geq 0$ be factored into $|C(\omega)|^2$?

2. (Practice) How is this spectral factorization actually done?

The answer to Question 1 is *yes*. This is the Féjer-Riesz Theorem. The answer to Question 2 is not so quick. There are many competing algorithms for spectral factorization. Short filters offer no serious difficulty, but with 100 or even 50 coefficients the weaker algorithms become slow and/or unreliable. When $C(\omega)$ is only approximate, the reconstruction is not perfect.

The trigonometric polynomials $P(\omega)$ and $C(\omega)$ are both of degree N:

$$\sum_{-N}^{N} \boldsymbol{p}(n)e^{-in\omega} = |C(e^{i\omega})|^2 = \left| \sum_{0}^{N} \boldsymbol{c}(n)e^{-in\omega} \right|^2 .$$

$P(z)$ has symmetric coefficients $\boldsymbol{p}(n) = \boldsymbol{p}(-n)$. There are $N+1$ independent coefficients in P and the same number in C. They are linked by quadratic equations, when we solve $P(\omega) = |C(\omega)|^2$. Those equations are solvable if and only if $P(\omega) \geq 0$ for all ω.

As an aside, note that *matrix* spectral factorization is also possible where $P(\omega)$ is symmetric positive definite. Both 1 and 2, theory and practice, are nontrivial. The Riccati equation is involved.

With real symmetric coefficients $\boldsymbol{p}(n)$, we have $P(z) = P(1/z)$. **If** z_i **is a root, so is** $1/z_i$. When z_i is inside the unit circle, $1/z_i$ is outside. The roots z_j *on* the unit circle must have even multiplicity, by the crucial assumption that $P(\omega) \geq 0$. Therefore the polynomial $z^N P(z)$ of degree $2N$, with leading coefficient $\boldsymbol{p}(N) \neq 0$, must have these $2N$ factors:

$$z^N P(z) = \boldsymbol{p}(N) \prod_{i=1}^{M} (z - z_i) \left(z - \frac{1}{z_i} \right) \prod_{j=1}^{N-M} (z - z_j)^2 . \tag{3.70}$$

This contains the key point, but we know more. Real coefficients ensure that the complex conjugate $\bar{z}$ is a root when z is a root. The complex roots off the unit circle actually come *four at a time*: z_i and $\bar{z}_i$ inside, $1/z_i$ and $1/\bar{z}_i$ outside. The complex roots on the circle also come four at a time: z_j twice and $\bar{z}_j$ twice. Real roots on the circle come two at a time (even multiplicity).

Now construct $C(z)$ by taking *all* the roots z_i (including $\bar{z}_i$) inside the circle, and also take one out of every double root z_j on the circle:

$$z^N C(z) = |\boldsymbol{p}(N)|^{1/2} \prod_{i=1}^{M} (z - z_i) \prod_{j=1}^{N-M} (z - z_j) . \tag{3.71}$$

This is the "minimum phase spectral factor." It has no roots outside the circle. The coefficients of $C(z)$ are still real, because the complex roots are automatically in conjugate pairs: $\bar{z}_i$ and $\bar{z}_j$ came with z_i and z_j.

Example 3. The 4-tap Daubechies filter has zeros at $z_i = 2 - \sqrt{3}$ and $z_i^{-1} = 2 + \sqrt{3}$. The other four roots of $z^3 P(z)$ are at $z_j = -1$ (on the unit circle and again real). Two of those roots go into the spectral factor (3.71). Thus $z^3 C(z)$ is a cubic polynomial with roots $2 - \sqrt{3}$, -1, and -1. It is minimum phase.

Every factorization of $P(z)$ into $F(z)\,H(z)$ must put some roots into $H(z)$ and the remaining roots into $F(z)$. The rules for this separation of roots of $P(z)$ are:

- For F and H to be **real** filters, z and $\bar{z}$ must stay together.
- For F and H to be **symmetric** filters, z and z^{-1} must stay together.
- For F to be the **transpose** of H, z and z^{-1} must go separately. This factorization $C(z)\,C(z^{-1})$ gives an orthogonal filter bank when P is halfband.

To achieve both properties at the same time, all zeros of $P(z)$ – not just the zeros on the unit circle – must be of even multiplicity. This gives another proof that orthogonality conflicts with symmetry.

Maxflat (Daubechies) Filters

This section is about an important family of filters, which lead to an outstanding family of wavelets. The same construction yields both.

1. These particular filters (and wavelets) are **orthogonal**.

2. The frequency responses have **maximum flatness** at $\omega = 0$ and $\omega = \pi$.

The lowpass filters will have $p = 1, 2, 3, 4, \ldots$ zeros at π. They have $2p = 2, 4, 6, 8, \ldots$ coefficients, so that $N = 2p - 1$. We use ***boldface p*** for the coefficients of $P(\omega) = |C(\omega)|^2$ and *lightface* p to count the zeros of $C(\omega)$ at $\omega = \pi$. The highpass coefficients $\boldsymbol{d}(k)$ come from an alternating flip. The first member of this family was $\boldsymbol{c}(0) = \boldsymbol{c}(1) = 1/\sqrt{2}$. Note the normalization $\boldsymbol{c}(0)^2 + \boldsymbol{c}(1)^2 = 1$. These numbers go into a *unitary matrix*. For each $p = 1, 2, 3, 4, \ldots$ the filter bank is orthonormal. The product filters have degree $2N = 4p - 2$:

$$P_0(z) = \left(\frac{1 + z^{-1}}{2}\right)^{2p} Q_{2p-2}(z) \quad \textit{will be halfband by special choice of} \quad Q.$$

In the literature on filters, this family is described as *maxflat*. The coefficients were given by Herrmann. They were already in formulas for interpolation, described below. In the history of wavelets, we are reproducing the great 1988 discovery by Ingrid Daubechies. The filters are FIR with $2p$ coefficients. The wavelets are supported on the interval $[0, N] = [0, 2p - 1]$. As p increases, the filters are increasingly "regular" and the wavelets are increasingly "smooth."

Condition O and Condition A$_p$

Before starting, it is helpful to count the requirements we must impose. There are $2p$ numbers to be chosen. These can be the coefficients $\boldsymbol{c}(0), \ldots, \boldsymbol{c}(2p - 1)$ in the lowpass filter, with frequency response $C(\omega)$. They could equally well be the coefficients $\boldsymbol{p}(0), \ldots, \boldsymbol{p}(2p - 1)$ of the centered (even) polynomial $P(\omega) = |C(\omega)|^2$. The $\boldsymbol{c}$'s come from the $\boldsymbol{p}$'s by spectral factorization. The nonnegative polynomial $P(\omega)$ is factored by the methods of the previous section:

$$P(\omega) = \sum_{1-2p}^{2p-1} \boldsymbol{p}(n) e^{-in\omega} \text{ equals } |C(\omega)|^2 = \left| \sum_{0}^{2p-1} \boldsymbol{c}(n) e^{-in\omega} \right|^2. \qquad (3.72)$$

Our formulas yield the numbers $\boldsymbol{p}(n) = \boldsymbol{p}(-n)$. Except for the first few filters in the family, there are no simple formulas for $\boldsymbol{c}(n)$.

These $2p$ numbers are determined by p conditions for orthogonality from Condition O, and p conditions for a flat response from Condition A. More precisely, the requirement is "Condition A_p" — the subscript indicates the order of flatness at $\omega = \pi$ (and $\omega = 0$). Here are the $p + p$ conditions:

Condition O $\quad P = |C|^2$ is a normalized halfband filter:

$$\boxed{\boldsymbol{p}(0) = 1 \text{ and } \boldsymbol{p}(2) = \boldsymbol{p}(4) = \cdots = \boldsymbol{p}(2p-2) = 0.} \tag{3.73}$$

Condition A_p $\quad C(\omega)$ has a zero of order p at $\omega = \pi$:

$$\boxed{C(\pi) = C'(\pi) = \cdots = C^{(p-1)}(\pi) = 0.} \tag{3.74}$$

The equation $C(\pi) = 0$ says that $\sum \boldsymbol{c}(n)(-1)^n = 0$. The odd-numbered coefficients have the same sum as the even-numbered coefficients:

$$\text{Condition } A_1 \text{ on } \boldsymbol{c}(n): \quad \sum_{\text{odd } n} \boldsymbol{c}(n) = \sum_{\text{even } n} \boldsymbol{c}(n). \tag{3.75}$$

This is the first of the "sum rules." Altogether we can impose the pth order zero in (3.74) as p sum rules on the coefficients:

$$\text{Condition } A_p \text{ on } \boldsymbol{c}(n): \quad \sum_{n=0}^{2p-1} (-1)^n n^k \boldsymbol{c}(n) = 0 \quad \text{for } k = 0, 1, \ldots, p-1. \tag{3.76}$$

The factor n^k comes from the kth derivative of $\sum \boldsymbol{c}(n)e^{-in\omega}$. Then $(-1)^n$ comes from substituting $\omega = \pi$. The convention for n^0 is 1.

The p zeros at π mean that $C(\omega)$ *has a factor* $\left(1 + e^{-i\omega}\right)^p$:

$$\text{Condition } A_p \text{ on } C(\omega): \quad C(\omega) = \left(\frac{1 + e^{-i\omega}}{2}\right)^p R(\omega). \tag{3.77}$$

$R(\omega)$ has degree $p-1$, to bring the total degree of $C(\omega)$ to $2p-1$. You could say that the pth order flatness is accounted for by $\left(1 + e^{-i\omega}\right)^p$. Then the p coefficients in $R(\omega)$ are chosen to satisfy the p equations of Condition O.

Formulas for $P(\omega)$

We intend to give two formulas for $P(\omega) = |C(\omega)|^2$. The one associated with Ingrid Daubechies has $(1+\cos\omega)^p$ times a sum of p terms. The formula associated with Yves Meyer gives the derivative of $P(\omega)$ as $-c(\sin\omega)^{2p-1}$. Then integration determines c and $P(\omega)$.

The best starting point is the ordinary polynomial $B_p(y)$. This has degree $p-1$, with p coefficients. It is the binomial series for $(1-y)^{-p}$, truncated after p terms:

$$B_p(y) = 1 + py + \frac{p(p+1)}{2}y^2 + \ldots + \binom{2p-2}{p-1} y^{p-1} = (1-y)^{-p} + O(y^p). \tag{3.78}$$

The coefficient of y^k is $\begin{pmatrix} p+k-1 \\ k \end{pmatrix}$. The remainder has order y^p because this is the first term to be dropped. The complex zeros of this polynomial $B_p(y)$ will be all-important for the Daubechies filters.

We combine $B_p(y)$ with the factor $(1-y)^p$ that has p zeros at $y=1$. The variable y on $[0,1]$ will correspond to the frequency ω on $[0,\pi]$. The product $\widetilde{P}(y) = 2(1-y)^p B_p(y)$ has exactly the flatness we want at $y=0$:

$$2(1-y)^p B_p(y) = 2(1-y)^p[(1-y)^{-p} + O(y^p)] = 2 + O(y^p). \tag{3.79}$$

This is a polynomial of degree $2p-1$. It is the unique polynomial with $2p$ coefficients that satisfies p conditions at each endpoint:

> $\widetilde{P}(y)$ **and its first** $p-1$ **derivatives are zero at** $y=0$ **and** $y=1$, **except** $\widetilde{P}(0)=2$.

Two more properties follow quickly. First, the derivative has $p-1$ zeros at both end points. It is a polynomial of degree $2p-2$ and with those zeros it must be

$$\widetilde{P}'(y) = -Cy^{p-1}(1-y)^{p-1} \quad \text{for some } C. \tag{3.80}$$

The second property comes when we add $\widetilde{P}(y)$ to $\widetilde{P}(1-y)$. The sum equals 2 at both ends and is still flat. Its $2p$ coefficients are uniquely determined — it must be the constant polynomial 2:

$$\widetilde{P}(y) + \widetilde{P}(1-y) \equiv 2. \tag{3.81}$$

At $y = \frac{1}{2}$ this gives $\widetilde{P}(\frac{1}{2}) = 1$. $\widetilde{P}(y)$ is odd around its middle value. This "Hermite interpolating polynomial" drops from 2 to 0 with flatness at the ends. Here are the polynomials for $p=2$ and $p=3$:

$$\begin{aligned} B_2(y) &= 1+2y \quad &&\text{and} \quad \widetilde{P}(y) = 2(1-y)^2(1+2y) = 2-6y^2+4y^3 \\ B_3(y) &= 1+3y+6y^2 \quad &&\text{and} \quad \widetilde{P}(y) = 2(1-y)^3 B_3(y) = 2-20y^3+30y^4-12y^5. \end{aligned}$$

Now we go from ordinary polynomials in y to trigonometric polynomials in ω. The degree stays at $2p-1$. The change that takes $0 \le y \le 1$ into $0 \le \omega \le \pi$ is

$$y = \frac{1-\cos\omega}{2} \quad \text{and} \quad 1-y = \frac{1+\cos\omega}{2}. \tag{3.82}$$

The polynomial $\widetilde{P}(y)$ *becomes our desired* $P(\omega)$. We summarize its properties.

Theorem 3.7 *The polynomial* $2(1-y)^p B_p(y)$ *becomes the halfband response*

$$\boxed{P(\omega) = 2\left(\tfrac{1+\cos\omega}{2}\right)^p \sum_{k=0}^{p-1} \begin{pmatrix} p+k-1 \\ k \end{pmatrix} \left(\tfrac{1-\cos\omega}{2}\right)^k.} \tag{3.83}$$

This satisfies Conditions O and A_p. Its Meyer form, by integrating $P'(\omega)$ and choosing c to give $P(\pi) = 0$, is

$$\boxed{P(\omega) = 2 - c\int_0^\omega (\sin\omega)^{2p-1} d\omega.} \tag{3.84}$$

For $p = 1, 2, 3$ the Daubechies and Meyer forms are

$$P(\omega) = \quad 1 + \cos\omega = 2 - \int_0^\omega \sin\omega\, d\omega$$

$$P(\omega) = \quad (1+\cos\omega)^2(1 - \tfrac{1}{2}\cos\omega) = 2 - \tfrac{3}{2}\int_0^\omega \sin^3\omega\, d\omega$$

$$P(\omega) = \quad (1+\cos\omega)^3(1 - \tfrac{9}{8}\cos\omega + \tfrac{3}{8}\cos^2\omega) = 2 - \tfrac{15}{4}\int_0^\omega \sin^5\omega\, d\omega$$

Most authors emphasize the Daubechies form, with its highly visible factor $(1+\cos\omega)^p$. That immediately ensures a pth order zero for the factors at $\omega = \pi$. Spectral factorization is speeded up, because only a lower-degree polynomial remains. It may not be so clear that (3.83) is a halfband filter. The even powers like $\cos^2\omega$ and $\cos^4\omega$ must disappear and they do. In the explicit formula for $p = 2$, multiplication produces $P(\omega) = 1 + \frac{3}{2}\cos\omega - \frac{1}{2}\cos^3\omega$.

The halfband property is $P(\omega) + P(\omega+\pi) \equiv 2$. This addition cancels the odd powers of $\cos\omega$, and the even powers are not present (except the constant term 1). This identity follows immediately from (3.81) because $1 - y = \frac{1+\cos\omega}{2} = \frac{1-\cos(\omega+\pi)}{2}$:

$$\widetilde{P}(y) + \widetilde{P}(1-y) \equiv 2 \;\text{ becomes }\; P(\omega) + P(\omega+\pi) \equiv 2. \tag{3.85}$$

The reader recognizes this "Condition O" as $|C(\omega)|^2 + |C(\omega+\pi)|^2 = 2$.

The halfband property is immediate in the Meyer form, with absolutely no calculations. Replace y by $(1-\cos\omega)/2$ in (3.80) to find $P'(\omega)d\omega$:

$$-Cy^{p-1}(1-y)^{p-1}dy = -C\left(\frac{1-\cos\omega}{2}\right)^{p-1}\left(\frac{1+\cos\omega}{2}\right)^{p-1}\frac{\sin\omega}{2}d\omega. \tag{3.86}$$

This is $-c(1-\cos^2\omega)^{p-1}\sin\omega d\omega$, which is also $-c(\sin\omega)^{2p-1}d\omega$. Its integral is

$$-c\int\left(1-\cos^2\omega\right)^{p-1}\sin\omega\, d\omega = \text{odd powers of } \cos\omega.$$

The only even frequency is a constant of integration. The filter is halfband.

The flatness condition requires first of all that $P(\pi) = 0$. The constant c makes this true. The derivative $P'(\omega) = -c(\sin\omega)^{2p-1}$ has a zero of order $2p-1$ at $\omega = \pi$. Then P itself has a zero of order $2p$. Its factor C has a zero of order p. Condition A_p is satisfied and Meyer's formula is confirmed.

Note that $P(\omega)$ *decreases monotonically* from $P(0) = 2$ to $P(\pi) = 0$. Then $P' = -c(\sin\omega)^{2p-1}$ is everywhere negative between 0 and π. There are no ripples. Therefore $P(\omega) \geq 0$ for all ω, and a factorization into $|C(\omega)|^2$ is assured.

The transition from passband (low frequencies) to stopband (high frequencies) becomes steeper and sharper as p increases. The slope at the midpoint $\omega = \frac{\pi}{2}$ is $-c\left(\sin\frac{\pi}{2}\right)^{2p-1}$, which is $-c$. We will show that c increases asymptotically like $\sqrt{p}$ as $p \to \infty$. Thus the transition band has width of order $1/\sqrt{p}$.

The Halfband Filter $P(z)$

Now we change from y and ω to the complex variable z. This will produce the filter coefficients in $P(z)$. That polynomial will be halfband and centered. The shifted polynomial $P_0(z) = z^{-N}P(z) = z^{1-2p}P(z)$ will be halfband and causal. The change of variables comes from $z = e^{i\omega}$:

$$\frac{z + z^{-1}}{2} = \cos\omega = 1 - 2y. \tag{3.87}$$

Thus $y = 0$ and $\omega = 0$ give $z = 1$. Similarly $y = 1$ and $\omega = \pi$ give $z = -1$.

Notice that the midpoints $y = \frac{1}{2}$ and $\omega = \frac{\pi}{2}$ give $z = \pm i$. **There are two z's for each y**, from $z + z^{-1} = 2 - 4y$. (This is a quadratic equation for z.) One z is inside the unit circle, while z^{-1} is outside. This "Joukowski transformation" is also central in fluid flow. The endpoints $z = 1$ and $z = -1$ are really double roots of $z + z^{-1} = 2$ and $z + z^{-1} = -2$.

The change of variable gives $1 - y$ and y in factored form:

$$1-y = \frac{1+\cos\omega}{2} = (\frac{1+z}{2})(\frac{1+z^{-1}}{2}) \text{ and } y = \frac{1-\cos\omega}{2} = (\frac{1-z}{2})(\frac{1-z^{-1}}{2}). \tag{3.88}$$

Substituting in $\widetilde{P}(y)$, the maxflat filter in the z-domain becomes $P(z)$:

$$P(z) = 2(\frac{1+z}{2})^p(\frac{1+z^{-1}}{2})^p \sum_{k=0}^{p-1} \binom{p+k-1}{k} (\frac{1-z}{2})^k(\frac{1-z^{-1}}{2})^k. \tag{3.89}$$

This factors into $P(z) = C(z)C(z^{-1})$ when $P(\omega)$ factors into $|C(\omega)|^2$. The p zeros at $y = 1$ and $\omega = \pi$ are now $2p$ zeros at $z = -1$. Half of them go into $C(z)$. The $p-1$ complex zeros of the other factor $B_p(y)$ become $2p-2$ zeros of $P(z)$. Half of those (the $p-1$ zeros inside the circle $|z| = 1$, if we want minimum phase) also go into $C(z)$. So the spectral factor $C(z)$ can be computed in two steps:

1. Find the $p - 1$ zeros of $B_p(y)$ and the $p - 1$ corresponding z's with $|z| < 1$.

2. Include p zeros at $z = -1$. Then $C(z)$ has these $2p - 1$ zeros.

Example: $p = 2$ gives Daubechies D_4 from $B_2(y) = 1 + 2y = \frac{1}{2}(-z + 4 - z^{-1})$.

The zero is at $y = -\frac{1}{2}$. Therefore $z + z^{-1} = 4$. This quadratic equation has roots $z = 2 \pm \sqrt{3}$. Then the $2p-1$ roots of $C(z)$ are $-1, -1, 2-\sqrt{3}$. The coefficients of D_4 are approximately .4830, .8365, .2241, and $-.1294$:

$$C(z) = \left[(1+\sqrt{3}) + (3+\sqrt{3})z^{-1} + (3-\sqrt{3})z^{-2} + (1-\sqrt{3})z^{-3}\right] / 4\sqrt{2}.$$

3.3 Eigenvalues of $(\downarrow 2)H$ and Convergence of the Cascade Algorithm

The key operator in multirate filtering is $(\downarrow 2)H$. The input signal x is filtered by H and then downsampled. We keep the even-numbered components of Hx. This is the lowpass channel in the analysis half of a filter bank. When $(\downarrow 2)H$ is iterated, either finitely often in practice or infinitely often in the passage to scaling functions and wavelets, its eigenvalues become all-important. These eigenvalues are intimately related to the number of "zeros at π" in the frequency response, and to the equal number of "vanishing moments" in the wavelets. This note studies the eigenvalues and eigenvectors (left as well as right). We also determine when the cascade algorithm converges to the scaling function.

The cascade algorithm is the iteration $\phi^{(i+1)} = M\phi^{(i)} = (\downarrow 2)2H\phi^{(i)}$. The extra factor 2 maintains constant area for the sequence of functions $\phi^{(i+1)}(t) = \sum 2h(k)\phi^{(i)}(2t-k)$, when the filter coefficients are normalized by $\sum h(k) = 1$. With double-shift orthogonality, $\sum 2h(k)h(k+2l) = \delta(l)$, convergence to $\phi(t)$ is almost (but not quite) certain. In a *biorthogonal* filter bank, with a more general lowpass filter H, this convergence is not at all assured. Nevertheless these non-orthogonal filters are giving the best results in compression, and their condition numbers are often quite moderate. We determine when they lead to wavelets.

This paper presents a simple proof of the necessary and sufficient condition for convergence to $\phi(t)$. Since we work in the L^2 norm (convergence in energy), inner products play a decisive part. They lead to the "transition operator" $T = (\downarrow 2)2HH^T$, whose eigenvalues control the convergence. Those eigenvalues also determine the smoothness of the scaling function and wavelets.

We summarize now the main conclusions. Some are already in the literature, with different proofs, and some are new. The (real) filter coefficients $h(0), ..., h(N)$ yield the transfer function $H(z) = \sum h(n)z^{-n}$. The input signal transforms to $X(z) = \sum x(n)z^{-n}$, and the filtered signal is $H(z)X(z)$. In the z-domain, the key operators $M = (\downarrow 2)2H$ and $T = (\downarrow 2)2HH^T$ involve multiplication by $H(z)$ from the filter and an aliasing term (identified by $-z$) from the downsampling:

$$(MX)(z^2) = H(z)X(z) + H(-z)X(-z) \tag{3.90}$$

$$(TX)(z^2) = H(z)X(z)H(z^{-1}) + H(-z)X(-z)H(-z^{-1}). \quad (3.91)$$

Note the argument z^2. We are dealing with the even part, because $(\downarrow 2)$ removes the odd terms.

In the time domain, the i, k entry of M is $2h(2i-k)$. It is $2i$ that reflects the double shift from $(\downarrow 2)$. The entries of T are $2p(2i-k)$, where $P(z) = H(z)H(z^{-1})$ corresponds to HH^T. The calculations involve finite matrices, in which i and k range from 0 to $N-1$ for M and from $1-N$ to $N-1$ for T. The frequency responses of interest have p zeros at π. Thus $H(z)$ has p zeros at $z = e^{j\pi} = -1$:

$$H(z) = \left(\frac{1+z^{-1}}{2}\right)^p Q(z) \quad \text{with} \quad Q(-1) \neq 0.$$

We state the conclusions in the time domain (for matrix eigenvalues), where they are easiest to check. We establish those conclusions in the z-domain, where they are easiest to prove.

Theorem 3.8 *Each time $H(z)$ is multiplied by $\frac{1+z^{-1}}{2}$, all the eigenvalues of M are multiplied by $\frac{1}{2}$ and a new eigenvalue $\lambda = 1$ is introduced. Thus the eigenvalues of M are*

$$1, \frac{1}{2}, \ldots, \left(\frac{1}{2}\right)^{p-1} \quad \textit{together with} \quad \frac{1}{2^p} \quad \textit{times the eigenvalues for} \quad (\downarrow 2)2Q. \quad (3.92)$$

Theorem 3.9 *When $H(z)$ is multiplied by $\left(\frac{1+z^{-1}}{2}\right)$, the new eigenvectors $\tilde{x}$ are the* differences *of the previous eigenvectors and the new left eigenvectors $\tilde{y}$ are the* sums *of the previous left eigenvectors:*

$$\tilde{X}(z) = \left(1 - z^{-1}\right) X(z) \quad \textit{and} \quad \tilde{Y}(z) = \frac{Y(z)}{(1-z)}. \quad (3.93)$$

The extra eigenvalue $\lambda = 1$ has left eigenvector $e = [1 \; 1 \; \cdots \; 1]$. The right eigenvector gives the new values of the scaling function at the integers.

Theorem 3.10 *If $H(-1) = \sum(-1)^k h(k) = 0$, the periodized functions $P^{(i)}(t) = \sum \phi^{(i)}(t-n)$ satisfy the identity*

$$P^{(i+1)}(t) = P^{(i)}(2t) \quad \textit{and thus} \quad P^{(i)}(t) = P^{(0)}(2^i t).$$

The cascade algorithm cannot converge to $\phi(t)$ unless $P^{(0)}(t) \equiv 1$. Otherwise $P^{(0)}(2^i t)$ will oscillate faster and faster. Thus Theorem 3.10 defines the acceptable class I of initial functions; they must satisfy $\sum \phi^{(0)}(t-n) \equiv 1$. This is equivalent to the so-called Strang-Fix condition on the Fourier transform: $\widehat{\phi}^{(0)}(2\pi n) = \delta(n)$. In that form, the requirement on $\phi^{(0)}(t)$ was discovered and proved necessary by Durand and by Meyer and Paiva. Our proof uses the identity $P^{(1)}(t) = P^{(0)}(2t)$.

The central question is convergence from these acceptable $\phi^{(0)}(t)$, and this is governed by the eigenvalues of T.

Theorem 3.11 *The cascade algorithm* $\phi^{(i+1)}(t) = \sum 2h(k)\phi^{(i)}(2t-k)$ *converges in* L^2 *for all* $\phi^{(0)}$ *in* I *if and only if the eigenvalues of* T *satisfy Condition* **E**:

$$\lambda = 1 \quad \textit{is a simple eigenvalue and all other eigenvalues have} \quad |\lambda| < 1. \tag{3.94}$$

Condition **E** is also the Cohen-Daubechies requirement for the translates $\phi(t-k)$ to be strongly independent. Jia has studied convergence and independence very carefully also in L^p.

The scaling function and wavelets are smoother by one more derivative, pointwise and in L^2, for every additional factor $(1+z^{-1})$ in $H(z)$. Splines come from the special choice $H(z) = \left(\frac{1+z^{-1}}{2}\right)^p$. They have no orthogonality, except in Haar's piecewise constant case $p=1$, but they have maximum smoothness. $H(z)$ has binomial coefficients $h(k)$ divided by 2^p. For $p=2$, 3, 4 we indicate the matrix $M = (\downarrow 2)2H$ and the eigenvalues predicted by Theorem 3.8:

$$\frac{1}{2}\begin{bmatrix} 1 & 0 \\ 1 & 2 \end{bmatrix} \qquad \frac{1}{4}\begin{bmatrix} 1 & 0 & 0 \\ 3 & 3 & 1 \\ 0 & 1 & 3 \end{bmatrix} \qquad \frac{1}{8}\begin{bmatrix} 1 & 0 & 0 & 0 \\ 6 & 4 & 1 & 0 \\ 1 & 4 & 6 & 4 \\ 0 & 0 & 1 & 4 \end{bmatrix}$$

$$\lambda = 1, \tfrac{1}{2} \qquad \lambda = 1, \tfrac{1}{2}, \tfrac{1}{4} \qquad \lambda = 1, \tfrac{1}{2}, \tfrac{1}{4}, \tfrac{1}{8}$$

The operator $(\downarrow 2)$ produces the double-shift between rows. The matrix on the right comes from the coefficients 1, 4, 6, 4, 1. This also illustrates the matrix T for the sequence $h(k) = 1, 2, 1$ because $h * h = 1, 4, 6, 4, 1$. (Actually the first row and column are dropped in T, so the eigenvalues are $1, \frac{1}{2}, \frac{1}{4}$.) Condition **E** in Theorem 3.11 is fully satisfied, and the cascade algorithm for the $\frac{1}{4}(1,2,1)$ filter converges quickly to the hat function—the linear spline.

In summary, the factor $\left(1+z^{-1}\right)^p$ gives the zeros at π that produce flatness of $H(z)$ and smoothness of $\phi(t)$. This pth order zero has important effects:

- p vanishing moments for the wavelets
- p sum rules for the coefficients $h(k)$
- pth order accuracy in approximation $f(t) \approx \sum_k a_k\phi(t-k)$
- pth order decay of wavelet coefficients for a smooth $f(t) = \sum b_{jk}w_{jk}(t)$
- All polynomials of degree $< p$ are combinations of the translates $\phi(t-k)$.

The smoothness of $\phi(t)$ is measured in the L^2 norm by using Parseval's equality. The scaling function has s derivatives when $|\omega|^s\hat{\phi}(\omega)$ has finite energy. The supremum $s_{\max}$ depends on p and the largest eigenvalue $|\lambda_{\max}|$ of $(\downarrow 2)2QQ^T$:

$$s_{\max} = p - \log_4 |\lambda_{\max}| \,. \tag{3.95}$$

Villemoes has given a particularly neat analysis of this smoothness formula. It is the eigenvalue $\lambda_{\max}$ from $Q(z)$ that has no simple expression (but is easily computed). Then Theorem 3.8 shows how the factor $(\frac{1+z^{-1}}{2})^{2p}$ in HH^T divides it by $2^{2p} = 4^p$. Smaller eigenvalues of T mean more smoothness of $\phi(t)$ and the wavelets.

Eigenvalues and Eigenvectors of $\boldsymbol{M}$

Theorems 3.8 and 3.9 will be proved together. By identifying the change in eigenvectors when $H(z)$ is multiplied by $\left(\frac{1+z^{-1}}{2}\right)$, we also confirm that the eigenvalues are cut in half. Starting from $Q(z)$ with no zeros at π, this multiplication occurs p times to reach the final $H(z)$ with p zeros at π. We go one step at a time, monitoring the eigenvectors. By equation (3.90), $(\downarrow 2)2Hx = \lambda x$ means

$$H(z)X(z) + H(-z)X(-z) = \lambda X(z^2). \tag{3.96}$$

Theorem 3.9 states that the step to $\left(\frac{1+z^{-1}}{2}\right) H(z)$ produces the new eigenfunction $\tilde{X}(z)= \left(1-z^{-1}\right) X(z)$ with eigenvalue $\tilde{\lambda} = \frac{1}{2}\lambda$. If this is true, then equation (3.96) will hold for $\tilde{H}(z)$ and $\tilde{X}(z)$ and $\tilde{\lambda}$:

$$\begin{aligned} &\left(\tfrac{1+z^{-1}}{2}\right) H(z)\left(1-z^{-1}\right) X(z) + \left(\tfrac{1-z^{-1}}{2}\right) H(-z)\left(1+z^{-1}\right) X(-z) \\ = \;& \frac{\lambda}{2}\left(1-z^{-2}\right) X\left(z^2\right). \end{aligned} \tag{3.97}$$

To verify (3.97), multiply (3.96) by $\frac{1}{2}\left(1-z^{-2}\right)$. That is the only step in the proof. Daubechies proved in a different way that $M = (\downarrow 2)2H$ has eigenvalues $1, \frac{1}{2}, \ldots, (\frac{1}{2})^{p-1}$.

Now consider the left eigenvectors. These are right eigenvectors of $M^T = 2H^T(\downarrow 2)^T$. In the z-domain this transposed operator takes $Y(z)$ into $2H(z^{-1})Y(z^2)$. Thus $yM = \lambda y$ means

$$2H(z^{-1})Y(z^2) = \lambda Y(z). \tag{3.98}$$

Theorem 3.9 says that equation (3.98) remains correct for $\tilde{H}(z) = \left(\frac{1+z^{-1}}{2}\right) H(z)$ and $\tilde{\lambda} = \frac{1}{2}\lambda$ when the eigenvector transforms to $\tilde{Y}(z) = \frac{Y(z)}{1-z}$. The left side is multiplied by $\frac{1+z}{2}$ and divided by $1-z^2$. The right side is divided by

$2(1-z)$. This agreement completes the proof of Theorem 3.9.

For Theorems 3.10 and 3.11 we refer to the full paper and to the textbook [SN]. The convergence of the cascade algorithm becomes convergence of the power method for the matrix T. This requires Condition **E** and it yields scaling functions and wavelets in L^2. The theory goes onward from there.

3.4 Zeros of The Daubechies Polynomials(*with Jianhong Shen*)

We study the asymptotics of the maximally flat Daubechies filters, as the number of zeros at π is steadily increased. The product filter of degree $4p-2$ with $2p$ zeros at π is factored into analysis times synthesis, $P(z) = F(z)H(z)$. When F is a time–reversal of H, the filter bank is orthogonal. Other factorizations yield linear phase filters by a different assignment of the zeros of $P(z)$. The zeros of this special $P(z)$ play an important role in the design of filters.

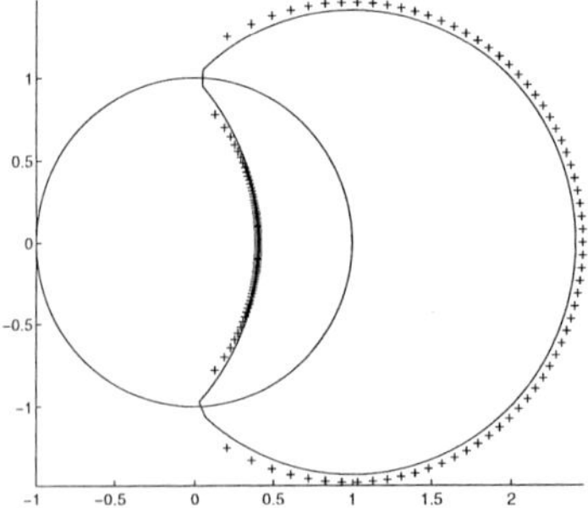

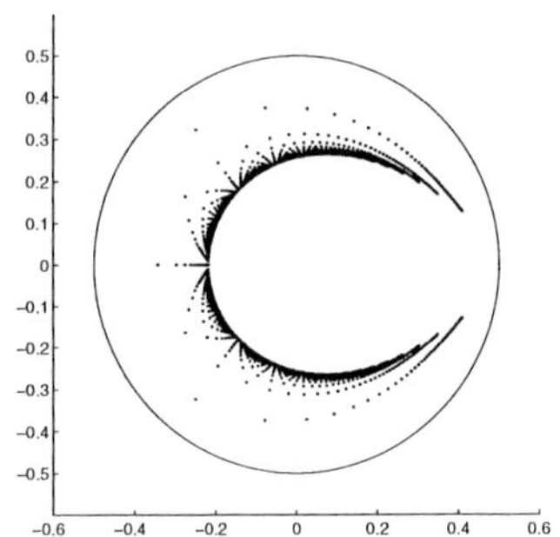

Left: The zeros are close to the limiting curve in the z–plane. Right: All zeros lie inside the circle of radius 1/2 in the y–plane, $p = 1:1:60$.

Matlab shows a remarkable plot for $p = 70$. The zeros are close to a limiting curve $|z - z^{-1}| = 2$. This is a union of two circular arcs of radius $\sqrt{2}$, around $z = 1$ and $z = -1$. We prove that the zeros do approach this limit and we find their distribution along the curve. For finite p, a simple approximation to all the zeros is based on their asymptotic equidistribution on the circle of radius $1+\log(4\pi p)/2p$ in the w–plane. The final figure shows asymptotic vs. actual zeros for $p = 70$, with $w = 4y(1-y) = ((z-z^{-1})/2)^2$.

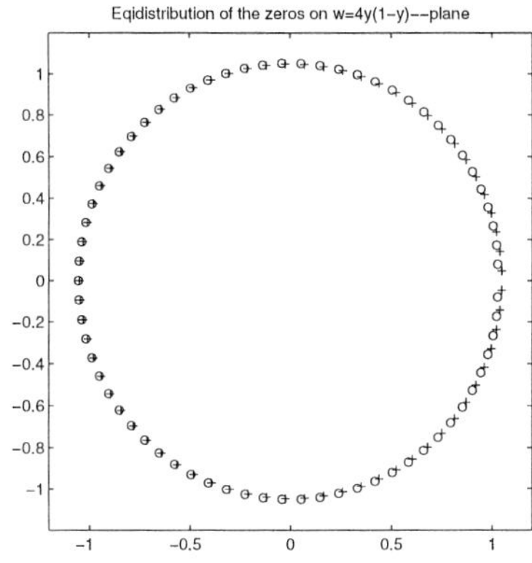

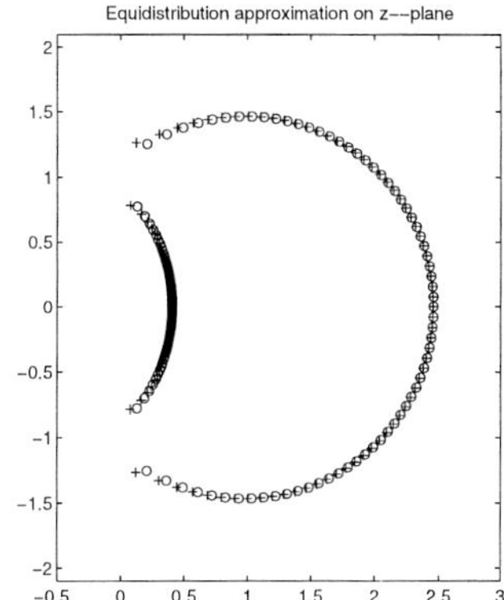

The wide dynamic range in the coefficients makes the zeros difficult to compute for large p. Rescaling y by 4 allows us to reach $p = 80$ by standard codes. This is "spectral factorization" of high order. The zeros at $z = -1$ stabilize the iteration of the lowpass filter in a wavelet filter bank.

Our starting point is the binomial series for $(1-y)^{-p}$, truncated after p terms. Its $p-1$ zeros give the $2p-2$ zeros of $P(z)$ inside and outside the unit circle, by the quadratic relation $z + z^{-1} = 2 - 4y$. The zeros in the complex y-plane approach the curve $|4y(1-y)| = 1$. All zeros have $|y| \leq 1/2$ and Re $z \geq 0$. The extreme zeros approach the singular point $y = 1/2$ (corresponding to $z = \pm i$) with speed $p^{-1/2}$. Then the $2p$ additional zeros at $z = -1$ yield the maximally flat $P(z)$.

Other important questions about the asymptotics of the Daubechies filters and scaling functions remain open. An important paper by Lemarié and Kateb will appear in Revista Matematica Iberoamericana.

3.5 References

[AH] A. N. Akansu and R. A. Haddad, *Multiresolution Signal Decomposition*, Academic Press (1992).

[AS] A. N. Akansu and M.J.T.Smith, eds.,*Subband and Wavelet Transforms*, Kluwer (1995).

[BF] J. Benedetto and M. Frazier, eds., *Wavelets: Mathematics and Applications*, CRC Press (1993).

[C] A. Cohen, *Wavelets and Multiscale Signal Processing*, Chapman and Hall (1995).

[CD] A. Cohen and I. Daubechies, A stability criterion for biorthogonal wavelet bases and their related subband coding scheme, *Duke Math. J.* **68** (1992) 313–335.

[CDF] A. Cohen, I. Daubechies, and J. C. Feauveau, Biorthogonal bases of compactly supported wavelets, *Comm. Pure Appl. Math.* **45** (1992) 485–560.

[CDM] A. S. Cavaretta, W. Dahmen, and C. A. Micchelli, *Stationary Subdivision*, Amer. Math. Soc. Memoirs 453 (1991).

[CGT] J. M. Combes, A. Grossmann, and Ph. Tchamitchian, eds., *Wavelets, Time-Frequency Methods, and Phase Space*, Springer (1990).

[CR] R. E. Crochiere and L. R. Rabiner, *Multirate Digital Signal Processing*, Prentice-Hall (1983).

[C1] C. K. Chui, *An Introduction to Wavelets*, Academic Press (1992).

[C2] C. K. Chui, ed., *Wavelets: A Tutorial in Theory and Applications*, Academic Press (1992).

[D] I. Daubechies, *Ten Lectures on Wavelets*, SIAM (1992).

[DL] I. Daubechies and J. Lagarias, Two-scale difference equations: I. Existence and global regularity of solutions, *SIAM J. Math. Anal.*, **22** (1991), 1388–1410; Two-scale difference equations: II. Local regularity, infinite products of matrices and fractals, *SIAM J. Math. Anal.*, **23** (1992), 1031–1079.

[E] T. Eirola, Sobolev characterization of solutions of dilation equations, *SIAM J. Math. Anal.* **23** (1992) 1015–1030.

[F] N.J. Fliege, Multirate Digital Signal Processing, John Wiley (1994).

[G] R. M. Gray, *Source Coding Theory*, Kluwer (1990).

[GG] A. Gersho and R. M. Gray, *Vector Quantization and Signal Compression*, Kluwer (1992).

[HC] C. Heil and D. Colella, *Dilation equations and the smoothness of compactly supported wavelets*, Wavelets: Mathematics and Applications, J. Benedetto and M. Frazier, eds., CRC Press (1993).

[IEEE1] Special Issue on Wavelets, *IEEE Transactions on Signal Processing*, **41** (12/93).

[IEEE2] Special Issue on Wavelet Transforms, *IEEE Transactions on Information Theory*, **38** (3/92).

[IEEE3] W.K.Chen, ed., The Circuits and Filters Handbook, chapters on filter banks and wavelets by T. Nguyen, I Djokovic, and P.P. Vaidyanathan, IEEE Press (1995).

[J] R.-Q. Jia, Subdivision schemes in L^p spaces, *Adv. in Comp. Math.* **3** (1995) 309-341.

[JN] N. J. Jayant and P. Noll, *Digital Coding of Waveforms*, Prentice-Hall (1984).

[K] G. Kaiser, *A Friendly Guide to Wavelets*, Birkhäuser (1994).

[L] W. Lawton, Necessary and sufficient conditions for constructing orthogonal wavelets, *J. Math. Phys.* **32** (1991) 52–61.

[LLS] W. Lawton, S. L. Lee, and Z. Shen, Convergence of multidimens ional cascade algorithm, preprint.

[Mr] H. S. Malvar, *Signal Processing with Lapped Transforms*, Artech House (1992).

[Mt] S. Mallat, *Wavelet Signal Processing*, Academic Press (1996).

[M1] Y. Meyer, *Wavelets and Operators*, Translation of *Ondelettes et Opérateurs* (Hermann, 1990), Cambridge University Press (1993).

[M2] Y. Meyer, *Wavelets: Algorithms and Applications*, SIAM (1993).

[M3] Y. Meyer, ed., *Proceedings of the Marseille Conference on Wavelets*, Masson (1993).

[OS] A. V. Oppenheim and R. W. Schafer, *Discrete-Time Signal Processing*, Prentice-Hall (1989).

[RBCDM] M. B. Ruskai et al., eds., *Wavelets and Their Applications*, Jones and Bartlett (1992).

[SN] G. Strang and T. Nguyen, *Wavelets and Filter Banks*, Wellesley-Cambridge Press (1996).

[SW] L. L. Schumaker and G. Webb, eds., *Recent Advances in Wavelet Analysis*, Academic Press (1994).

[V] P. P. Vaidyanathan, *Multirate Systems and Filter Banks*, Prentice-Hall (1992).

[VK] M. Vetterli and J. Kovacevic, *Wavelets and Subband Coding*, Prentice-Hall (1995).

[Vi] L. F. Villemoes, Energy moments in time and frequency for two-scale difference equation solutions and wavelets, *SIAM J. Math. Anal.* **23** (1992) 1519–1543; Wavelet analysis of refinement equations, *SIAM J. Math. Anal.* **25** (1994) 1433–1460.

[Wa] G. G. Walter, *Wavelets and Other Orthogonal Systems with Applications*, CRC Press (1994).

[Wi] M. V. Wickerhauser, *Adapted Wavelet Analysis from Theory to Software*, AK Peters (1994).

[Wo] J. W. Woods, ed., *Subband Image Coding*, Kluwer (1991).

4

Constructing Linear Algebra Software Libraries for High-Performance Computers

Jack J. Dongarra*
David W. Walker†

This chapter discusses the design of linear algebra libraries for high performance computers. Particular emphasis is placed on the development of scalable algorithms for MIMD distributed memory concurrent computers. A brief description of the EISPACK, LINPACK, and LAPACK libraries is given, followed by an outline of ScaLAPACK, which is a distributed memory version of LAPACK currently under development. The importance of block-partitioned algorithms in reducing the frequency of data movement between different levels of hierarchical memory is stressed. The use of such algorithms helps reduce the message startup costs on distributed memory concurrent computers. Other key ideas in our approach are the use of distributed versions of the Level 3 Basic Linear Algebra Subprograms (BLAS) as computational building blocks, and the use of Basic Linear Algebra Communication Subprograms (BLACS) as communication building blocks. Together the distributed BLAS and the BLACS can be used to construct higher-level algorithms, and hide many details of the parallelism from the application developer.

The block-cyclic data distribution is described, and adopted as a good way of distributing block-partitioned matrices. Block-partitioned versions of the Cholesky and LU factorizations are presented, and optimization issues associated with the implementation of the LU factorization algorithm

*Department of Computer Science, University of Tennessee, 107 Ayres Hall, Knoxville, TN 37996-1301. This work was supported in part by ARPA under contract number DAAL03-91-C-0047 administered by ARO, and in part by DOE under contract number DE-AC05-84OR21400, and by National Science Foundation Grant number ASC-ASC-9005933.

†Mathematical Sciences Section, Oak Ridge National Laboratory, P. O. Box 2008, Bldg. 6012, Oak Ridge, TN 37831-6367

on distributed memory concurrent computers are discussed, together with its performance on the Intel Delta system. Finally, approaches to the design of library interfaces are reviewed.

4.1 Introduction

The increasing availability of advanced-architecture computers is having a very significant effect on all spheres of scientific computation, including algorithm research and software development in numerical linear algebra. Linear algebra—in particular, the solution of linear systems of equations—lies at the heart of most calculations in scientific computing. This chapter discusses some of the recent developments in linear algebra designed to exploit these advanced-architecture computers. Particular attention will be paid to dense factorization routines, such as the Cholesky and LU factorizations, and these will be used as examples to highlight the most important factors that must be considered in designing linear algebra software for advanced-architecture computers. We use these factorization routines for illustrative purposes not only because they are relatively simple, but also because of their importance in several scientific and engineering applications that make use of boundary element methods. These applications include electromagnetic scattering and computational fluid dynamics problems, as discussed in more detail in Section 4.4.1.

Much of the work in developing linear algebra software for advanced-architecture computers is motivated by the need to solve large problems on the fastest computers available. In this chapter, we focus on four basic issues: (1) the motivation for the work; (2) the development of standards for use in linear algebra and the building blocks for a library; (3) aspects of algorithm design and parallel implementation; and (4) future directions for research.

For the past 15 years or so, there has been a great deal of activity in the area of algorithms and software for solving linear algebra problems. The linear algebra community has long recognized the need for help in developing algorithms into software libraries, and several years ago, as a community effort, put together a *de facto* standard for identifying basic operations required in linear algebra algorithms and software. The hope was that the routines making up this standard, known collectively as the Basic Linear Algebra Subprograms (BLAS), would be efficiently implemented on advanced-architecture computers by many manufacturers, making it possible to reap the portability benefits of having them efficiently implemented on a wide range of machines. This goal has been largely realized.

The key insight of our approach to designing linear algebra algorithms for advanced architecture computers is that the frequency with which data are moved between different levels of the memory hierarchy must be minimized in order to attain high performance. Thus, our main algorithmic approach for exploiting both vectorization and parallelism in our implementations is the use of block-partitioned algorithms, particularly in conjunction with highly-tuned kernels for performing matrix-vector and matrix-matrix operations (the Level 2 and 3 BLAS). In general, the use of block-partitioned algorithms requires data to be moved as blocks, rather than as vectors or scalars, so that although the total amount of data moved is unchanged, the latency (or startup cost) associated with the movement is greatly reduced because fewer messages are needed to move the data.

A second key idea is that the performance of an algorithm can be tuned by a user by varying the parameters that specify the data layout. On shared memory machines, this is controlled by the block size, while on distributed memory machines it is controlled by the block size and the configuration of the logical process mesh, as described in more detail in Section 4.5.

In Section 4.1, we first give an overview of some of the major software projects aimed at solving dense linear algebra problems. Next, we describe the types of machine that benefit most from the use of block-partitioned algorithms, and discuss what is meant by high-quality, reusable software for advanced-architecture computers. Section 4.2 discusses the role of the BLAS in portability and performance on high-performance computers. We discuss the design of these building blocks, and their use in block-partitioned algorithms, in Section 4.3. Section 4.4 focuses on the design of a block-partitioned algorithm for LU factorization, and Sections 4.5, 4.6, and 4.7 use this example to illustrate the most important factors in implementing dense linear algebra routines on MIMD, distributed memory, concurrent computers. Section 4.5 deals with the issue of mapping the data onto the hierarchical memory of a concurrent computer. The layout of an application's data is crucial in determining the performance and scalability of the parallel code. In Sections 4.6 and 4.7, details of the parallel implementation and optimization issues are discussed. Section 4.8 presents some future directions for investigation.

4.1.1 Dense Linear Algebra Libraries

Over the past twenty-five years, the first author has been directly involved in the development of several important packages of dense linear algebra software: EISPACK, LINPACK, LAPACK, and the BLAS. In addition, both authors are currently involved in the development of ScaLAPACK, a scalable version of LAPACK for distributed memory concurrent computers. In this section, we give a brief review of these packages—their history, their advantages, and their limitations on high-performance computers.

EISPACK

EISPACK is a collection of Fortran subroutines that compute the eigenvalues and eigenvectors of nine classes of matrices: complex general, complex Hermitian, real general, real symmetric, real symmetric banded, real symmetric tridiagonal, special real tridiagonal, generalized real, and generalized real symmetric matrices. In addition, two routines are included that use singular value decomposition to solve certain least-squares problems.

EISPACK is primarily based on a collection of Algol procedures developed in the 1960s and collected by J. H. Wilkinson and C. Reinsch in a volume entitled *Linear Algebra* in the *Handbook for Automatic Computation* [59] series. This volume was not designed to cover every possible method of solution; rather, algorithms were chosen on the basis of their generality, elegance, accuracy, speed, or economy of storage.

Since the release of EISPACK in 1972, over ten thousand copies of the collection have been distributed worldwide.

LINPACK

LINPACK is a collection of Fortran subroutines that analyze and solve linear equations and linear least-squares problems. The package solves linear systems whose matrices are general, banded, symmetric indefinite, symmetric positive definite, triangular, and tridiagonal square. In addition, the package computes the QR and singular value decompositions of rectangular matrices and applies them to least-squares problems.

LINPACK is organized around four matrix factorizations: LU factorization, pivoted Cholesky factorization, QR factorization, and singular value decomposition. The term LU factorization is used here in a very general sense to mean the factorization of a square matrix into a lower triangular part and an upper triangular part, perhaps with pivoting. These factorizations will be treated at greater length later, when the actual LINPACK subroutines are discussed. But first a digression on organization and factors influencing LINPACK's efficiency is necessary.

LINPACK uses column-oriented algorithms to increase efficiency by preserving locality of reference. This means that if a program references an item in a particular block, the next reference is likely to be in the same block. By column orientation we mean that the LINPACK codes always reference arrays down columns, not across rows. This works because Fortran stores arrays in column major order. Thus, as one proceeds down a column of an array, the memory references proceed sequentially in memory. On the other hand, as one proceeds across a row, the memory references jump across memory, the length of the jump being proportional to the length of a column. The effects of column orientation are quite dramatic: on systems with virtual or cache memories, the LINPACK codes will significantly outperform codes that are not column oriented. We note, however, that textbook examples of matrix algorithms are seldom column oriented.

Another important factor influencing the efficiency of LINPACK is the use of the Level 1 BLAS; there are three effects.

First, the overhead entailed in calling the BLAS reduces the efficiency of the code. This reduction is negligible for large matrices, but it can be quite significant for small matrices. The matrix size at which it becomes unimportant varies from system to system; for square matrices it is typically between $n = 25$ and $n = 100$. If this seems like an unacceptably large overhead, remember that on many modern systems the solution of a system of order 25 or less is itself a negligible calculation. Nonetheless, it cannot be denied that a person whose programs depend critically on solving small matrix problems in inner loops will be better off with BLAS-less versions of the LINPACK codes. Fortunately, the BLAS can be removed from the smaller, more frequently used program in a short editing session.

Second, the BLAS improve the efficiency of programs when they are run on nonoptimizing compilers. This is because doubly subscripted array references in the inner loop of the algorithm are replaced by singly subscripted array references in the appropriate BLAS. The effect can be seen for matrices of quite small order, and for large orders the savings are quite significant.

Finally, improved efficiency can be achieved by coding a set of BLAS [19] to take advantage of the special features of the computers on which LINPACK is being run. For most computers, this simply means producing machine-language versions. However, the code can also take advantage of more exotic architectural features, such as vector operations.

Further details about the BLAS are presented in Section 4.2.

LAPACK

LAPACK [16] provides routines for solving systems of simultaneous linear equations, least-squares solutions of linear systems of equations, eigenvalue problems, and singular value problems. The associated matrix factorizations (LU, Cholesky, QR, SVD, Schur, generalized Schur) are also provided, as are related computations such as reordering of the Schur factorizations and estimating condition numbers. Dense and banded matrices are handled, but not general sparse matrices. In all areas, similar functionality is provided for real and complex matrices, in both single and double precision.

The original goal of the LAPACK project was to make the widely used EISPACK and LINPACK libraries run efficiently on shared-memory vector and parallel processors. On these machines, LINPACK and EISPACK are inefficient because their memory access patterns disregard the multilayered memory hierarchies of the machines, thereby spending too much time moving data instead of doing useful floating-point operations. LAPACK addresses this problem by reorganizing the algorithms to use block matrix operations, such as matrix multiplication, in the innermost loops [4, 16]. These block operations can be optimized for each architecture to account for the memory hierarchy [3], and so provide a transportable way to achieve high efficiency on diverse modern machines. Here we use the term "transportable" instead of "portable" because, for fastest possible performance, LAPACK requires that highly optimized block matrix operations be already implemented on each machine. In other words, the correctness of the code is portable, but high performance is not—if we limit ourselves to a single Fortran source code.

LAPACK can be regarded as a successor to LINPACK and EISPACK. It has virtually all the capabilities of these two packages and much more besides. LAPACK improves on LINPACK and EISPACK in four main respects: speed, accuracy, robustness and functionality. While LINPACK and EISPACK are based on the vector operation kernels of the Level 1 BLAS, LAPACK was designed at the outset to exploit the Level 3 BLAS —a set of specifications for Fortran subprograms that do various types of matrix multiplication and the solution of triangular systems with multiple right-hand sides. Because of the coarse granularity of the Level 3 BLAS operations, their use tends to promote high efficiency on many high-performance computers, particularly if specially coded implementations are provided by the manufacturer.

ScaLAPACK

The ScaLAPACK software library, scheduled for completion by the end of 1994, will extend the LAPACK library to run scalably on MIMD, distributed memory, concurrent computers [12, 13]. For such machines the memory hierarchy includes the off-processor memory of other processors, in addition to the hierarchy of registers, cache, and local memory on each processor. Like LAPACK, the ScaLAPACK routines are based on block-partitioned algorithms in order to minimize the frequency of data movement between different levels of the memory hierarchy. The fundamental building blocks of the ScaLAPACK library are distributed memory versions of the Level 2 and Level 3 BLAS, and a set of Basic Linear Algebra Communication Subprograms (BLACS) [18, 28] for communication tasks that arise frequently in parallel linear algebra computations. In the ScaLAPACK routines, all interprocessor communication occurs within the distributed BLAS and the BLACS, so the source code of the top software layer of ScaLAPACK looks very similar to that of LAPACK.

We envisage a number of user interfaces to ScaLAPACK. Initially, the interface will be similar to that of LAPACK, with some additional arguments passed to each routine to specify the data layout. Once this is in place, we intend to modify the interface so the arguments to each ScaLAPACK routine are the same as in LAPACK. This will require information about the data distribution of each matrix and vector to be hidden from the user. This may be done by means of a ScaLAPACK initialization routine. This interface will be fully compatible with LAPACK. Provided "dummy" versions of the ScaLAPACK initialization routine and the BLACS are added to LAPACK, there will be no distinction between LAPACK and ScaLAPACK at the application level, though each will link to different versions of the BLAS and BLACS. Following on from this, we will experiment with object-based interfaces for LAPACK and ScaLAPACK, with the goal of developing interfaces compatible with Fortran 90 [12] and C++ [26].

4.1.2 Target Architectures

The EISPACK and LINPACK software libraries were designed for supercomputers used in the 1970s and early 1980s, such as the CDC-7600, Cyber 205, and Cray-1. These machines featured multiple functional units pipelined for good performance [45]. The CDC-7600 was basically a high-performance scalar computer, while the Cyber 205 and Cray-1 were early vector computers.

The development of LAPACK in the late 1980s was intended to make the EISPACK and LINPACK libraries run efficiently on shared memory, vector supercomputers. The ScaLAPACK software library will extend the use of LAPACK to distributed memory concurrent supercomputers. The development of ScaLAPACK began in 1991 and is expected to be completed by the end of 1994.

The underlying concept of both the LAPACK and ScaLAPACK libraries is the use of block-partitioned algorithms to minimize data movement between different levels in hierarchical memory. Thus, the ideas discussed in this chapter for developing a library for dense linear algebra computations are applicable to any computer with a hierarchical memory that (1) imposes a sufficiently large startup cost on the movement of data between different levels in the hierarchy, and for which (2) the cost of a context switch is too great to make fine grain size multithreading worthwhile. Our target machines are, therefore, medium and large grain size advanced-architecture computers. These include "traditional" shared memory, vector supercomputers, such as the Cray Y-MP and C90, and MIMD distributed memory concurrent supercomputers, such as the Intel Paragon, and Thinking Machines' CM-5, and the more recently announced IBM SP1 and Cray T3D concurrent systems. Since these machines have only very recently become available, most of the ongoing development of the ScaLAPACK library is being done on a 128-node Intel iPSC/860 hypercube and on the 520-node Intel Delta system.

The Intel Paragon supercomputer can have up to 2000 nodes, each consisting of an i860 processor and a communications processor. The nodes each have at least 16 Mbytes of memory, and are connected by a high-speed network with the topology of a two-dimensional mesh. The CM-5 from Thinking Machines Corporation [55] supports both SIMD and MIMD programming models, and may have up to 16k processors, though the largest CM-5 currently installed has 1024 processors. Each CM-5 node is a Sparc processor and up to 4 associated vector processors. Point-to-point communication between nodes is supported by a data network with the topology of a "fat tree" [48]. Global communication operations, such as synchronization and reduction, are supported by a separate control network. The IBM SP1 system is based on the same RISC chip used in the IBM RS/6000 workstations and uses a multistage switch to connect processors. The Cray T3D uses the Alpha chip from Digital Equipment Corporation, and connects the processors in a three-dimensional torus.

Future advances in compiler and hardware technologies in the mid to late 1990s are expected to make multithreading a viable approach for masking communication costs. Since the blocks in a block-partitioned algorithm can be regarded as separate threads, our approach will still be applicable on machines that exploit medium and coarse grain size multithreading.

4.1.3 High-Quality, Reusable, Mathematical Software

In developing a library of high-quality subroutines for dense linear algebra computations the design goals fall into three broad classes:

- performance
- ease-of-use

- range-of-use

Performance

Two important performance metrics are *concurrent efficiency* and *scalability*. We seek good performance characteristics in our algorithms by eliminating, as much as possible, overhead due to load imbalance, data movement, and algorithm restructuring. The way the data are distributed (or decomposed) over the memory hierarchy of a computer is of fundamental importance to these factors. Concurrent efficiency, ϵ, is defined as the concurrent speedup per processor [34], where the concurrent speedup is the execution time, T_{seq}, for the best sequential algorithm running on one processor of the concurrent computer, divided by the execution time, T, of the parallel algorithm running on N_p processors. When direct methods are used, as in LU factorization, the concurrent efficiency depends on the problem size and the number of processors, so on a given parallel computer and for a fixed number of processors, the running time should not vary greatly for problems of the same size. Thus, we may write,

$$\epsilon(N, N_p) = \frac{1}{N_p} \frac{T_{\mathrm{seq}}(N)}{T(N, N_p)} \tag{4.1}$$

where N represents the problem size. In dense linear algebra computations, the execution time is usually dominated by the floating-point operation count, so the concurrent efficiency is related to the performance, G, measured in floating-point operations per second by,

$$G(N, N_p) = \frac{N_p}{t_{\mathrm{calc}}} \epsilon(N, N_p) \tag{4.2}$$

where t_{calc} is the time for one floating-point operation. For iterative routines, such as eigensolvers, the number of iterations, and hence the execution time, depends not only on the problem size, but also on other characteristics of the input data, such as condition number. A parallel algorithm is said to be scalable [39] if the concurrent efficiency depends on the problem size and number of processors only through their ratio. This ratio is simply the problem size per processor, often referred to as the granularity. Thus, for a scalable algorithm, the concurrent efficiency is constant as the number of processors increases while keeping the granularity fixed. Alternatively, Eq. 4.2 shows that this is equivalent to saying that, for a scalable algorithm, the performance depends linearly on the number of processors for fixed granularity.

Ease-Of-Use

Ease-of-use is concerned with factors such as portability and the user interface to the library. Portability, in its most inclusive sense, means that the code is written in a standard language, such as Fortran, and that the source code can be compiled on an arbitrary machine to produce a program that will run correctly. We call this the "mail-order software" model of portability, since it reflects the model used by software servers such as *netlib* [22]. This notion of portability is quite demanding. It requires that all relevant properties of the computer's arithmetic and architecture be discovered at runtime within the confines of a Fortran code. For example, if it is important to know the overflow threshold for scaling purposes, it must be determined at runtime *without overflowing*, since overflow is generally fatal. Such demands have resulted in quite large and sophisticated programs [30, 46] which must be modified frequently to deal with new architectures and software releases. This "mail-order" notion of software portability also means that codes generally must be written for the worst possible machine expected to be used, thereby often degrading performance on all others. Ease-of-use is also enhanced if implementation details are largely hidden from the user, for example, through the use of an object-based interface to the library [26].

Range-Of-Use

Range-of-use may be gauged by how numerically stable the algorithms are over a range of input problems, and the range of data structures the library will support. For example, LINPACK and EISPACK deal with dense matrices stored in a rectangular array, packed matrices where only the upper or lower half of a symmetric matrix is stored, and banded matrices where only the nonzero bands are stored. In addition, some special formats such as Householder vectors are used internally to represent orthogonal matrices. There are also sparse matrices, which may be stored in many different ways; but in this chapter we focus on dense and banded matrices, the mathematical types addressed by LINPACK, EISPACK, and LAPACK.

4.2 The BLAS as the Key to Portability

At least three factors affect the performance of portable Fortran code.

1. **Vectorization.** Designing vectorizable algorithms in linear algebra is usually straightforward. Indeed, for many computations there are

several variants, all vectorizable, but with different characteristics in performance (see, for example, [17]). Linear algebra algorithms can approach the peak performance of many machines—principally because peak performance depends on some form of chaining of vector addition and multiplication operations, and this is just what the algorithms require. However, when the algorithms are realized in straightforward Fortran 77 code, the performance may fall well short of the expected level, usually because vectorizing Fortran compilers fail to minimize the number of memory references—that is, the number of vector load and store operations.

2. **Data movement.** What often limits the actual performance of a vector, or scalar, floating-point unit is the rate of transfer of data between different levels of memory in the machine. Examples include the transfer of vector operands in and out of vector registers, the transfer of scalar operands in and out of a high-speed scalar processor, the movement of data between main memory and a high-speed cache or local memory, paging between actual memory and disk storage in a virtual memory system, and interprocessor communication on a distributed memory concurrent computer.

3. **Parallelism.** The nested loop structure of most linear algebra algorithms offers considerable scope for loop-based parallelism. This is the principal type of parallelism that LAPACK and ScaLAPACK presently aim to exploit. On shared memory concurrent computers, this type of parallelism can sometimes be generated automatically by a compiler, but often requires the insertion of compiler directives. On distributed memory concurrent computers, data must be moved between processors. This is usually done by explicit calls to message passing routines, although parallel language extensions such as Coherent Parallel C [33] and Split-C [15] do the message passing implicitly.

The question arises, "How can we achieve sufficient control over these three factors to obtain the levels of performance that machines can offer?" The answer is through use of the BLAS.

There are now three levels of BLAS:

Level 1 BLAS [47]: for vector operations, such as $y \leftarrow \alpha x + y$

Level 2 BLAS [20]: for matrix-vector operations, such as $y \leftarrow \alpha Ax + \beta y$

Level 3 BLAS [19]: for matrix-matrix operations, such as $C \leftarrow \alpha AB + \beta C$.

TABLE 4.1. Speed (Megaflops) of Level 2 and Level 3 BLAS Operations on a CRAY Y-MP. All matrices are of order 500; U is upper triangular.

Number of processors:	1	2	4	8
Level 2: $y \leftarrow \alpha Ax + \beta y$	311	611	1197	2285
Level 3: $C \leftarrow \alpha AB + \beta C$	312	623	1247	2425
Level 2: $x \leftarrow Ux$	293	544	898	1613
Level 3: $B \leftarrow UB$	310	620	1240	2425
Level 2: $x \leftarrow U^{-1}x$	272	374	479	584
Level 3: $B \leftarrow U^{-1}B$	309	618	1235	2398

Here, A, B and C are matrices, x and y are vectors, and α and β are scalars.

The Level 1 BLAS are used in LAPACK, but for convenience rather than for performance: they perform an insignificant fraction of the computation, and they cannot achieve high efficiency on most modern supercomputers.

The Level 2 BLAS can achieve near-peak performance on many vector processors, such as a single processor of a CRAY X-MP or Y-MP, or Convex C-2 machine. However, on other vector processors such as a CRAY-2 or an IBM 3090 VF, the performance of the Level 2 BLAS is limited by the rate of data movement between different levels of memory.

The Level 3 BLAS overcome this limitation. This third level of BLAS performs $O(n^3)$ floating-point operations on $O(n^2)$ data, whereas the Level 2 BLAS perform only $O(n^2)$ operations on $O(n^2)$ data. The Level 3 BLAS also allow us to exploit parallelism in a way that is transparent to the software that calls them. While the Level 2 BLAS offer some scope for exploiting parallelism, greater scope is provided by the Level 3 BLAS, as Table 4.1 illustrates.

4.3 Block Algorithms and Their Derivation

It is comparatively straightforward to recode many of the algorithms in LINPACK and EISPACK so that they call Level 2 BLAS. Indeed, in the simplest cases the same floating-point operations are done, possibly even in the same order: it is just a matter of reorganizing the software. To illustrate this point, we consider the Cholesky factorization algorithm used in the LINPACK routine SPOFA, which factorizes a symmetric positive definite matrix as $A = U^TU$. We consider Cholesky factorization because the algorithm is simple, and no pivoting is required. In Section 4.4 we shall consider the slightly more complicated example of LU factorization.

Suppose that after $j-1$ steps the block A_{00} in the upper lefthand corner of A has been factored as $A_{00} = U_{00}^T U_{00}$. The next row and column of the factorization can then be computed by writing $A = U^T U$ as

$$\begin{pmatrix} A_{00} & b_j & A_{02} \\ . & a_{jj} & c_j^T \\ . & . & A_{22} \end{pmatrix} = \begin{pmatrix} U_{00}^T & 0 & 0 \\ v_j^T & u_{jj} & 0 \\ U_{02}^T & w_j & U_{22}^T \end{pmatrix} \begin{pmatrix} U_{00} & v_j & U_{02} \\ 0 & u_{jj} & w_j^T \\ 0 & 0 & U_{22} \end{pmatrix}$$

where b_j, c_j, v_j, and w_j are column vectors of length $j-1$, and a_{jj} and u_{jj} are scalars. Equating coefficients of the jth column, we obtain

$$\begin{aligned} b_j &= U_{00}^T v_j \\ a_{jj} &= v_j^T v_j + u_{jj}^2. \end{aligned}$$

Since U_{00} has already been computed, we can compute v_j and u_{jj} from the equations

$$\begin{aligned} U_{00}^T v_j &= b_j \\ u_{jj}^2 &= a_{jj} - v_j^T v_j. \end{aligned}$$

The body of the code of the LINPACK routine SPOFA that implements the above method is shown in Figure 4.1. The same computation recoded in "LAPACK-style" to use the Level 2 BLAS routine STRSV (which solves a triangular system of equations) is shown in Figure 4.2. The call to STRSV has replaced the loop over K which made several calls to the Level 1 BLAS routine SDOT. (For reasons given below, this is not the actual code used in LAPACK — hence the term "LAPACK-style".)

This change by itself is sufficient to result in large gains in performance on a number of machines—for example, from 72 to 251 megaflops for a matrix of order 500 on one processor of a CRAY Y-MP. Since this is 81% of the peak speed of matrix-matrix multiplication on this processor, we cannot hope to do very much better by using Level 3 BLAS.

We can, however, restructure the algorithm at a deeper level to exploit the faster speed of the Level 3 BLAS. This restructuring involves recasting the algorithm as a **block algorithm**—that is, an algorithm that operates on **blocks** or submatrices of the original matrix.

4.3.1 Deriving a Block Algorithm

To derive a block form of Cholesky factorization, we partition the matrices as shown in Figure 4.4, in which the diagonal blocks of A and U are square, but of differing sizes. We assume that the first block has already been factored as $A_{00} = U_{00}^T U_{00}$, and that we now want to determine the second block column of U consisting of the blocks U_{01} and U_{11}. Equating submatrices in the second block of columns, we obtain

$$\begin{aligned} A_{01} &= U_{00}^T U_{01} \\ A_{11} &= U_{01}^T U_{01} + U_{11}^T U_{11}. \end{aligned}$$

Hence, since U_{00} has already been computed, we can compute U_{01} as the solution to the equation

$$U_{00}^T U_{01} = A_{01}$$

by a call to the Level 3 BLAS routine STRSM; and then we can compute U_{11} from

$$U_{11}^T U_{11} = A_{11} - U_{01}^T U_{01}.$$

This involves first updating the symmetric submatrix A_{11} by a call to the Level 3 BLAS routine SSYRK, and then computing its Cholesky factorization. Since Fortran does not allow recursion, a separate routine must be called (using Level 2 BLAS rather than Level 3), named SPOTF2 in Figure 4.3. In this way, successive blocks of columns of U are computed. The LAPACK-style code for the block algorithm is shown in Figure 4.3. This code runs at 49 megaflops on an IBM 3090, more than double the speed of the LINPACK code. On a CRAY Y-MP, the use of Level 3 BLAS squeezes a little more performance out of one processor, but makes a large improvement when using all 8 processors.

But that is not the end of the story, and the code given above is not the code actually used in the LAPACK routine SPOTRF. We mentioned earlier that for many linear algebra computations there are several algorithmic variants, often referred to as i-, j-, and k-variants, according to a convention introduced in [17] and used in [38]. The same is true of the corresponding block algorithms.

It turns out that the j-variant chosen for LINPACK, and used in the above examples, is not the fastest on many machines, because it performs most of the work in solving triangular systems of equations, which can be significantly slower than matrix-matrix multiplication. The variant actually used in LAPACK is the i-variant, which relies on matrix-matrix multiplication for most of the work.

Table 4.2 summarizes the results.

```
do j = 0, n-1
  info = j + 1
  s = 0.0e0
  jm1 = j
  if (jm1 .ge. 1) then
    do k = 0, jm1 - 1
      t = a(k,j) - sdot(k,a(0,k),1,a(0,j),1)
      t = t/a(k,k)
      a(k,j) = t
      s = s + t*t
    end do
  end if
  s = a(j,j) - s
  if (s .le. 0.0e0) go to 40
  a(j,j) = sqrt(s)
  end do
```

FIGURE 4.1. The body of the LINPACK routine SPOFA for Cholesky factorization.

```
do j = 0, n - 1
  call strsv( 'upper', 'transpose', 'non-unit', j, a,
              lda, a(0,j), 1 )
  s = a(j,j) - sdot( j, a(0,j), 1, a(0,j), 1 )
  if ( s .le. zero ) go to 20
  a(j,j) = sqrt( s )
end do
```

FIGURE 4.2. The body of the "LAPACK-style" routine SPOFA for Cholesky factorization.

```
do j = 0, n-1, nb
  jb = min( nb, n-j )
  call strsm( 'left', 'upper', 'transpose', 'non-unit',
              j, jb, one, a, lda, a(0,j), lda )
  call ssyrk( 'upper', 'transpose', jb, j, -one,
              a(0,j), lda, one, a(j,j), lda )
  call spotf2( 'upper', jb, a(j,j), lda, info )
  if( info .ne. 0 ) go to 20
end do
```

FIGURE 4.3. The body of the "LAPACK-style" routine SPOFA for block Cholesky factorization. In this code fragment, `nb` denotes the width of the blocks.

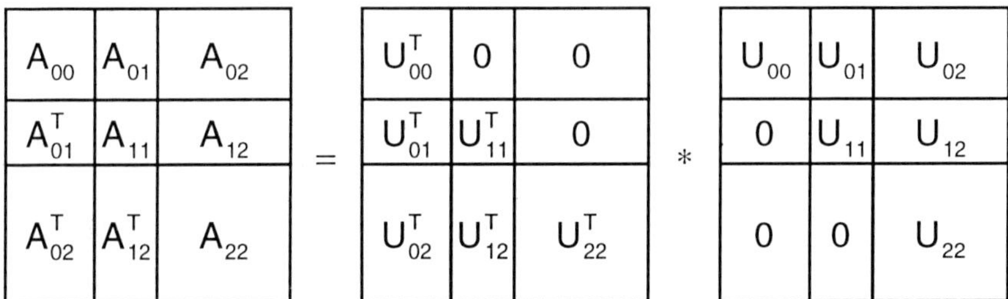

FIGURE 4.4. Partitioning of A, U^T, and U into blocks. It is assumed that the first block has already been factored as $A_{00} = U_{00}^T U_{00}$, and we next want to determine the block column consisting of U_{01} and U_{11}. Note that the diagonal blocks of A and U are square matrices.

TABLE 4.2. Speed (Megaflops) of Cholesky Factorization $A = U^T U$ for $n = 500$

	IBM 3090 VF, 1 proc.	CRAY Y-MP, 1 proc.	CRAY Y-MP, 8 proc.
j-variant: LINPACK	23	72	72
j-variant: using Level 2 BLAS	24	251	378
j-variant: using Level 3 BLAS	49	287	1225
i-variant: using Level 3 BLAS	50	290	1414

TABLE 4.3. Speed (Megaflops) of SGETRF/DGETRF for Square Matrices of Order n

Machine	No. of processors	Block size	Values of n				
			100	200	300	400	500
IBM RISC/6000-530	1	32	19	25	29	31	33
Alliant FX/8	8	16	9	26	32	46	57
IBM 3090J VF	1	64	23	41	52	58	63
Convex C-240	4	64	31	60	82	100	112
CRAY Y-MP	1	1	132	219	254	272	283
CRAY-2	1	64	110	211	292	318	358
Siemens/Fujitsu VP 400-EX	1	64	46	132	222	309	397
NEC SX2	1	1	118	274	412	504	577
CRAY Y-MP	8	64	195	556	920	1188	1408

4.3.2 Examples of Block Algorithms in LAPACK

Having discussed in detail the derivation of one particular block algorithm, we now describe examples of the performance achieved with two well-known block algorithms: LU and Cholesky factorizations. No extra floating-point operations nor extra working storage are required for either of these simple block algorithms. (See Gallivan et al. [35] and Dongarra et al. [21] for surveys of algorithms for dense linear algebra on high-performance computers.)

Table 4.3 illustrates the speed of the LAPACK routine for LU factorization of a real matrix, SGETRF in single precision on CRAY machines, and DGETRF in double precision on all other machines. Thus, 64-bit floating-point arithmetic is used on all machines tested. A block size of 1 means that the unblocked algorithm is used, since it is faster than – or at least as fast as – a block algorithm.

LAPACK is designed to give high efficiency on vector processors, high-performance "superscalar" workstations, and shared memory multiprocessors. LAPACK in its present form is less likely to give good performance on other types of parallel architectures (for example, massively parallel SIMD machines, or MIMD distributed memory machines), but the ScaLAPACK project, described in Section 4.1.1, is intended to adapt LAPACK to these new architectures. LAPACK can also be used satisfactorily on all types of scalar machines (PCs, workstations, mainframes).

Table 4.4 gives similar results for Cholesky factorization, extending the results given in Table 4.2.

LAPACK, like LINPACK, provides LU and Cholesky factorizations of

TABLE 4.4. Speed (Megaflops) of SPOTRF/DPOTRF for Matrices of Order n. Here UPLO = 'U', so the factorization is of the form $A = U^T U$.

Machine	No. of processors	Block size	Values of n				
			100	200	300	400	500
IBM RISC/6000-530	1	32	21	29	34	36	38
Alliant FX/8	8	16	10	27	40	49	52
IBM 3090J VF	1	48	26	43	56	62	67
Convex C-240	4	64	32	63	82	96	103
CRAY Y-MP	1	1	126	219	257	275	285
CRAY-2	1	64	109	213	294	318	362
Siemens/Fujitsu VP 400-EX	1	64	53	145	237	312	369
NEC SX2	1	1	155	387	589	719	819
CRAY Y-MP	8	32	146	479	845	1164	1393

band matrices. The LINPACK algorithms can easily be restructured to use Level 2 BLAS, though restructuring has little effect on performance for matrices of very narrow bandwidth. It is also possible to use Level 3 BLAS, at the price of doing some extra work with zero elements outside the band [24]. This process becomes worthwhile for large matrices and semi-bandwidth greater than 100 or so.

4.4 LU Factorization

In this section, we first discuss the uses of dense LU factorization in several fields. We next develop a block-partitioned version of the k, or right-looking, variant of the LU factorization algorithm. In subsequent sections, the parallelization of this algorithm is described in detail in order to highlight the issues and considerations that must be taken into account in developing an efficient, scalable, and transportable dense linear algebra library for MIMD, distributed memory, concurrent computers.

4.4.1 Uses of LU Factorization in Science and Engineering

A major source of large dense linear systems is problems involving the solution of boundary integral equations. These are integral equations defined on the boundary of a region of interest. All examples of practical interest compute some intermediate quantity on a two-dimensional boundary and then use this information to compute the final desired quantity in three-dimensional space. The price one pays for replacing three dimensions with two is that what started as a sparse problem in $O(n^3)$ variables is replaced by a dense problem in $O(n^2)$.

Dense systems of linear equations are found in numerous applications, including:

- airplane wing design;
- radar cross-section studies;
- flow around ships and other off-shore constructions;
- diffusion of solid bodies in a liquid;
- noise reduction; and
- diffusion of light through small particles.

The electromagnetics community is a major user of dense linear systems solvers. Of particular interest to this community is the solution of the so-called radar cross-section problem. In this problem, a signal of fixed frequency bounces off an object; the goal is to determine the intensity of the reflected signal in all possible directions. The underlying differential equation may vary, depending on the specific problem. In the design of stealth aircraft, the principal equation is the Helmholtz equation. To solve this equation, researchers use the *method of moments* [40, 58]. In the case of fluid flow, the problem often involves solving the Laplace or Poisson equation. Here, the boundary integral solution is known as the *panel method* [42, 43], so named from the quadrilaterals that discretize and approximate a structure such as an airplane. Generally, these methods are called *boundary element methods.*

Use of these methods produces a dense linear system of size $O(N)$ by $O(N)$, where N is the number of boundary points (or panels) being used. It is not unusual to see size $3N$ by $3N$, because of three physical quantities of interest at every boundary element.

A typical approach to solving such systems is to use LU factorization. Each entry of the matrix is computed as an interaction of two boundary elements. Often, many integrals must be computed. In many instances, the time required to compute the matrix is considerably larger than the time for solution.

Only the builders of stealth technology who are interested in radar cross-sections are considering using direct Gaussian elimination methods for solving dense linear systems. These systems are always symmetric and complex, but not Hermitian.

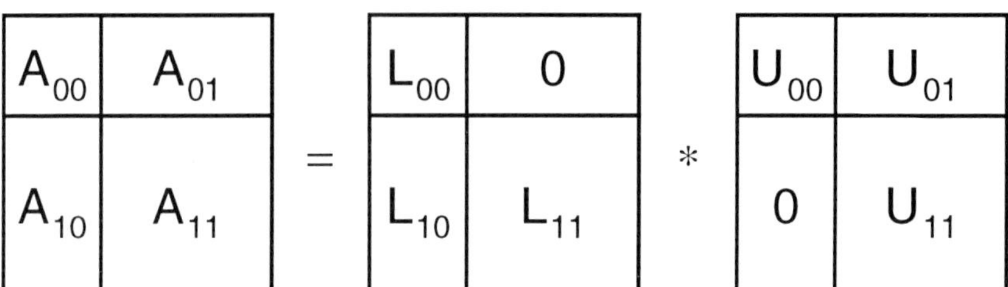

FIGURE 4.5. Block LU factorization of the partitioned matrix A. A_{00} is $r \times r$, A_{01} is $r \times (N-r)$, A_{10} is $(M-r) \times r$, and A_{11} is $(M-r) \times (N-r)$. L_{00} and L_{11} are lower triangular matrices with 1's on the main diagonal, and U_{00} and U_{11} are upper triangular matrices.

For further information on various methods for solving large dense linear algebra problems that arise in computational fluid dynamics, see the report by Alan Edelman [32].

4.4.2 Derivation of a Block Algorithm for LU Factorization

Suppose the $M \times N$ matrix A is partitioned as shown in Figure 4.5, and we seek a factorization $A = LU$, where the partitioning of L and U is also shown in Figure 4.5. Then we may write,

$$\begin{aligned} L_{00}U_{00} &= A_{00} && (4.3)\\ L_{10}U_{00} &= A_{10} && (4.4)\\ L_{00}U_{01} &= A_{01} && (4.5)\\ L_{10}U_{01} + L_{11}U_{11} &= A_{11} && (4.6) \end{aligned}$$

where A_{00} is $r \times r$, A_{01} is $r \times (N-r)$, A_{10} is $(M-r) \times r$, and A_{11} is $(M-r) \times (N-r)$. L_{00} and L_{11} are lower triangular matrices with 1s on the main diagonal, and U_{00} and U_{11} are upper triangular matrices.

Equations 4.3 and 4.4 taken together perform an LU factorization on the first $M \times r$ panel of A (i.e., A_{00} and A_{10}). Once this is completed, the matrices L_{00}, L_{10}, and U_{00} are known, and the lower triangular system in Eq. 4.5 can be solved to give U_{01}. Finally, we rearrange Eq. 4.6 as,

$$A'_{11} = A_{11} - L_{10}U_{01} = L_{11}U_{11} \qquad (4.7)$$

From this equation we see that the problem of finding L_{11} and U_{11} reduces to finding the LU factorization of the $(M-r)\times(N-r)$ matrix A'_{11}. This can be done by applying the steps outlined above to A'_{11} instead of to A. Repeating these steps K times, where

$$K = \min\left(\lceil M/r \rceil, \lceil N/r \rceil\right) \tag{4.8}$$

we obtain the LU factorization of the original $M \times N$ matrix A. For an in-place algorithm, A is overwritten by L and U – the 1s on the diagonal of L do not need to be stored explicitly. Similarly, when A is updated by Eq. 4.7 this may also be done in place.

After k of these K steps, the first kr columns of L and the first kr rows of U have been evaluated, and matrix A has been updated to the form shown in Figure 4.6, in which panel B is $(M-kr)\times r$ and C is $r\times(N-(k-1)r)$. Step $k+1$ then proceeds as follows,

1. factor B to form the next panel of L, performing partial pivoting over rows if necessary (see Figure 4.14). This evaluates the matrices L_0, L_1, and U_0 in Figure 4.6.

2. solve the triangular system $L_0 U_1 = C$ to get the next row of blocks of U.

3. do a rank-r update on the trailing submatrix E, replacing it with $E' = E - L_1 U_1$.

The LAPACK implementation of this form of LU factorization uses the Level 3 BLAS routines xTRSM and xGEMM to perform the triangular solve and rank-r update. We can regard the algorithm as acting on matrices that have been partitioned into blocks of $r\times r$ elements, as shown in Figure 4.7.

4.5 Data Distribution

The fundamental data object in the LU factorization algorithm presented in Section 4.4.2 is a block-partitioned matrix. In this section, we describe the block-cyclic method for distributing such a matrix over a two-dimensional mesh of processes, or template. In general, each process has an independent thread of control, and with each process is associated some local memory directly accessible only by that process. The assignment of these processes to physical processors is a machine-dependent optimization issue, and will be considered later in Section 4.7.

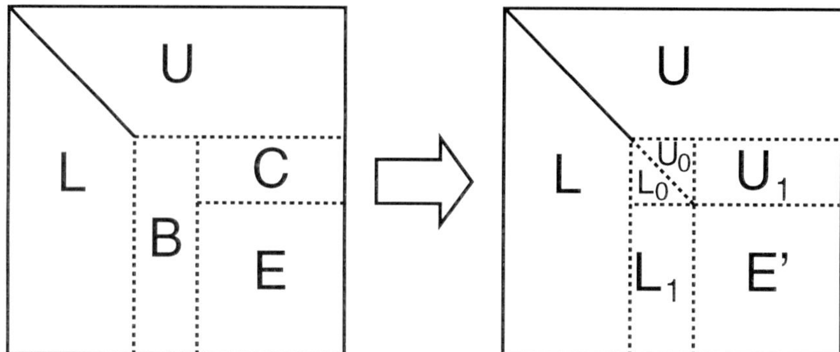

FIGURE 4.6. Stage $k+1$ of the block LU factorization algorithm showing how the panels B and C, and the trailing submatrix E are updated. The trapezoidal submatrices L and U have already been factored in previous steps. L has kr columns, and U has kr rows. In the step shown another r columns of L and r rows of U are evaluated.

$A_{0,0}$	$A_{0,1}$	$A_{0,2}$	$A_{0,3}$	$A_{0,4}$	$A_{0,5}$
$A_{1,0}$	$A_{1,1}$	$A_{1,2}$	$A_{1,3}$	$A_{1,4}$	$A_{1,5}$
$A_{2,0}$	$A_{2,1}$	$A_{2,2}$	$A_{2,3}$	$A_{2,4}$	$A_{2,5}$
$A_{3,0}$	$A_{3,1}$	$A_{3,2}$	$A_{3,3}$	$A_{3,4}$	$A_{3,5}$
$A_{4,0}$	$A_{4,1}$	$A_{4,2}$	$A_{4,3}$	$A_{4,4}$	$A_{4,5}$
$A_{5,0}$	$A_{5,1}$	$A_{5,2}$	$A_{5,3}$	$A_{5,4}$	$A_{5,5}$

FIGURE 4.7. Block-partitioned matrix A. Each block $A_{i,j}$ consists of $r \times r$ matrix elements.

An important property of the class of data distribution we shall use is that independent decompositions are applied over rows and columns. We shall, therefore, begin by considering the distribution of a vector of M data objects over P processes. This can be described by a mapping of the global index, m, of a data object to an index pair (p, i), where p specifies the process to which the data object is assigned, and i specifies the location in the local memory of p at which it is stored. We shall assume $0 \leq m < M$ and $0 \leq p < P$.

Two common decompositions are the *block* and the *cyclic* decompositions [57, 34]. The block decomposition, that is often used when the computational load is distributed homogeneously over a regular data structure such as a Cartesian grid, assigns contiguous entries in the global vector to the processes in blocks.

$$m \mapsto (\lfloor m/L \rfloor, m \bmod L), \tag{4.9}$$

where $L = \lceil M/P \rceil$. The cyclic decomposition (also known as the wrapped or scattered decomposition) is commonly used to improve load balance when the computational load is distributed inhomogeneously over a regular data structure. The cyclic decomposition assigns consecutive entries in the global vector to successive different processes,

$$m \mapsto (m \bmod P, \lfloor m/P \rfloor) \tag{4.10}$$

Examples of the block and cyclic decompositions are shown in Figure 4.8.

m	0	1	2	3	4	5	6	7	8	9
p	0	0	0	0	1	1	1	1	2	2
i	0	1	2	3	0	1	2	3	0	1

(a) Block

m	0	1	2	3	4	5	6	7	8	9
p	0	1	2	0	1	2	0	1	2	0
i	0	0	0	1	1	1	2	2	2	3

(b) Cyclic

FIGURE 4.8. Examples of block and cyclic decompositions of $M = 10$ data objects over $P = 3$ processes.

The block cyclic decomposition is a generalization of the block and cyclic decompositions in which blocks of consecutive data objects are distributed cyclically over the processes. In the block cyclic decomposition the mapping of the global index, m, can be expressed as $m \mapsto (p, b, i)$, where p is the process number, b is the block number in process p, and i is the index within block b to which m is mapped. Thus, if the number of data objects in a block is r, the block cyclic decomposition may be written,

$$m \mapsto \left(\left\lfloor \frac{m \bmod T}{r} \right\rfloor, \left\lfloor \frac{m}{T} \right\rfloor, m \bmod r \right) \tag{4.11}$$

m	0	1	2	3	4	5	6	7	8	9	10	11	12	13	14	15	16	17	18	19	20	21	22
p	0	0	1	1	2	2	0	0	1	1	2	2	0	0	1	1	2	2	0	0	1	1	2
b	0	0	0	0	0	0	1	1	1	1	1	1	2	2	2	2	2	2	3	3	3	3	3
i	0	1	0	1	0	1	0	1	0	1	0	1	0	1	0	1	0	1	0	1	0	1	0

(a) $m \mapsto (p, b, i)$

p	0	0	0	0	0	0	0	0	1	1	1	1	1	1	1	1	2	2	2	2	2	2	2
b	0	0	1	1	2	2	3	3	0	0	1	1	2	2	3	3	0	0	1	1	2	2	3
i	0	1	0	1	0	1	0	1	0	1	0	1	0	1	0	1	0	1	0	1	0	1	0
m	0	1	6	7	12	13	18	19	2	3	8	9	14	15	20	21	4	5	10	11	16	17	22

(b) $(p, b, i) \mapsto m$

FIGURE 4.9. An example of the block cyclic decomposition of $M = 23$ data objects over $P = 3$ processes for a block size of $r = 2$. (a) shows the mapping from global index, m, to the triplet (p, b, i), and (b) shows the inverse mapping.

where $T = rP$. It should be noted that this reverts to the cyclic decomposition when $r = 1$, with local index $i = 0$ for all blocks. A block decomposition is recovered when $r = L$, in which case there is a single block in each process with block number $b = 0$. The inverse mapping of the triplet (p, b, i) to a global index is given by,

$$(p, b, i) \mapsto Br + i = pr + bT + i \tag{4.12}$$

where $B = p + bP$ is the global block number. The block cyclic decomposition is one of the data distributions supported by High Performance Fortran (HPF) [44], and has been previously used, in one form or another, by several researchers (see [2, 5, 6, 11, 25, 29, 52, 54, 56] for examples of its use). The block cyclic decomposition is illustrated with an example in Figure 4.9.

The form of the block cyclic decomposition given by Eq. 4.11 ensures that the block with global index 0 is placed in process 0, the next block is placed in process 1, and so on. However, it is sometimes necessary to offset the processes relative to the global block index so that, in general, the first block is placed in process p_0, the next in process $p_0 + 1$, and so on. We, therefore, generalize the block cyclic decomposition by replacing m on the righthand side of Eq. 4.11 by $m' = m + rp_0$ to give,

$$\begin{aligned} m \quad &\mapsto \quad \left(\left\lfloor \frac{m' \bmod T}{r} \right\rfloor, \left\lfloor \frac{m'}{T} \right\rfloor, m' \bmod r \right) \\ &= \quad \left(\left(\left\lfloor \frac{m \bmod T}{r} \right\rfloor + p_0 \right) \bmod P, \left\lfloor \frac{m + rp_0}{T} \right\rfloor, m \bmod r \right) \end{aligned} \tag{4.13}$$

Equation 4.12 may also be generalized to,

$$(p, b, i) \mapsto Br + i = (p - p_0)r + bT + i \tag{4.14}$$

where now the global block number is given by $B = (p - p_0) + bP$. It should be noted that in processes with $p < p_0$, block 0 is not within the range of the block cyclic mapping and it is, therefore, an error to reference it in any way.

In decomposing an $M \times N$ matrix we apply independent block cyclic decompositions in the row and column directions. Thus, suppose the matrix rows are distributed with block size r and offset p_0 over P processes by the block cyclic mapping $\mu_{r,p_0,P}$, and the matrix columns are distributed with block size s and offset q_0 over Q processes by the block cyclic mapping $\nu_{s,q_0,Q}$. Then the matrix element indexed globally by (m, n) is mapped as follows,

$$\begin{aligned} m &\stackrel{\mu}{\longmapsto} (p, b, i) \\ n &\stackrel{\nu}{\longmapsto} (q, d, j). \end{aligned} \tag{4.15}$$

The decomposition of the matrix can be regarded as the tensor product of the row and column decompositions, and we can write,

$$(m, n) \mapsto \big((p, q),\ (b, d),\ (i, j) \big). \tag{4.16}$$

The block cyclic matrix decomposition given by Eqs. 4.15 and 4.16 distributes blocks of size $r \times s$ to a mesh of $P \times Q$ processes. We shall refer to this mesh as the *process template*, and refer to processes by their position in the template. Equation 4.16 says that global index (m, n) is mapped to process (p, q), where it is stored in the block at location (b, d) in a two-dimensional array of blocks. Within this block it is stored at location (i, j). The decomposition is completely specified by the parameters r, s, p_0, q_0, P, and Q. In Figure 4.10 an example is given of the block cyclic decomposition of a 36×80 matrix for block size 4×5, a process template 3×4, and a template offset $(p_0, q_0) = (0, 0)$. Figure 4.11 shows the same example but for a template offset of $(1, 2)$.

The block cyclic decomposition can reproduce most of the data distributions commonly used in linear algebra computations on parallel computers. For example, if $Q = 1$ and $r = \lceil M/P \rceil$ the block row decomposition is obtained. Similarly, $P = 1$ and $s = \lceil N/Q \rceil$ gives a block column decomposition. These decompositions, together with row and column cyclic decompositions, are shown in Figure 4.12. Other commonly used block cyclic matrix decompositions are shown in Figure 4.13.

4.6 Parallel Implementation

In this section we describe the parallel implementation of LU factorization, with partial pivoting over rows, for a block-partitioned matrix. The matrix, A, to be factored is assumed to have a block cyclic decomposition, and at the end of the computation is overwritten by the lower and upper triangular factors, L and U. This implicitly determines the decomposition of L and U. Quite a high-level description is given here since the details of the parallel implementation involve optimization issues that will be addressed in Section 4.7.

The sequential LU factorization algorithm described in Section 4.4.2 uses square blocks. Although in the parallel algorithm we could choose to decompose the matrix using nonsquare blocks, this would result in a more complicated code, and additional sources of concurrent overhead. For LU factorization we, therefore, restrict the decomposition to use only square blocks, so that the blocks used to decompose the matrix are the same as those used to partition the computation. If the block size is $r \times r$, then an $M \times N$ matrix consists of $M_b \times N_b$ blocks, where $M_b = \lceil M/r \rceil$ and $N_b = \lceil N/r \rceil$.

As discussed in Section 4.4.2, LU factorization proceeds in a series of sequential steps indexed by $k = 0, \min(M_b, N_b) - 1$, in each of which the following three tasks are performed,

1. factor the kth column of blocks, performing pivoting if necessary. This evaluates the matrices L_0, L_1, and U_0 in Figure 4.6.
2. evaluate the kth block row of U by solving the lower triangular system $L_0 U_1 = C$.
3. do a rank-r update on the trailing submatrix E, replacing it with $E' = E - L_1 U_1$.

We now consider the parallel implementation of each of these tasks. The computation in the factorization step involves a single column of blocks, and these lie in a single column of the process template. In the kth factorization step, each of the r columns in block column k is processed in turn. Consider the ith column in block column k. The pivot is selected by finding the element with largest absolute value in this column between row $kr + i$ and the last row, inclusive. The elements involved in the pivot search at this stage are shown shaded in Figure 4.14. Having selected the pivot, the value of the pivot and its row are broadcast to all other processors. Next, pivoting is performed by exchanging the entire row $kr + i$ with the row containing the pivot. We exchange entire rows, rather than just the part to the right

B \ D (p,q)	0	1	2	3	4	5	6	7	8	9	10	11	12	13	14	15
0	0,0	0,1	0,2	0,3	0,0	0,1	0,2	0,3	0,0	0,1	0,2	0,3	0,0	0,1	0,2	0,3
1	1,0	1,1	1,2	1,3	1,0	1,1	1,2	1,3	1,0	1,1	1,2	1,3	1,0	1,1	1,2	1,3
2	2,0	2,1	2,2	2,3	2,0	2,1	2,2	2,3	2,0	2,1	2,2	2,3	2,0	2,1	2,2	2,3
3	0,0	0,1	0,2	0,3	0,0	0,1	0,2	0,3	0,0	0,1	0,2	0,3	0,0	0,1	0,2	0,3
4	1,0	1,1	1,2	1,3	1,0	1,1	1,2	1,3	1,0	1,1	1,2	1,3	1,0	1,1	1,2	1,3
5	2,0	2,1	2,2	2,3	2,0	2,1	2,2	2,3	2,0	2,1	2,2	2,3	2,0	2,1	2,2	2,3
6	0,0	0,1	0,2	0,3	0,0	0,1	0,2	0,3	0,0	0,1	0,2	0,3	0,0	0,1	0,2	0,3
7	1,0	1,1	1,2	1,3	1,0	1,1	1,2	1,3	1,0	1,1	1,2	1,3	1,0	1,1	1,2	1,3
8	2,0	2,1	2,2	2,3	2,0	2,1	2,2	2,3	2,0	2,1	2,2	2,3	2,0	2,1	2,2	2,3
9	0,0	0,1	0,2	0,3	0,0	0,1	0,2	0,3	0,0	0,1	0,2	0,3	0,0	0,1	0,2	0,3
10	1,0	1,1	1,2	1,3	1,0	1,1	1,2	1,3	1,0	1,1	1,2	1,3	1,0	1,1	1,2	1,3
11	2,0	2,1	2,2	2,3	2,0	2,1	2,2	2,3	2,0	2,1	2,2	2,3	2,0	2,1	2,2	2,3

(a) Assignment of global block indices, (B, D), to processes, (p, q).

p \ q (B,D)	0				1				2				3			
0	0,0	0,4	0,8	0,12	0,1	0,5	0,9	0,13	0,2	0,6	0,10	0,14	0,3	0,7	0,11	0,15
	3,0	3,4	3,8	3,12	3,1	3,5	3,9	3,13	3,2	3,6	3,10	3,14	3,3	3,7	3,11	3,15
	6,0	6,4	6,8	6,12	6,1	6,5	6,9	6,13	6,2	6,6	6,10	6,14	6,3	6,7	6,11	6,15
	9,0	9,4	9,8	9,12	9,1	9,5	9,9	9,13	9,2	9,6	9,10	9,14	9,3	9,7	9,11	9,15
1	1,0	1,4	1,8	1,12	1,1	1,5	1,9	1,13	1,2	1,6	1,10	1,14	1,3	1,7	1,11	1,15
	4,0	4,4	4,8	4,12	4,1	4,5	4,9	4,13	4,2	4,6	4,10	4,14	4,3	4,7	4,11	4,15
	7,0	7,4	7,8	7,12	7,1	7,5	7,9	7,13	7,2	7,6	7,10	7,14	7,3	7,7	7,11	7,15
	10,0	10,4	10,8	10,12	10,1	10,5	10,9	10,13	10,2	10,6	10,10	10,14	10,3	10,7	10,11	10,15
2	2,0	2,4	2,8	2,12	2,1	2,5	2,9	2,13	2,2	2,6	2,10	2,14	2,3	2,7	2,11	2,15
	5,0	5,4	5,8	5,12	5,1	5,5	5,9	5,13	5,2	5,6	5,10	5,14	5,3	5,7	5,11	5,15
	8,0	8,4	8,8	8,12	8,1	8,5	8,9	8,13	8,2	8,6	8,10	8,14	8,3	8,7	8,11	8,15
	11,0	11,4	11,8	11,12	11,1	11,5	11,9	11,13	11,2	11,6	11,10	11,14	11,3	11,7	11,11	11,15

(b) Global blocks, (B, D), in each process, (p, q).

FIGURE 4.10. Block cyclic decomposition of a 36×80 matrix with a block size of 4×5, onto a 3×4 process template. Each small rectangle represents one matrix block – individual matrix elements are not shown. In (a), shading is used to emphasize the process template that is periodically stamped over the matrix, and each block is labeled with the process to which it is assigned. In (b), each shaded region shows the blocks in one process, and is labeled with the corresponding global block indices. In both figures, the black rectangles indicate the blocks assigned to process $(0, 0)$.

B \ p,q (D)	0	1	2	3	4	5	6	7	8	9	10	11	12	13	14	15
0	1,2	1,3	1,0	1,1	1,2	1,3	1,0	1,1	1,2	1,3	1,0	1,1	1,2	1,3	1,0	1,1
1	2,2	2,3	2,0	2,1	2,2	2,3	2,0	2,1	2,2	2,3	2,0	2,1	2,2	2,3	2,0	2,1
2	0,2	0,3	0,0	0,1	0,2	0,3	0,0	0,1	0,2	0,3	0,0	0,1	0,2	0,3	0,0	0,1
3	1,2	1,3	1,0	1,1	1,2	1,3	1,0	1,1	1,2	1,3	1,0	1,1	1,2	1,3	1,0	1,1
4	2,2	2,3	2,0	2,1	2,2	2,3	2,0	2,1	2,2	2,3	2,0	2,1	2,2	2,3	2,0	2,1
5	0,2	0,3	0,0	0,1	0,2	0,3	0,0	0,1	0,2	0,3	0,0	0,1	0,2	0,3	0,0	0,1
6	1,2	1,3	1,0	1,1	1,2	1,3	1,0	1,1	1,2	1,3	1,0	1,1	1,2	1,3	1,0	1,1
7	2,2	2,3	2,0	2,1	2,2	2,3	2,0	2,1	2,2	2,3	2,0	2,1	2,2	2,3	2,0	2,1
8	0,2	0,3	0,0	0,1	0,2	0,3	0,0	0,1	0,2	0,3	0,0	0,1	0,2	0,3	0,0	0,1
9	1,2	1,3	1,0	1,1	1,2	1,3	1,0	1,1	1,2	1,3	1,0	1,1	1,2	1,3	1,0	1,1
10	2,2	2,3	2,0	2,1	2,2	2,3	2,0	2,1	2,2	2,3	2,0	2,1	2,2	2,3	2,0	2,1
11	0,2	0,3	0,0	0,1	0,2	0,3	0,0	0,1	0,2	0,3	0,0	0,1	0,2	0,3	0,0	0,1

(a) Assignment of global block indices, (B, D), to processes, (p, q).

B,D (p \ q)			0					1		q			2					3		
	—	—	—	—	—	—	—	—	—	—	—	—	—	—	—	—	—	—	—	—
	—	2,2	2,6	2,10	2,14	—	2,3	2,7	2,11	2,15	2,0	2,4	2,8	2,12	—	2,1	2,5	2,9	2,13	—
0	—	5,2	5,6	5,10	5,14	—	5,3	5,7	5,11	5,15	5,0	5,4	5,8	5,12	—	5,1	5,5	5,9	5,13	—
	—	8,2	8,6	8,10	8,14	—	8,3	8,7	8,11	8,15	8,0	8,4	8,8	8,12	—	8,1	8,5	8,9	8,13	—
	—	11,2	11,6	11,10	11,14	—	11,3	11,7	11,11	11,15	11,0	11,4	11,8	11,12	—	11,1	11,5	11,9	11,13	—
	—	0,2	0,6	0,10	0,14	—	0,3	0,7	0,11	0,15	0,0	0,4	0,8	0,12	—	0,1	0,5	0,9	0,13	—
	—	3,2	3,6	3,10	3,14	—	3,3	3,7	3,11	3,15	3,0	3,4	3,8	3,12	—	3,1	3,5	3,9	3,13	—
p 1	—	6,2	6,6	6,10	6,14	—	6,3	6,7	6,11	6,15	6,0	6,4	6,8	6,12	—	6,1	6,5	6,9	6,13	—
	—	9,2	9,6	9,10	9,14	—	9,3	9,7	9,11	9,15	9,0	9,4	9,8	9,12	—	9,1	9,5	9,9	9,13	—
	—	—	—	—	—	—	—	—	—	—	—	—	—	—	—	—	—	—	—	—
	—	1,2	1,6	1,10	1,14	—	1,3	1,7	1,11	1,15	1,0	1,4	1,8	1,12	—	1,1	1,5	1,9	1,13	—
	—	4,2	4,6	4,10	4,14	—	4,3	4,7	4,11	4,15	4,0	4,4	4,8	4,12	—	4,1	4,5	4,9	4,13	—
2	—	7,2	7,6	7,10	7,14	—	7,3	7,7	7,11	7,15	7,0	7,4	7,8	7,12	—	7,1	7,5	7,9	7,13	—
	—	10,2	10,6	10,10	10,14	—	10,3	10,7	10,11	10,15	10,0	10,4	10,8	10,12	—	10,1	10,5	10,9	10,13	—
	—	—	—	—	—	—	—	—	—	—	—	—	—	—	—	—	—	—	—	—

(b) Global blocks, (B, D), in each process, (p, q).

FIGURE 4.11. The same matrix decomposition as shown in Figure 4.10, but for a template offset of $(p_0, q_0) = (1, 2)$. Dashed entries in (b) indicate that the block does not contain any data. In both figures, the black rectangles indicate the blocks assigned to process $(0, 0)$.

0,0	0,0	0,0	0,0	0,0	0,0	0,0	0,0	0,0	0,0
0,0	0,0	0,0	0,0	0,0	0,0	0,0	0,0	0,0	0,0
0,0	0,0	0,0	0,0	0,0	0,0	0,0	0,0	0,0	0,0
1,0	1,0	1,0	1,0	1,0	1,0	1,0	1,0	1,0	1,0
1,0	1,0	1,0	1,0	1,0	1,0	1,0	1,0	1,0	1,0
1,0	1,0	1,0	1,0	1,0	1,0	1,0	1,0	1,0	1,0
2,0	2,0	2,0	2,0	2,0	2,0	2,0	2,0	2,0	2,0
2,0	2,0	2,0	2,0	2,0	2,0	2,0	2,0	2,0	2,0
2,0	2,0	2,0	2,0	2,0	2,0	2,0	2,0	2,0	2,0
3,0	3,0	3,0	3,0	3,0	3,0	3,0	3,0	3,0	3,0

(a) $r = 3, s = 10, P = 4, Q = 1$

0,0	0,0	0,0	0,0	0,0	0,0	0,0	0,0	0,0	0,0
1,0	1,0	1,0	1,0	1,0	1,0	1,0	1,0	1,0	1,0
2,0	2,0	2,0	2,0	2,0	2,0	2,0	2,0	2,0	2,0
3,0	3,0	3,0	3,0	3,0	3,0	3,0	3,0	3,0	3,0
0,0	0,0	0,0	0,0	0,0	0,0	0,0	0,0	0,0	0,0
1,0	1,0	1,0	1,0	1,0	1,0	1,0	1,0	1,0	1,0
2,0	2,0	2,0	2,0	2,0	2,0	2,0	2,0	2,0	2,0
3,0	3,0	3,0	3,0	3,0	3,0	3,0	3,0	3,0	3,0
0,0	0,0	0,0	0,0	0,0	0,0	0,0	0,0	0,0	0,0
1,0	1,0	1,0	1,0	1,0	1,0	1,0	1,0	1,0	1,0

(b) $r = 1, s = 10, P = 4, Q = 1$

0,0	0,0	0,0	0,1	0,1	0,1	0,2	0,2	0,2	0,3
0,0	0,0	0,0	0,1	0,1	0,1	0,2	0,2	0,2	0,3
0,0	0,0	0,0	0,1	0,1	0,1	0,2	0,2	0,2	0,3
0,0	0,0	0,0	0,1	0,1	0,1	0,2	0,2	0,2	0,3
0,0	0,0	0,0	0,1	0,1	0,1	0,2	0,2	0,2	0,3
0,0	0,0	0,0	0,1	0,1	0,1	0,2	0,2	0,2	0,3
0,0	0,0	0,0	0,1	0,1	0,1	0,2	0,2	0,2	0,3
0,0	0,0	0,0	0,1	0,1	0,1	0,2	0,2	0,2	0,3
0,0	0,0	0,0	0,1	0,1	0,1	0,2	0,2	0,2	0,3
0,0	0,0	0,0	0,1	0,1	0,1	0,2	0,2	0,2	0,3

(c) $r = 10, s = 3, P = 1, Q = 4$

0,0	0,1	0,2	0,3	0,0	0,1	0,2	0,3	0,0	0,1
0,0	0,1	0,2	0,3	0,0	0,1	0,2	0,3	0,0	0,1
0,0	0,1	0,2	0,3	0,0	0,1	0,2	0,3	0,0	0,1
0,0	0,1	0,2	0,3	0,0	0,1	0,2	0,3	0,0	0,1
0,0	0,1	0,2	0,3	0,0	0,1	0,2	0,3	0,0	0,1
0,0	0,1	0,2	0,3	0,0	0,1	0,2	0,3	0,0	0,1
0,0	0,1	0,2	0,3	0,0	0,1	0,2	0,3	0,0	0,1
0,0	0,1	0,2	0,3	0,0	0,1	0,2	0,3	0,0	0,1
0,0	0,1	0,2	0,3	0,0	0,1	0,2	0,3	0,0	0,1
0,0	0,1	0,2	0,3	0,0	0,1	0,2	0,3	0,0	0,1

(d) $r = 10, s = 1, P = 1, Q = 4$

FIGURE 4.12. These 4 figures show different ways of decomposing a 10×10 matrix. Each cell represents a matrix element, and is labeled by the position, (p, q), in the template of the process to which it is assigned. To emphasize the pattern of decomposition, the matrix entries assigned to the process in the first row and column of the template are shown shaded, and each separate shaded region represents a matrix block. Figures (a) and (b) show block and cyclic row-oriented decompositions, respectively, for 4 nodes. In figures (c) and (d) the corresponding column-oriented decompositions are shown. Below each figure we give the values of r, s, P, and Q corresponding to the decomposition. In all cases $p_0 = q_0 = 0$.

0,0	0,0	0,0	0,1	0,1	0,1	0,2	0,2	0,2	0,3
0,0	0,0	0,0	0,1	0,1	0,1	0,2	0,2	0,2	0,3
0,0	0,0	0,0	0,1	0,1	0,1	0,2	0,2	0,2	0,3
1,0	1,0	1,0	1,1	1,1	1,1	1,2	1,2	1,2	1,3
1,0	1,0	1,0	1,1	1,1	1,1	1,2	1,2	1,2	1,3
1,0	1,0	1,0	1,1	1,1	1,1	1,2	1,2	1,2	1,3
2,0	2,0	2,0	2,1	2,1	2,1	2,2	2,2	2,2	2,3
2,0	2,0	2,0	2,1	2,1	2,1	2,2	2,2	2,2	2,3
2,0	2,0	2,0	2,1	2,1	2,1	2,2	2,2	2,2	2,3
3,0	3,0	3,0	3,1	3,1	3,1	3,2	3,2	3,2	3,3

(a) $r = 3, s = 3, P = 4, Q = 4$

0,0	0,1	0,2	0,3	0,0	0,1	0,2	0,3	0,0	0,1
0,0	0,1	0,2	0,3	0,0	0,1	0,2	0,3	0,0	0,1
0,0	0,1	0,2	0,3	0,0	0,1	0,2	0,3	0,0	0,1
1,0	1,1	1,2	1,3	1,0	1,1	1,2	1,3	1,0	1,1
1,0	1,1	1,2	1,3	1,0	1,1	1,2	1,3	1,0	1,1
1,0	1,1	1,2	1,3	1,0	1,1	1,2	1,3	1,0	1,1
2,0	2,1	2,2	2,3	2,0	2,1	2,2	2,3	2,0	2,1
2,0	2,1	2,2	2,3	2,0	2,1	2,2	2,3	2,0	2,1
2,0	2,1	2,2	2,3	2,0	2,1	2,2	2,3	2,0	2,1
3,0	3,1	3,2	3,3	3,0	3,1	3,2	3,3	3,0	3,1

(b) $r = 3, s = 1, P = 4, Q = 4$

0,0	0,0	0,0	0,1	0,1	0,1	0,2	0,2	0,2	0,3
1,0	1,0	1,0	1,1	1,1	1,1	1,2	1,2	1,2	1,3
2,0	2,0	2,0	2,1	2,1	2,1	2,2	2,2	2,2	2,3
3,0	3,0	3,0	3,1	3,1	3,1	3,2	3,2	3,2	3,3
0,0	0,0	0,0	0,1	0,1	0,1	0,2	0,2	0,2	0,3
1,0	1,0	1,0	1,1	1,1	1,1	1,2	1,2	1,2	1,3
2,0	2,0	2,0	2,1	2,1	2,1	2,2	2,2	2,2	2,3
3,0	3,0	3,0	3,1	3,1	3,1	3,2	3,2	3,2	3,3
0,0	0,0	0,0	0,1	0,1	0,1	0,2	0,2	0,2	0,3
1,0	1,0	1,0	1,1	1,1	1,1	1,2	1,2	1,2	1,3

(c) $r = 1, s = 3, P = 4, Q = 4$

0,0	0,1	0,2	0,3	0,0	0,1	0,2	0,3	0,0	0,1
1,0	1,1	1,2	1,3	1,0	1,1	1,2	1,3	1,0	1,1
2,0	2,1	2,2	2,3	2,0	2,1	2,2	2,3	2,0	2,1
3,0	3,1	3,2	3,3	3,0	3,1	3,2	3,3	3,0	3,1
0,0	0,1	0,2	0,3	0,0	0,1	0,2	0,3	0,0	0,1
1,0	1,1	1,2	1,3	1,0	1,1	1,2	1,3	1,0	1,1
2,0	2,1	2,2	2,3	2,0	2,1	2,2	2,3	2,0	2,1
3,0	3,1	3,2	3,3	3,0	3,1	3,2	3,3	3,0	3,1
0,0	0,1	0,2	0,3	0,0	0,1	0,2	0,3	0,0	0,1
1,0	1,1	1,2	1,3	1,0	1,1	1,2	1,3	1,0	1,1

(d) $r = 1, s = 1, P = 4, Q = 4$

FIGURE 4.13. These 4 figures show different ways of decomposing a 10×10 matrix over 16 processes arranged as a 4×4 template. Below each figure we give the values of r, s, P, and Q corresponding to the decomposition. In all cases $p_0 = q_0 = 0$.

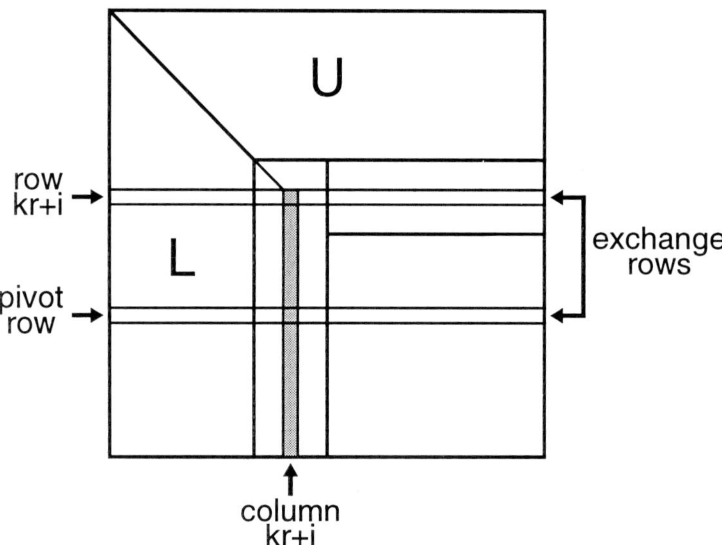

FIGURE 4.14. This figure shows pivoting for step i of the kth stage of LU factorization. The element with largest absolute value in the gray shaded part of column $kr+i$ is found, and the row containing it is exchanged with row $kr+i$. If the rows exchanged lie in different processes, communication may be necessary.

of the columns already factored, in order to simplify the application of the pivots to the righthand side in any subsequent solve phase. Finally, each value in the column below the pivot is divided by the pivot. If a cyclic column decomposition is used, like that shown in Figure 4.12(d), only one processor is involved in the factorization of the block column, and no communication is necessary between the processes. However, in general P processes are involved, and communication is necessary in selecting the pivot, and exchanging the pivot rows.

The solution of the lower triangular system $L_0U_1 = C$ to evaluate the kth block row of U involves a single row of blocks, and these lie in a single row of the process template. If a cyclic row decomposition is used, like that shown in Figure 4.12(b), only one processor is involved in the triangular solve, and no communication is necessary between the processes. However, in general Q processes are involved, and communication is necessary to broadcast the lower triangular matrix, L_0, to all processes in the row. Once this has been done, each process in the row independently performs a lower triangular solve for the blocks of C that it holds.

The communication necessary to update the trailing submatrix at step k takes place in two steps. First, each process holding part of L_1 broadcasts these blocks to the other processes in the same row of the template. This may be done in conjunction with the broadcast of L_0, mentioned in the preceding paragraph, so that all of the factored panel is broadcast together. Next, each process holding part of U_1 broadcasts these blocks to the other processes in the same column of the template. Each process can then complete the update of the blocks that it holds with no further communication.

A pseudocode outline of the parallel LU factorization algorithm is given in Figure 4.15. There are two points worth noting in Figure 4.15. First, the triangular solve and update phases operate on matrix blocks and may, therefore, be done with parallel versions of the Level 3 BLAS (specifically, xTRSM and xGEMM, respectively). The factorization of the column of blocks, however, involves a loop over matrix columns. Hence, is it not a block-oriented computation, and cannot be performed using the Level 3 BLAS. The second point to note is that most of the parallelism in the code comes from updating the trailing submatrix since this is the only phase in which all the processes are busy.

Figure 4.15 also shows quite clearly where communication is required; namely, in finding the pivot, exchanging pivot rows, and performing various types of broadcast. The exact way in which these communications are done and interleaved with computation generally has an important effect on performance, and will be discussed in more detail in Section 4.7.

Figure 4.15 refers to broadcasting data to all processes in the same row or column of the template. This is a common operation in parallel linear algebra algorithms, so the idea will be described here in a little more detail. Consider, for example, the task of broadcasting the lower triangular block, L_0, to all processes in the same row of the template, as required before solving $L_0U_1 = C$. If L_0 is in process (p, q), then it will be broadcast to all processes in row p of the process template. As a second example, consider the broadcast of L_1 to all processes in the same template row, as required before updating the trailing submatrix. This type of "rowcast" is shown schematically in Figure 4.16(a). If L_1 is in column q of the template, then each process (p, q) broadcasts its blocks of L_1 to the other processes in row p of the template. Loosely speaking, we can say that L_0 and L_1 are broadcast along the rows of the template. This type of data movement is the same as that performed by the Fortran 90 routine SPREAD [9]. The broadcast of U_1 to all processes in the same template column is very similar. This type of communication is sometimes referred to as a "colcast", and is shown in Figure 4.16(b).

4.7 Optimization, Tuning, and Trade-offs

In this section, we shall examine techniques for optimizing the basic LU factorization code presented in Section 4.4.2. Among the issues to be considered are the assignment of processes to physical processors, the arrangement of the data in the local memory of each process, the trade-off between load imbalance and communication latency, the potential for overlapping communication and calculation, and the type of algorithm used to broadcast data. Many of these issues are interdependent, and in addition the portability and ease of code maintenance and use must be considered. For further details of the optimization of parallel LU factorization algorithms for specific concurrent machines, together with timing results, the reader is referred to the work of Chu and George [14], Geist and Heath [36], Geist and Romine [37], Van de Velde [57], Brent [10], Hendrickson and Womble [41], Lichtenstein and Johnsson [49], and Dongarra and co-workers [12, 27].

4.7.1 Mapping Logical Memory to Physical Memory

In Section 4.5, a logical (or virtual) matrix decomposition was described in which the global index (m, n) is mapped to a position, (p, q), in a logical process template, a position, (b, d), in a logical array of blocks local to the process, and a position, (i, j), in a logical array of matrix elements local to the block. Thus, the block cyclic decomposition is hierarchical, and attempts to represent the hierarchical memory of advanced-architecture computers. Although the parallel LU factorization algorithm can be specified solely in terms of this logical hierarchical memory, its performance depends on how the logical memory is mapped to physical memory.

Assignment of Processes to Processors

Consider, first, the assignment of processes, (p, q), to physical processors. In general, more than one process may be assigned to a processor, so the problem may be overdecomposed. To avoid load imbalance the same number of processes should be assigned to each processor as nearly as possible. If this condition is satisfied, the assignment of processes to processors can still affect performance by influencing the communication overhead. On recent distributed memory machines, such as the Intel Delta and CM-5, the time to send a single message between two processors is largely independent of their physical location [31, 50, 51], and hence the assignment of processes to processors does not have much direct effect on performance. However, when a collective communication task, such as a broadcast, is being done, contention for physical resources can degrade performance.

Thus, the way in which processes are assigned to processors can affect performance if some assignments result in differing amounts of contention. Logarithmic contention-free broadcast algorithms have been developed for processors connected as a two-dimensional mesh [7, 53], so on such machines process (p, q) is usually mapped to the processor at position (p, q) in the mesh of processors. Such an assignment also ensures that the multiple one-dimensional broadcasts of L_1 and U_1 along the rows and columns of the template, respectively, do not give rise to contention.

Layout of Local Process Memory

The layout of matrix blocks in the local memory of a process, and the arrangement of matrix elements within each block, can also affect performance. Here, tradeoffs among several factors need to be taken into account. When communicating matrix blocks, for example in the broadcasts of L_1 and U_1, we would like the data in each block to be contiguous in physical memory so there is no need to pack them into a communication buffer before sending them. On the other hand, when updating the trailing submatrix, E, each process multiplies a column of blocks by a row of blocks, to do a rank-r update on the part of E that it contains. If this were done as a series of separate block-block matrix multiplications, as shown in Figure 4.18(a), the performance would be poor except for sufficiently large block sizes, r, since the vector and/or pipeline units on most processors would not be fully utilized, as may be seen in Figure 4.17 for the i860 processor. Instead, we arrange the loops of the computation as shown in Figure 4.18(b). Now, if the data are laid out in physical memory first by running over the i index and then over the d index the inner two loops can be merged, so that the length of the inner loop is now $rd_{\max}$. This generally results in much better vector/pipeline performance. The b and j loops in Figure 4.18(b) can also be merged, giving the algorithm shown in Figure 4.18(c). This is just the outer product form of the multiplication of an $rd_{\max} \times r$ by an $r \times rb_{\max}$ matrix, and would usually be done by a call to the Level 3 BLAS routine xGEMM of which an assembly coded sequential version is available on most machines. Note that in Figure 4.18(c) the order of the inner two loops is appropriate for a Fortran implementation – for the C language this order should be reversed, and the data should be stored in each process by rows instead of by columns.

We have found in our work on the Intel iPSC/860 hypercube and the Delta system that it is better to optimize for the sequential matrix multiplication with an (i, d, j, b) ordering of memory in each process, rather than adopting an (i, j, d, b) ordering to avoid buffer copies when communicating blocks. However, there is another reason for doing this. On most distributed memory computers the message startup cost is sufficiently large that it is preferable wherever possible to send data as one large message rather than as several smaller messages. Thus, when communicating L_1 and U_1 the blocks to be broadcast would be amalgamated into a single message, which requires a buffer copy. The emerging Message Passing Interface (MPI) standard [23] provides support for noncontiguous messages, so in the future the need to avoid buffer copies will not be of such concern to the application developer.

4.7.2 Tradeoffs between Load Balance and Communication Latency

We have discussed the mapping of the logical hierarchical memory to physical memory. In addition, we have pointed out the importance of maintaining long inner loops to get good sequential performance for each process, and the desirability of sending a few large messages rather than many smaller ones. We next consider load balance issues. Assuming that equal numbers of processes have been assigned to each processor, load imbalance arises in two phases of the parallel LU factorization algorithm; namely, in factoring each column block, which involves only P processes, and in solving the lower triangular system to evaluate each row block of U, which involves only Q processes. If the time for data movement is negligible, the aspect ratio of the template that minimizes load imbalance in step k of the algorithm is,

$$\begin{aligned} \frac{P}{Q} &= \frac{\text{Sequential time to factor column block}}{\text{Sequential time for triangular solve}} \\ &= \frac{M_b - k - 1/3 + O(1/r^2)}{N_b - k - 1 + O(1/r^2)} \end{aligned} \tag{4.17}$$

where $M_b \times N_b$ is the matrix size in blocks, and r the block size. Thus, the optimal aspect ratio of the template should be the same as the aspect ratio of the matrix, i.e., M_b/N_b in blocks, or M/N in elements. If the effect of communication time is included then we must take into account the relative times taken to locate and broadcast the pivot information, and the time to broadcast the lower triangular matrix, L_0, along a row of the template. For both tasks the communication time increases with the number of processes involved, and since the communication time associated with the pivoting is greater than that associated with the triangular solve, we would expect the optimum aspect ratio of the template to be less than M/N. In fact, for our runs on the Intel Delta system we found an aspect ratio, P/Q, of between

1/4 and 1/8 to be optimal for most problems with square matrices, and that performance depends rather weakly on the aspect ratio, particularly for large grain sizes. Some typical results are shown in Figure 4.19 for 256 processors, which show a variation of less than 20% in performance as P/Q varies between 1/16 and 1 for the largest problem.

The block size, r, also affects load balance. Here the tradeoff is between the load imbalance that arises as rows and columns of the matrix are eliminated as the algorithm progresses, and communication startup costs. The block cyclic decomposition seeks to maintain good load balance by cyclically assigning blocks to processes, and the load balance is best if the blocks are small. On the other hand, cumulative communication startup costs are less if the block size is large since, in this case, fewer messages must be sent (although the total volume of data sent is independent of the block size). Thus, there is a block size that optimally balances the load imbalance and communication startup costs.

4.7.3 Optimality and Pipelining Tradeoffs

The communication algorithms used also influence performance. In the LU factorization algorithm, all the communication can be done by moving data along rows and/or columns of the process template. This type of communication can be done by passing from one process to the next along the row or column. We shall call this a "ring" algorithm, although the ring may, or may not, be closed. An alternative is to use a spanning tree algorithm, of which there are several varieties. The complexity of the ring algorithm is linear in the number of processes involved, whereas that of spanning tree algorithms is logarithmic (for example, see [7]). Thus, considered in isolation, the spanning tree algorithms are preferable to a ring algorithm. However, in a spanning tree algorithm, a process may take part in several of the logarithmic steps, and in some implementations these algorithms act as a barrier. In a ring algorithm, each process needs to communicate only once, and can then continue to compute, in effect overlapping the communication with computation. An algorithm that interleaves communication and calculation in this way is often referred to as a pipelined algorithm. In a pipelined LU factorization algorithm with no pivoting, communication and calculation would flow in waves across the matrix. Pivoting tends to inhibit this advantage of pipelining.

In the pseudocode in Figure 4.15, we do not specify how the pivot information should be broadcast. In an optimized implementation, we need to finish with the pivot phase, and the triangular solve phase, as soon as possible in order to begin the update phase which is richest in parallelism.

Thus, it is not a good idea to broadcast the pivot information from a single source process using a spanning tree algorithm, since this may occupy some of the processes involved in the panel factorization for too long. It is important to get the pivot information to the other processes in this template column as soon as possible, so the pivot information is first sent to these processes which subsequently broadcast it along the template rows to the other processes not involved in the panel factorization. In addition, the exchange of the parts of the pivot rows lying within the panel is done separately from that of the parts outside the pivot panel. Another factor to consider here is when the pivot information should be broadcast along the template columns. In Figure 4.15, the information is broadcast, and rows exchanged, immediately after the pivot is found. An alternative is to store up the sequence of r pivots for a panel and to broadcast them along the template rows when panel factorization is complete. This defers the exchange of pivot rows for the parts outside the panel until the panel factorization has been done, as shown in the pseudocode fragment in Figure 4.20. An advantage of this second approach is that only one message is used to send the pivot information for the panel along the template rows, instead of r messages.

In our implementation of LU factorization on the Intel Delta system, we used a spanning tree algorithm to locate the pivot and to broadcast it within the column of the process template performing the panel factorization. This ensures that pivoting, which involves only P processes, is completed as quickly as possible. A ring broadcast is used to pipeline the pivot information and the factored panel along the template rows. Finally, after the triangular solve phase has completed, a spanning tree broadcast is used to send the newly-formed block row of U along the template columns. Results for square matrices from runs on the Intel Delta system are shown in Figure 4.21. For each curve the results for the best process template configuration are shown. Recalling that for a scalable algorithm the performance should depend linearly on the number of processors for fixed granularity (see Eq. 4.2), it is apparent that scalability may be assessed by the extent to which isogranularity curves differ from linearity. An isogranularity curve is a plot of performance against number of processors for a fixed granularity. The results in Figure 4.21 can be used to generate the isogranularity curves shown in Figure 4.22 which show that on the Delta system the LU factorization routine starts to lose scalability when the granularity falls below about 0.2×10^6. This corresponds to a matrix size of about $M = 10000$ on 512 processors, or about 13% of the memory available to applications on the Delta, indicating that LU factorization scales rather well on the Intel

Delta system.

4.8 Conclusions and Future Research Directions

Portability of programs has always been an important consideration. Portability was easy to achieve when there was a single architectural paradigm (the serial von Neumann machine) and a single programming language for scientific programming (Fortran) embodying that common model of computation. Architectural and linguistic diversity have made portability much more difficult, but no less important, to attain. Users simply do not wish to invest significant amounts of time to create large-scale application codes for each new machine. Our answer is to develop portable software libraries that hide machine-specific details.

4.8.1 Portability, Scalability, and Standards

In order to be truly portable, parallel software libraries must be *standardized.* In a parallel computing environment in which the higher-level routines and/or abstractions are built upon lower-level computation and message-passing routines, the benefits of standardization are particularly apparent. Furthermore, the definition of computational and message-passing standards provides vendors with a clearly defined base set of routines that they can implement efficiently.

From the user's point of view, portability means that, as new machines are developed, they are simply added to the network, supplying cycles where they are most appropriate.

From the mathematical software developer's point of view, portability may require significant effort. Economy in development and maintenance of mathematical software demands that such development effort be leveraged over as many different computer systems as possible. Given the great diversity of parallel architectures, this type of portability is attainable to only a limited degree, but machine dependences can at least be isolated.

LAPACK is an example of a mathematical software package whose highest-level components are portable, while machine dependences are hidden in lower-level modules. Such a hierarchical approach is probably the closest one can come to software portability across diverse parallel architectures. And the BLAS that are used so heavily in LAPACK provide a portable, efficient, and flexible standard for applications programmers.

Like portability, *scalability* demands that a program be reasonably effective over a wide range of number of processors. The scalability of parallel algorithms, and software libraries based on them, over a wide range of architectural designs and numbers of processors will likely require that the fundamental granularity of computation be adjustable to suit the particular circumstances in which the software may happen to execute. Our approach to this problem is block algorithms with adjustable block size. In many cases, however, polyalgorithms[1] may be required to deal with the full range of architectures and processor multiplicity likely to be available in the future.

Scalable parallel architectures of the future are likely to be based on a distributed memory architectural paradigm. In the longer term, progress in hardware development, operating systems, languages, compilers, and communications may make it possible for users to view such distributed architectures (without significant loss of efficiency) as having a shared memory with a global address space. For the near term, however, the distributed nature of the underlying hardware will continue to be visible at the programming level; therefore, efficient procedures for explicit communication will continue to be necessary. Given this fact, standards for basic message passing (send/receive), as well as higher-level communication constructs (global summation, broadcast, etc.), become essential to the development of scalable libraries that have any degree of portability. In addition to standardizing general communication primitives, it may also be advantageous to establish standards for problem-specific constructs in commonly occurring areas such as linear algebra.

The BLACS (Basic Linear Algebra Communication Subprograms) [18, 28] is a package that provides the same ease of use and portability for MIMD message-passing linear algebra communication that the BLAS [19, 20, 47] provide for linear algebra computation. Therefore, we recommend that future software for dense linear algebra on MIMD platforms consist of calls to the BLAS for computation and calls to the BLACS for communication. Since both packages will have been optimized for a particular platform, good performance should be achieved with relatively little effort. Also, since both packages will be available on a wide variety of machines, code modifications required to change platforms should be minimal.

---

[1] In a polyalgorithm the actual algorithm used depends on the computing environment and the input data. The optimal algorithm in a particular instance is automatically selected at runtime.

4.8.2 Alternative Approaches

Traditionally, large, general-purpose mathematical software libraries have required users to write their own programs that call library routines to solve specific subproblems that arise during a computation. Adapted to a shared-memory parallel environment, this conventional interface still offers some potential for hiding underlying complexity. For example, the LAPACK project incorporates parallelism in the Level 3 BLAS, where it is not directly visible to the user.

But when going from shared-memory systems to the more readily scalable distributed memory systems, the complexity of the distributed data structures required is more difficult to hide from the user. Not only must the problem decomposition and data layout be specified, but different phases of the user's problem may require transformations between different distributed data structures.

These deficiencies in the conventional user interface have prompted extensive discussion of alternative approaches for scalable parallel software libraries of the future. Possibilities include:

1. Traditional function library (i.e., minimum possible change to the status quo in going from serial to parallel environment). This will allow one to protect the programming investment that has been made.

2. Reactive servers on the network. A user would be able to send a computational problem to a server that was specialized in dealing with the problem. This fits well with the concepts of a networked, heterogeneous computing environment with various specialized hardware resources (or even the heterogeneous partitioning of a single homogeneous parallel machine).

3. General interactive environments like Matlab or Mathematica, perhaps with "expert" drivers (i.e., knowledge-based systems). With the growing popularity of the many integrated packages based on this idea, this approach would provide an interactive, graphical interface for specifying and solving scientific problems. Both the algorithms and data structures are hidden from the user, because the package itself is responsible for storing and retrieving the problem data in an efficient, distributed manner. In a heterogeneous networked environment, such interfaces could provide seamless access to computational engines that would be invoked selectively for different parts of the user's computation according to which machine is most appropriate for a particular subproblem.

4. Domain-specific problem solving environments, such as those for structural analysis. Environments like Matlab and Mathematica have proven to be especially attractive for rapid prototyping of new algorithms and systems that may subsequently be implemented in a more customized manner for higher performance.

5. Reusable templates (i.e., users adapt "source code" to their particular applications). A template is a description of a general algorithm rather than the executable object code or the source code more commonly found in a conventional software library. Nevertheless, although templates are general descriptions of key data structures, they offer whatever degree of customization the user may desire.

Novel user interfaces that hide the complexity of scalable parallelism will require new concepts and mechanisms for representing scientific computational problems and for specifying how those problems relate to each other. Very high level languages and systems, perhaps graphically based, not only would facilitate the use of mathematical software from the user's point of view, but also would help to automate the determination of effective partitioning, mapping, granularity, data structures, etc. However, new concepts in problem specification and representation may also require new mathematical research on the analytic, algebraic, and topological properties of problems (e.g., existence and uniqueness).

One of our goals as software designers is to communicate to the high-performance computing community algorithms and methods for the solution of system of linear equations. In the past we have provided black-box software in the form of a mathematical software library, such as LAPACK, LINPACK, NAG, and IMSL. These software libraries provide for:

- Easy interface with hidden details
- Reliability; the code should fail as rarely as possible
- Speed.

The high-performance computing community, on the other hand, which wants to solve the largest, hardest problems as quickly as possible, seems to want

- Speed
- Access to internal details to tune data structures to the application
- Algorithms which are fast for the particular application even if not reliable as general methods.

These differing priorities make for different approaches to algorithms and software. The first set of priorities lead us to produce "black boxes" for general problem classes. The second set of priorities seems to lead us to want to produce "template codes" or "toolboxes" where the users can assemble, modify and tune building blocks starting from well-documented subparts. This leads to software which is not going to be reliable on all problems, and requires extensive user tuning to make work. This is just what the block-box users do not want.

In scientific high-performance computing we see three different computational platforms emerging, each with a distinct set of users. The first group of computers contains the traditional supercomputer. Computers in this group exploit vector and modest parallel computing. They are general purpose computers that can accommodate a large cross section of applications while providing a high percentage of their peak computing rate. They are the computers such as the Cray, IBM, and NEC; the so called general purpose vector supercomputers.

The second group of computers are the highly parallel computers. These machines often contain hundreds or even thousands of processors, usually RISC in design. The machines are usually loosely coupled having a switching network and relatively long communication times compared to the computational times. These computers are suitable for fine grain and coarse grain parallelism. As a system the cost is usually less than the traditional supercomputer and the programming environment is very poor and primitive. Portability between different manufacturer's machines is a problem since user's programs depend heavily on a particular architecture and on a particular software environment.

The third group of computers are the clusters of workstations. Each workstation usually contains single very fast RISC processor. Each workstation is connected through a Local Area Network or LAN and as such the communication time is very slow making this setup not very suitable for fine grain parallelism. They usually have a rich software environment and operating system, usually UNIX. This solution is usually view as a very cost-effective solution compared to the vector supercomputers and highly-parallel computers.

Users are in general not a monolithic entity, but in fact represent a wide diversity of needs. Some are the sophisticated computational scientists who eagerly moves to the newest architecture in search of ever higher performance. Others want only to solve their problem with the least change to their computational approach.

We hope to satisfy the high-performance computing community's needs by the use of reusable software templates. With the templates we describe the basic features of the algorithms. These templates offer the opportunity for whatever degree of customization the user may desire, and would also potentially serve a valuable pedagogical role in teaching parallel programming and installing a better understanding of the algorithms employed and results obtained. While providing the reusable software templates we hope to retain the delicate numerical details in many algorithms.

We believe it is important for users to have trust in the algorithms, and hope this approach conveys the spirit of the algorithm and provides a clear path for implementation where the appropriate data structures can be integrated into the implementation. We believe that this approach of templates allows for easy modification to suit various needs. More specifically, each template should have:

1. Working software for as general a matrix as the method allows.

2. Mathematical description of the flow of the iteration.

3. Algorithms described in a Fortran-77 program with calls to BLAS [2] [47], [20],[19] and LAPACK routines [1].

4. Discussion of convergence and stopping criteria.

5. Suggestions for extending a method to more specific matrix types (for example, banded systems).

6. Suggestions for tuning (for example, which preconditioners are applicable and which are not).

7. Performance: when to use a method and why.

8. Reliability: for what class of problems the method is appropriate.

9. Accuracy: suggestions for measuring the accuracy of the solution, or the stability of the method.

An area where this will work well is with sparse matrix computations. Many important practical problems give rise to large sparse systems of linear equations. One reason for the great interest in sparse linear equations solvers and iterative methods is the importance of being able to obtain numerical solutions to partial differential equations. Such systems appear in studies of electrical networks, economic-system models, and physical processes such as diffusion, radiation, and elasticity. Iterative methods work

[2]For a discussion of BLAS as building blocks, see [21]

by continually refining an initial approximate solution so that it becomes closer and closer to the correct solution. With an iterative method a sequence of approximate solutions $\{x^{(k)}\}$ is constructed which essentially involve the matrix A only in the context of matrix-vector multiplication. Thus the sparsity can be taken advantage of so that each iteration requires $O(n)$ operations.

Many basic methods exist for iteratively solving linear systems and finding eigenvalues. The trick is finding the most effective method for the problem at hand. The method that works well for one problem type may not work as well for another. Or it may not work at all.

There are many iterative methods that have been constructed. Indeed this field of research is rich and active, thus making it impossible to cover all methods and approaches. Some of the methods of interest are listed below:

1. Jacobi
2. Gauss-Seidel
3. Successive Over-Relaxation (SOR)
4. Conjugate Gradient (CG)
5. Preconditioned Conjugate Gradient Method (PCG)
6. Adaptive Chebyshev
7. Biconjugate Gradient (BCG)
8. Conjugate Gradient Squared (CGS)
9. Biconjugate Gradient Stabilized (BCGSTAB)
10. Generalized Minimum Residual (GMRES)
11. Lanczos
12. Quasi-Minimal Residual (QMR)

These methods have been summarized in the monograph on on iterative methods [8].

4.8.3 In Conclusion

We hope the insight we gained from our work will influence future developers of hardware, compilers and systems software so that they provide tools to facilitate development of high quality portable numerical software.

The EISPACK, LINPACK, and LAPACK linear algebra libraries are in the public domain, and are available from *netlib*. For example, for more information on how to obtain LAPACK, send the following one-line email message to `netlib@ornl.gov`:

```
send index from lapack
```

Information for EISPACK and LINPACK can be similarly obtained. The ScaLAPACK library is available from *netlib*.

Acknowledgments

This research was performed in part using the Intel Touchstone Delta System operated by the California Institute of Technology on behalf of the Concurrent Supercomputing Consortium. Access to this facility was provided through the Center for Research on Parallel Computing.

4.9 References

[1] E. Anderson, Z. Bai, C. H. Bischof, J. Demmel, J. J. Dongarra, J. Du Croz, A. Greenbaum, S. Hammarling, A. McKenney, S. Ostrouchov and D. C. Sorensen. *LAPACK Users' Guide Release 2.0.* SIAM, Philadelphia, 1995.

[2] E. Anderson, A. Benzoni, J. J. Dongarra, S. Moulton, S. Ostrouchov, B. Tourancheau, and R. van de Geijn. LAPACK for distributed memory architectures: Progress report. In *Parallel Processing for Scientific Computing, Fifth SIAM Conference.* SIAM, 1991.

[3] E. Anderson and J. Dongarra. Results from the initial release of LAPACK. Technical Report LAPACK working note 16, Computer Science Department, University of Tennessee, Knoxville, TN, 1989.

[4] E. Anderson and J. Dongarra. Evaluating block algorithm variants in LAPACK. Technical Report LAPACK working note 19, Computer Science Department, University of Tennessee, Knoxville, TN, 1990.

[5] C. C. Ashcraft. The distributed solution of linear systems using the torus wrap data mapping. Engineering Computing and Analysis Technical Report ECA-TR-147, Boeing Computer Services, 1990.

[6] C. C. Ashcraft. A taxonamy of distributed dense LU factorization methods. Engineering Computing and Analysis Technical Report ECA-TR-161, Boeing Computer Services, 1991.

[7] M. Barnett, D. G. Payne, and R. van de Geijn. Broadcasting on meshes with worm-hole routing. Technical report, Department of Computer Science, University of Texas at Austin, April 1993. Submitted to Supercomputing '93.

[8] R. Barrett, M. Berry, T. F. Chan, J. Demmel, J. Donato, J. Dongarra, V. Eijkhout, R. Pozo, C. Romine, and H. van der Vorst. *Templates for the Solution of Linear Systems: Building Blocks for Iterative Methods.* SIAM, Philadelphia, 1994.

[9] W. S. Brainerd, C. H. Goldbergs, and J. C. Adams. *Programmers Guide to Fortran 90.* McGraw-Hill, New York, 1990.

[10] R. P. Brent. The LINPACK benchmark for the Fujitsu AP 1000. In *Proceedings of the Fourth Symposium on the Frontiers of Massively Parallel Computation*, pages 128–135. IEEE Computer Society Press, 1992.

[11] R. P. Brent. The LINPACK benchmark on the AP 1000: Preliminary report. In *Proceedings of the 2nd CAP Workshop*, NOV 1991.

[12] J. Choi, J. J. Dongarra, R. Pozo, and D. W. Walker. Scalapack: A scalable linear algebra library for distributed memory concurrent computers. In *Proceedings of the Fourth Symposium on the Frontiers of Massively Parallel Computation*, pages 120–127. IEEE Computer Society Press, 1992.

[13] J. Choi, J. J. Dongarra, and D. W. Walker. The design of scalable software libraries for distributed memory concurrent computers. In J. J. Dongarra and B. Tourancheau, editors, *Environments and Tools for Parallel Scientific Computing.* Elsevier Science Publishers, 1993.

[14] E. Chu and A. George. Gaussian elimination with partial pivoting and load balancing on a multiprocessor. *Parallel Computing*, 5:65–74, 1987.

[15] D. E. Culler, A. Dusseau, S. C. Goldstein, A. Krishnamurthy, S. Lumetta, T. von Eicken, and K. Yelick. Introduction to Split-C: Version 0.9. Technical report, Computer Science Division – EECS, University of California, Berkeley, CA 94720, February 1993.

[16] J. Demmel. LAPACK: A portable linear algebra library for supercomputers. In *Proceedings of the 1989 IEEE Control Systems Society Workshop on Computer-Aided Control System Design*, December 1989.

[17] J. J. Dongarra. Increasing the performance of mathematical software through high-level modularity. In *Proc. Sixth Int. Symp. Comp. Methods in Eng. & Applied Sciences, Versailles, France*, pages 239–248. North-Holland, 1984.

[18] J. J. Dongarra. LAPACK Working Note 34: Workshop on the BLACS. Computer Science Dept. Technical Report CS-91-134, University of Tennessee, Knoxville, TN, May 1991. (LAPACK Working Note #34).

[19] J. J. Dongarra, J. Du Croz, S. Hammarling, and I. Duff. A set of level 3 basic linear algebra subprograms. *ACM Transactions on Mathematical Software*, 16(1):1–17, 1990.

[20] J. J. Dongarra, J. Du Croz, S. Hammarling, and R. Hanson. An extended set of Fortran basic linear algebra subroutines. *ACM Transactions on Mathematical Software*, 14(1):1–17, March 1988.

[21] J. J. Dongarra, I. S. Duff, D. C. Sorensen, and H. A. Van der Vorst. *Solving Linear Systems on Vector and Shared Memory Computers*. SIAM Publications, Philadelphia, PA, 1991.

[22] J. J. Dongarra and E. Grosse. Distribution of mathematical software via electronic mail. *Communications of the ACM*, 30(5):403–407, July 1987.

[23] J. J. Dongarra, R. Hempel, A. J. G. Hey, and D. W. Walker. A proposal for a user-level message passing interface in a distributed memory environment. Technical Report TM-12231, Oak Ridge National Laboratory, February 1993.

[24] J. J. Dongarra, Peter Mayes, and Giuseppe Radicati di Brozolo. The IBM RISC System/6000 and linear algebra operations. *Supercomputer*, 44(VIII-4):15–30, 1991.

[25] J. J. Dongarra and S. Ostrouchov. LAPACK block factorization algorithms on the Intel iPSC/860. Technical Report CS-90-115, University of Tennessee at Knoxville, Computer Science Department, October 1990.

[26] J. J. Dongarra, R. Pozo, and D. W. Walker. An object oriented design for high performance linear algebra on distributed memory architectures. In *Proceedings of the Object Oriented Numerics Conference*, 1993.

[27] J. J. Dongarra, R. van de Geijn, and D. W. Walker. A look at scalable dense linear algebra libraries. In IEEE, editor, *Proceedings of the Scalable High-Performance Computing Conference*, pages 372–379. IEEE Publishers, 1992.

[28] J. J. Dongarra and R. A. van de Geijn. Two-dimensional basic linear algebra communication subprograms. Technical Report LAPACK working note 37, Computer Science Department, University of Tennessee, Knoxville, TN, October 1991.

[29] J. J. Dongarra and R. A. van de Geijn. Reduction to condensed form for the eigenvalue problem on distributed memory architectures. *Parallel Computing*, 18:973–982, 1992.

[30] J. Du Croz and M. Pont. The development of a floating-point validation package. In M. J. Irwin and R. Stefanelli, editors, *Proceedings of the 8th Symposium on Computer Arithmetic, Como, Italy, May 19-21, 1987.* IEEE Computer Society Press, 1987.

[31] T. H. Dunigan. Communication performance of the Intel Touchstone Delta mesh. Technical Report TM-11983, Oak Ridge National Laboratory, January 1992.

[32] A. Edelman. Large dense numerical linear algebra in 1993: The parallel computing influence. *International Journal Supercomputer Applications*, 1993. Accepted for publication.

[33] E. W. Felten and S. W. Otto. Coherent parallel C. In G. C. Fox, editor, *Proceedings of the Third Conference on Hypercube Concurrent Computers and Applications*, pages 440–450. ACM Press, 1988.

[34] G. C. Fox, M. A. Johnson, G. A. Lyzenga, S. W. Otto, J. K. Salmon, and D. W. Walker. *Solving Problems on Concurrent Processors*, volume 1. Prentice Hall, Englewood Cliffs, N.J., 1988.

[35] K. Gallivan, R. Plemmons, and A. Sameh. Parallel algorithms for dense linear algebra computations. *SIAM Review*, 32(1):54–135, 1990.

[36] A. Geist and M. Heath. Matrix factorization on a hypercube multiprocessor. In M. Heath, editor, *Hypercube Multiprocessors, 1986*, pages

161–180, Philadelphia, PA, 1986. Society for Industrial and Applied Mathematics.

[37] A. Geist and C. Romine. LU factorization algorithms on distributed-memory multiprocessor architectures. *SIAM J. Sci. Statist. Comput.*, 9(4):639–649, July 1988.

[38] G. H. Golub and C. F. Van Loan. *Matrix Computations.* The Johns Hopkins Press, Baltimore, Maryland, 2nd edition, 1989.

[39] A. Gupta and V. Kumar. On the scalability of FFT on parallel computers. In *Proceedings of the Frontiers 90 Conference on Massively Parallel Computation.* IEEE Computer Society Press, 1990. Also available as technical report TR 90-20 from the Computer Science Department, University of Minnesota, Minneapolis, MN 55455.

[40] R. Harrington. Origin and development of the method of moments for field computation. *IEEE Antennas and Propagation Magazine*, June 1990.

[41] B. Hendrickson and D. Womble. The torus-wrap mapping for dense matrix computations on massively parallel computers. Technical Report SAND92-0792, Sandia National Laboratories, April 1992.

[42] J. L. Hess. Panel methods in computational fluid dynamics. *Annual Reviews of Fluid Mechanics*, 22:255–274, 1990.

[43] J. L. Hess and M. O. Smith. Calculation of potential flows about arbitrary bodies. In D. Küchemann, editor, *Progress in Aeronautical Sciences, Volume 8.* Pergamon Press, 1967.

[44] High Performance Fortran Forum. *High Performance Fortran Language Specification, Version 1.0*, January 1993.

[45] R. W. Hockney and C. R. Jesshope. *Parallel Computers.* Adam Hilger Ltd., Bristol, UK, 1981.

[46] W. Kahan. Paranoia. Available from netlib [22].

[47] C. Lawson, R. Hanson, D. Kincaid, and F. Krogh. Basic linear algebra subprograms for Fortran usage. *ACM Trans. Math. Softw.*, 5:308–323, 1979.

[48] C. Leiserson. Fat trees: Universal networks for hardware-efficient supercomputing. *IEEE Transactions on Computers*, C-34(10):892–901, 1985.

[49] W. Lichtenstein and S. L. Johnsson. Block-cyclic dense linear algebra. Technical Report TR-04-92, Harvard University, Center for Research in Computing Technology, January 1992.

[50] M. Lin, D. Du, A. E. Klietz, and S. Saroff. Performance evaluation of the CM-5 interconnection network. Technical report, Department of Computer Science, University of Minnesota, 1992.

[51] R. Ponnusamy, A. Choudhary, and G. Fox. Communication overhead on CM-5: An experimental performance evaluation. In *Proceedings of the Fourth Symposium on the Frontiers of Massively Parallel Computation*, pages 108–115. IEEE Computer Society Press, 1992.

[52] Y. Saad and M. H. Schultz. Parallel direct methods for solving banded linear systems. Technical Report YALEU/DCS/RR-387, Department of Computer Science, Yale University, 1985.

[53] S. R. Seidel. Broadcasting on linear arrays and meshes. Technical Report TM-12356, Oak Ridge National Laboratory, April 1993.

[54] A. Skjellum and A. Leung. LU factorization of sparse, unsymmetric, Jacobian matrices on multicomputers. In D. W. Walker and Q. F. Stout, editors, *Proceedings of the Fifth Distributed Memory Concurrent Computing Conference*, pages 328–337. IEEE Press, 1990.

[55] Thinking Machines Corporation, Cambridge, MA. *CM-5 Technical Summary*, 1991.

[56] R. A. van de Geijn. Massively parallel LINPACK benchmark on the Intel Touchstone Delta and iPSC/860 systems. Computer Science report TR-91-28, Univ. of Texas, 1991.

[57] E. F. Van de Velde. Data redistribution and concurrency. *Parallel Computing*, 16, December 1990.

[58] J. J. H. Wang. *Generalized Moment Methods in Electromagnetics.* John Wiley & Sons, New York, 1991.

[59] J. Wilkinson and C. Reinsch. *Handbook for Automatic Computation: Volume II - Linear Algebra.* Springer-Verlag, New York, 1971.

```
pcol= q_0
prow= p_0
do k= 0, min(M_b, N_b) - 1
    do i= 0, r - 1
        if (q =pcol) find pivot value and location
        broadcast pivot value and location to all processes
        exchange pivot rows
        if (q =pcol) divide column r below diagonal by pivot
    end do

    if (p =prow) then
        broadcast L_0 to all process in same template row
        solve L_0 U_1 = C
    end if

    broadcast L_1 to all processes in same template row
    broadcast U_1 to all processes in same template column
    update E <- E - L_1 U_1

    pcol= (pcol + 1) mod Q
    prow= (prow + 1) mod P
end do
```

FIGURE 4.15. Pseudocode for the basic parallel block-partitioned LU factorization algorithm. This code is executed by each process. The first box inside the k loop factors the kth column of blocks. The second box solves a lower triangular system to evaluate the kth row of blocks of U, and the third box updates the trailing submatrix. The template offset is given by (p_0, q_0), and (p, q) is position of a process in the template.

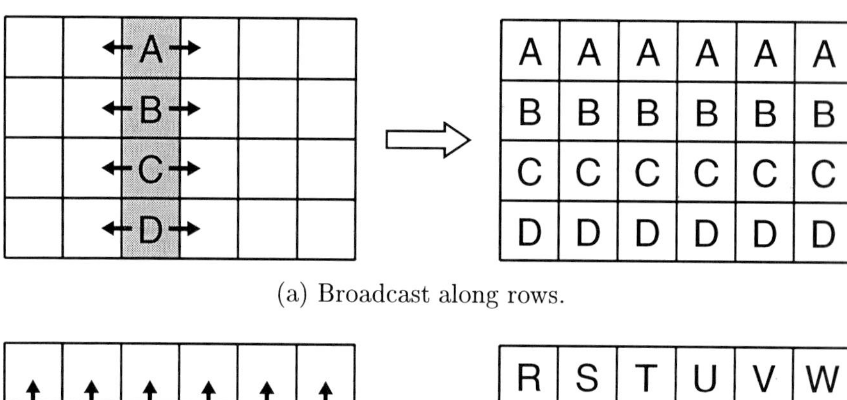

(a) Broadcast along rows.

(b) Broadcast along columns.

FIGURE 4.16. Schematic representation of broadcast along rows and columns of a 4×6 process template. In (a), each shaded process broadcasts to the processes in the same row of the process template. In (b), each shaded process broadcasts to the processes in the same column of the process template.

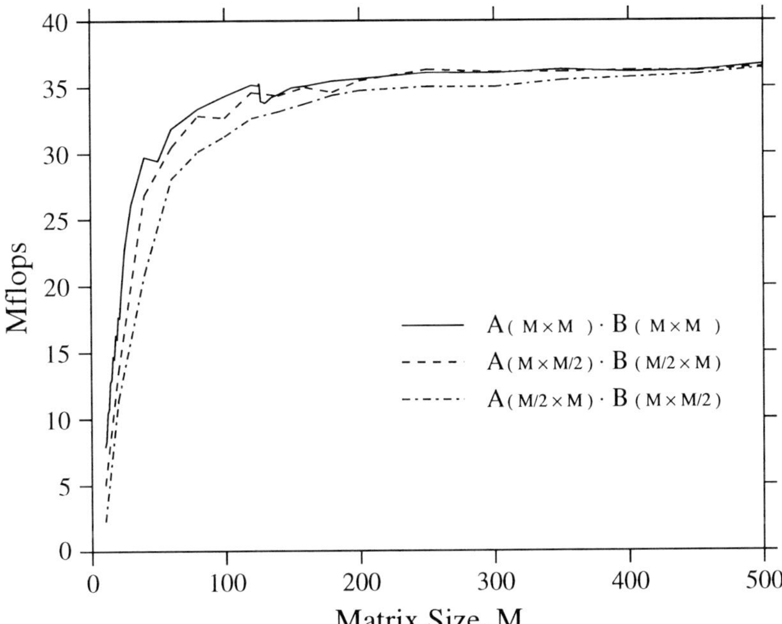

FIGURE 4.17. Performance of the assembly-coded Level 3 BLAS matrix multiplication routine DGEMM on one i860 processor of the Intel Delta system. Results for square and rectangular matrices are shown. Note that the peak performance of about 35 Mflops is attained only for matrices whose smallest dimension exceeds 100. Thus, performance is improved if a few large matrices are multiplied by each process, rather than many small ones.

```
do b = 0, b_max - 1
   do d = 0, d_max - 1
      do i = 0, r - 1
         do j = 0, r - 1
            do k = 0, r - 1
               E(b,d;i,j) = E(b,d;i,j) - L1(b,d;i,k) U1(b,d;k,j)
end all do loops
```

(a) Block-block multiplication

```
do k = 0, r - 1
   do b = 0, b_max - 1
      do j = 0, r - 1
         do d = 0, d_max - 1
            do i = 0, r - 1
               E(b,d;i,j) = E(b,d;i,j) - L1(b,d;i,k) U1(b,d;k,j)
end all do loops
```

(b) Intermediate form of algorithm

```
do k = 0, r - 1
   do x = 0, r*b_max - 1
      do y = 0, r*d_max - 1
         E(x,y) = E(x,y) - L1(x,k) U1(k,y)
end all do loops
```

(c) Outer product form of algorithm

FIGURE 4.18. Pseudocode for different versions of the rank-r update, $E \leftarrow E - L_1U_1$, for one process. The number of row and column blocks per process is given by $b_{\max}$ and $d_{\max}$, respectively; r is the block size. Blocks are indexed by (b, d), and elements within a block by (i, j). In version (a) the $r \times r$ blocks are multiplied one at a time, giving an inner loop of length r. (b) shows the loops rearranged before merging the i and d loops, and the j and b loops. This leads to the outer product form of the algorithm shown in (c) in which the inner loop is now of length $rd_{\max}$.

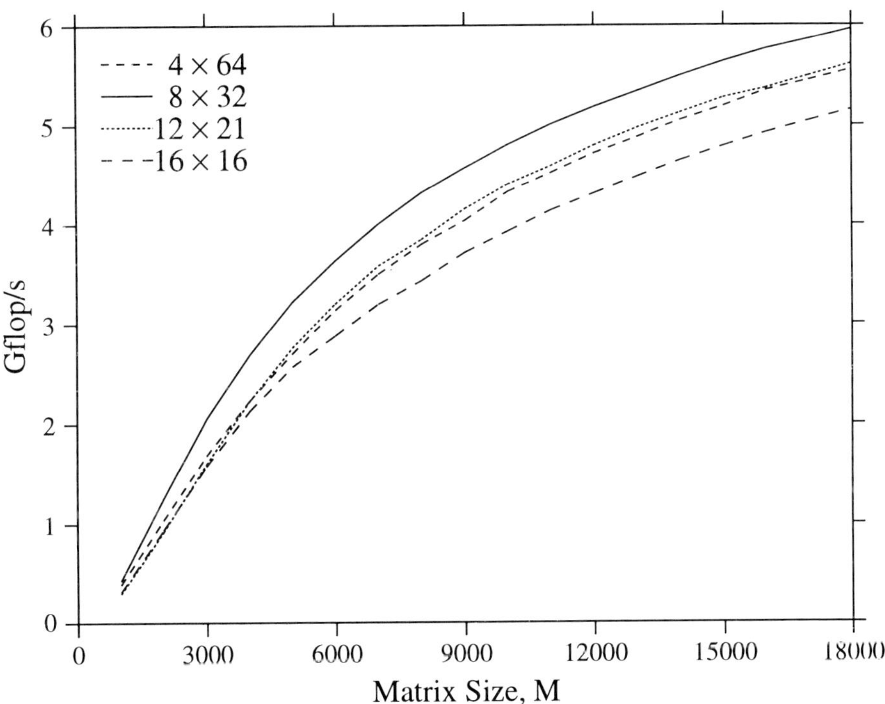

FIGURE 4.19. Performance of LU factorization on the Intel Delta as a function of square matrix size for different processor templates containing approximately 256 processors. The best performance is for an aspect ratio of 1/4, though the dependence on aspect ratio is rather weak.

```
if (q =pcol) then
   do i= 0, r − 1
      find pivot value and location
      exchange pivot rows lying within panel
      divide column r below diagonal by pivot
   end do
end if
broadcast pivot information for r pivots along template rows
exchange pivot rows lying outside the panel for each of r pivots
```

FIGURE 4.20. Pseudocode fragment for partial pivoting over rows. This may be regarded as replacing the first box inside the k loop in Figure 4.15. In the above code pivot information is first disseminated within the template column doing the panel factorization. The pivoting of the parts of the rows lying outside the panel is deferred until the panel factorization has been completed.

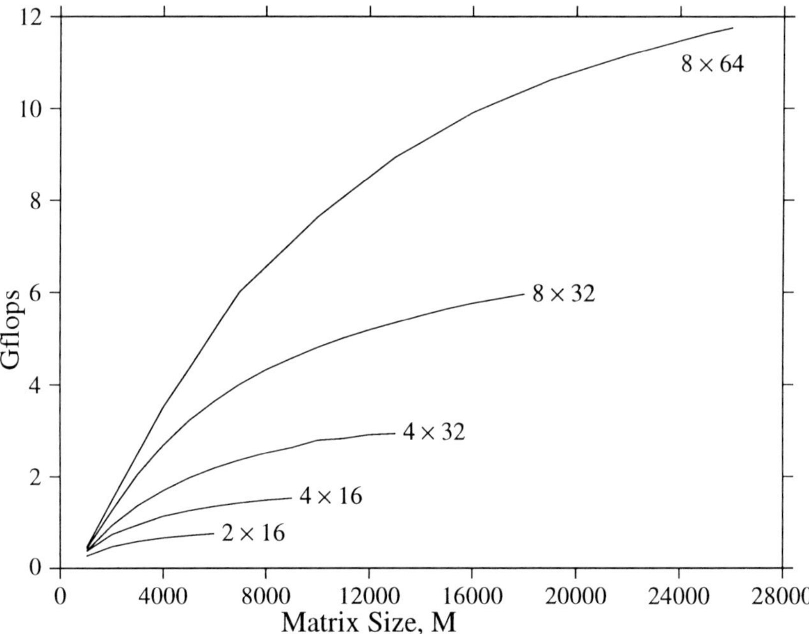

FIGURE 4.21. Performance of LU factorization on the Intel Delta as a function of square matrix size for different numbers of processors. For each curve, results are shown for the process template configuration that gave the best performance for that number of processors.

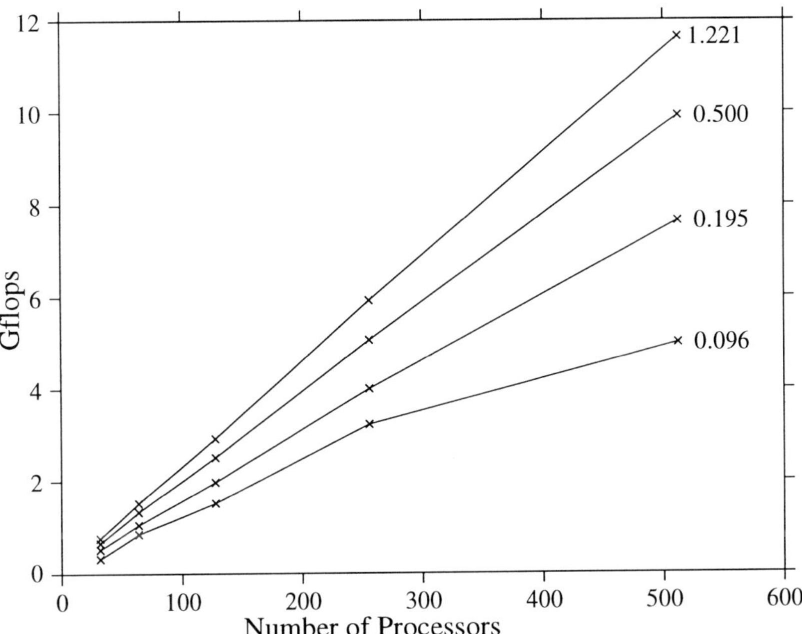

FIGURE 4.22. Isogranularity curves in the (N_p, G) plane for the LU factorization of square matrices on the Intel Delta system. The curves are labeled by the granularity in units of 10^6 matrix elements per processor. The linearity of the plots for granularities exceeding about 0.2×10^6 indicates that the LU factorization algorithm scales well on the Delta.

5

An Introduction to Multigrid Convergence Theory

Jinchao Xu *

An introduction is given in this paper to the basic idea and the convergence theory of multigrid methods. Brief discussions are first given to some basic properties of some elementary linear iterative methods such as Jacobi and Gauss-Seidel iterations and preconditioned conjugate gradient methods, and then more detailed discussions are given to a general framework of subspace correction method that can be applied to, among many other things, multigrid methods. A framework of auxiliary space method is also briefly presented for the construction of preconditioners. The multigrid method is introduced with a model elliptic boundary value problem of second order. Convergence estimates are obtained for basic multigrid methods such as backslash (\) cycle, V-cycle and W-cycle. Two different approaches are used in the convergence analysis. The first approach is the more traditional one that makes crucial use of elliptic regularity, while the second approach is based on the subspace correction framework that very weakly depends on the elliptic regularity. The first approach gives more precise estimates for simpler problems, while the second approach can be applied to more complex problems such as locally refined meshes and interface problems with large discontinuous jumps. As some more advanced topics, a general framework is briefly described on multigrid methods for nonnested multilevel subspaces and varying bilinear forms, and an optimal multigrid peconditioning technique is given for general unstructured grids using the auxiliary space framework. In addition to the aforementioned theoretical analysis, some discussions are also given to the implementation of some basic multigrid algorithms.

---

*Department of Mathematics, Penn State University. Email: `xu@math.psu.edu`.

The author was partially supported by NSF DMS94-03915-1 through Penn State and by NSF ASC-92-01266 and ONR-N00014-92-J-1890 through UCLA. This work was carried out while the author was visiting Department of Mathematics of UCLA. Special thanks go to the Applied Math Group at UCLA and especially to Professor Tony Chan.

5.1 Introduction

Multigrid methods are among the most efficient modern techniques for solving large scale algebraic systems arising from the discretization of partial differential equations. In this paper, we shall give a introduction to these methods and their convergence properties by considering their applications to a model elliptic boundary value problem of second order.

Multigrid methods have been most efficiently used in solving the linear algebraic system arising from the finite element discretizations of partial differential equations. The theory of the methods is an elegant combination of linear algebra, theory of finite element approximation and of partial differential equations. In this paper, we shall explore all these three aspects of the multigrid theory. We shall devote § 2, § 3 and § 4 to the technical materials for the theory of multigrid methods.

§ 2 is on the basic linear iterative methods and preconditioning concepts. Many elementary iterative methods, such as Jacobi and Gauss-Seidel iterations, are often the major components in a multigrid procedure, and also a multigrid method is often used in conjunction with a preconditioned conjugate gradient method. Therefore the materials in § 2 are fundamental to our multigrid algorithms and theory.

§ 3 is on an algebraic framework of subspace correction method (following Xu [32]) that can be in general used for construction and analysis of linear iterative methods. This framework will be a main technical tool in the analysis of multigrid methods in § 5.

The most technical materials in this paper are perhaps those in §4 for finite element approximation theory. In this section, some basic materials in finite elements are reviewed and some approximation results concerning multiple level of finite element spaces are presented. Some of these results depend crucially on the regularity theory for elliptic boundary value problems.

The core of this paper is §5 in which many major multigrid algorithms are introduced and analyzed. An attempt is made to explain the basic ideas behind multigrid methods and also to describe the implementation issues. But the major concern here is to present the multigrid convergence theory. The multigrid methods are analyzed with two different approaches. The first approach is the more traditional one which makes a crucial use of regularity theory of partial differential equations. The second approach is the subspace correction framework in §3.

The multigrid algorithms and their convergence analysis presented in §5 are only for the case that the underlying multilevel spaces are nested in the sense that the coarse spaces are subspaces of finer spaces. To give readers an idea on how multigrid methods can be applied to more complicated situations, we devote §6 to a general framework of nonnested multigrid methods which can be applied to cases like unstructured grids and nonconforming elements, and §7 to a special technical for construction optimal multigrid preconditioning technique for unstrutured grid using the framework of auxiliary space method.

Multigrid methods have been extensively studied in a vast literature by researchers in many different areas, a short article like this can only give a glimpse of small part of the whole subject. For further details, we refer to the tutorial book Briggs [15], research monographes Hackbusch [21, 22], McCormick [24], Wesseling [30], Bramble [5], review articles Xu [32] and Yserentant [37].

For convenience, following [32], the symbols $\lesssim, \gtrsim$ and $\eqsim$ will be used in this paper. That $x_1 \lesssim y_1, x_2 \gtrsim y_2$ and $x_3 \eqsim y_3$, mean that $x_1 \leq C_1 y_1$, $x_2 \geq c_2 y_2$ and $c_3 x_3 \leq y_3 \leq C_3 x_3$ for some constants C_1, c_2, c_3 and C_3 that are independent of mesh parameters.

5.2 Iterative and Preconditioning Methods

Assume $\mathcal{V}$ is a finite dimensional vector space. The goal of this section is to study iterative methods and preconditioning techniques for solving the following kind of equation:

$$Au = f. \tag{5.2.1}$$

Here $A : \mathcal{V} \mapsto \mathcal{V}$ is an SPD linear operator over $\mathcal{V}$ and $f \in \mathcal{V}$ is given.

5.2.1 Elementary linear iterative methods

A single step linear iterative method which uses an old approximation, u^{old}, of the solution u of (5.2.1), to produce a new approximation, u^{new}, usually consists of three steps:

1. Form $r^{\text{old}} = f - Au^{\text{old}}$;
2. Solve $Ae = r^{\text{old}}$ approximately: $\hat{e} = Br^{\text{old}}$;
3. Update $u^{new} = u^{\text{old}} + \hat{e}$,

where B is a linear operator on $\mathcal{V}$ and can be thought of as an approximate inverse of A.

As a result, we have the following iterative algorithm.

Algorithm 5.2.2 *Given* $u^0 \in \mathcal{V}$,

$$u^{k+1} = u^k + B(f - Au^k), \quad k = 0, 1, 2, \cdots.$$

The core of the above iterate scheme is the operator B. Notice that if $B = A^{-1}$, after one iteration, u^1 is then the exact solution. B will be called an iterator of A.

We say that an iterative scheme like (5.2.2) converges if $\lim_{k\to\infty} u_k = u$ for any $u_0 \in \mathcal{V}$. Assume that u and u^k are solutions of (5.2.1) and (5.2.2) respectively, then

$$u - u^k = (I - BA)^k(u - u_0).$$

Therefore the iterative scheme (5.2.2) converges iff $\rho(I - BA) < 1$.

(5.2.3) Symmetrization. Sometimes it is more desirable that the iterator B is symmetric. If B is not symmetric, there is a natural way to symmetrize it. Consider the following iteration

$$\begin{aligned} u^{k+1/2} &= u^k + B(f - Au^k) \\ u^{k+1} &= u^{k+1/2} + B^t(f - Au^{k+1/2}) \end{aligned}$$

where "t" denotes the adjoint operator with respect to $(\cdot,\cdot)$. Eliminating the intermediate $u^{k+1/2}$ gives

$$u - u^{k+1} = (I - B^tA)(I - BA)(u - u^k)$$

or

$$u^{k+1} = u^k + \bar{B}(f - Au^k) \tag{5.2.4}$$

where, with "$*$" denoting the adjoint operator with respect to $(\cdot,\cdot)_A$,

$$\bar{B} = (I - (I - BA)^*(I - BA))A^{-1} = B^t + B - B^tAB \tag{5.2.5}$$

or

$$I - \bar{B}A = (I - BA)^*(I - BA). \tag{5.2.6}$$

Obviously $\bar{B}$ is symmetric and will be called the *symmetrization* of iterator B. The following identities obviously hold

$$(\bar{B}Av, v)_A = ((2I - BA)v, BAv)_A \quad \forall\, v \in \mathcal{V}. \tag{5.2.7}$$

and

$$\|v\|_A^2 - \|(I - BA)v\|_A^2 = (\bar{B}Av, v)_A \quad \forall\, v \in \mathcal{V}. \tag{5.2.8}$$

A simple consequence of (5.2.8) is that

$$\lambda_{\max}(\bar{B}A) \le 1.$$

Theorem 1 *The following are equivalent:*

1. The symmetrized scheme (5.2.4) is convergent.

2. The operator $\bar{B}$ given by (5.2.5) is SPD.

3. The matrix $B^{-t} + B^{-1} - A$ is SPD

4. There exists a constant $\omega_1 \in (0, 2)$ such that any one of the following is satisfied for any $v \in \mathcal{V}$:

$$(BAv, BAv)_A \le \omega_1(BAv, v)_A; \tag{5.2.9}$$

$$(Av, v) \leq \omega_1(B^{-1}v, v); \tag{5.2.10}$$

$$(\frac{2}{\omega_1} - 1)(Av, v) \leq ((B^{-1} + B^{-t} - A)v, v); \tag{5.2.11}$$

$$(2 - \omega_1)(Bv, v) \leq (\bar{B}v, v). \tag{5.2.12}$$

Furthermore, the scheme 5.2.2 converges if (and only if, when B is symmetric) its symmetrized scheme 5.2.4 converges.

The above results can be proved easily by definition. We further notice that

$$(2 - \omega_1)B \leq \bar{B} \leq 2B. \tag{5.2.13}$$

(5.2.14) Richardson iterative methods. Richardson iteration is perhaps the simplest iterative method which correspond to (5.2.2) with $B = \frac{\omega}{\rho(A)}I$. Namely,

$$u^{k+1} = u^k + \frac{\omega}{\rho(A)}(f - Au^k), \quad k = 0, 1, 2, \cdots, \tag{5.2.15}$$

One can imagine that Richardson method is not very efficient method, but it is theoretically a very important one. One of the most important property of this method is its "smoothing property" that will be discussed now.

Let $A\phi_i = \lambda_i\phi_i$ with $\lambda_1 < \lambda_2 \leq \ldots \lambda_n$, $(\phi_i, \phi_j) = \delta_{ij}$, and $u - u^0 = \sum \alpha_i\phi_i$, then

$$u - u^k = \sum_i \alpha_i(1 - \omega\lambda_i/\lambda_h)^k\phi_i.$$

For a fixed $\omega \in (0, 2)$, it is clear that $(1 - \omega\lambda_i/\lambda_h)^k$ converges to zero very fast as $k \to \infty$ if λ_i is close to λ_h. This exactly means that the high frequency modes in the error get damped out very quickly.

An iterative method (5.2.2) is said to be Richardson like if there exists an $\omega \in (0, 2)$ such that

$$\|(I - BA)v\|_A \leq \|(I - \frac{\omega}{\rho_A}A)v\|_A \quad \forall\, v \in \mathcal{V}. \tag{5.2.16}$$

Lemma 1 *For the iterative method* (5.2.2), *the followings are equivalent*

1. *The inequality* (5.2.16) *satisfies with* $\omega = C_0^{-1}$.

2. $(C_0\rho_A)^{-1}\|v\|^2 \leq (\bar{B}v, v) \quad \forall\, v \in \mathcal{V}$.

3. $(C_0\rho_A)^{-1}\|Av\|^2 \leq \|v\|_A^2 - \|(I - BA)v\|_A^2 \quad \forall\, v \in \mathcal{V}$.

5.2.2 Jacobi and Gauss-Seidel Methods

Assume $\mathcal{V} = \mathbb{R}^n$ and $A = (a_{ij}) \in \mathbb{R}^{n\times n}$ is the usual SPD matrix. We write $A = D - L - U$ with D being the diagonal of A and $-L$ and $-U$ the lower and upper triangular part of A respectively. The easiest approximate inverse of A are perhaps

$$B = D^{-1} \quad \text{or} \quad B = (D-L)^{-1}.$$

As we shall see that these two choices of B result in the well-known Jacobi and Gauss-Seidel methods. More generally, we have the following choice of B that result in various different iterative methods:

$$B = \begin{cases} \omega & \text{Richardson;} \\ D^{-1} & \text{Jacobi;} \\ \omega D^{-1} & \text{Damped Jacobi;} \\ (D-L)^{-1} & \text{Gauss-Seidel;} \\ \omega(D-\omega L)^{-1} & \text{SOR.} \end{cases} \tag{5.2.17}$$

The symmetrization of the aforementioned Gauss-Seidel method is called the symmetric Gauss-Seideil method.

Theorem 2 *Assume A is SPD. Then*

- *Richardson method converges iff* $0 < \omega < 2/\rho(A)$;
- *Jacobi method converges iff* $2D - A$ *is SPD;*
- *Damped Jacobi method converges iff* $0 < \omega < 2/\rho(D^{-1}A)$;
- *Gauss-Seidel method always converges;*
- *SOR method converges iff* $0 < \omega < 2$.

The proof of the above results follow directly from Theorem 1 by (5.2.17) to compute $B^{-t} + B^{-1} - A$. For example, for SOR method, $B^{-t} + B^{-1} - A = \frac{2-\omega}{\omega} D$

5.2.3 Alternative formulations of iterative schemes

Assume that $\mathcal{V}$ and $\mathcal{W}$ are two vector spaces and $A \in L(\mathcal{V}, \mathcal{W})$. By convention, the matrix representation of A with respect to a basis $(\phi_1, \cdots, \phi_n)$ of $\mathcal{V}$ and a basis $(\psi_1, \cdots, \psi_m)$ of $\mathcal{W}$ is the matrix $\widetilde{A} \in \mathbb{R}^{m\times n}$ satisfying

$$(A\phi_1, \cdots, A\phi_n) = (\psi_1 \cdots, \psi_m)\widetilde{A}.$$

Given any $v \in \mathcal{V}$, there exists a unique $\nu = (\nu_i) \in \mathbb{R}^n$ such that $v = \sum_{i=1}^n \nu_i \phi_i$. The vector ν can be regarded as the matrix representation of v, denoted by $\nu = \widetilde{v}$.

By definition, we have, for any two operators A, B and a vector v

$$\widetilde{AB} = \widetilde{A}\widetilde{B} \quad \text{and} \quad \widetilde{Av} = \widetilde{A}\widetilde{v}. \tag{5.2.18}$$

Under the basis (ϕ_k), we define the so-called mass matrix and stiffness matrix as follows

$$\mathcal{M} = ((\phi_i, \phi_j))_{n\times n} \quad \text{and} \quad \mathcal{A} = ((A\phi_i, \phi_j))_{n\times n},$$

respectively. It can be easily shown that

$$\mathcal{A} = \mathcal{M}\widetilde{A}.$$

and that $\mathcal{M}$ is the matrix representation of the operator defined by

$$Rv = \sum_{i=1}^{n} (v, \phi_i)\phi_i, \quad \forall\, v \in \mathcal{V}. \tag{5.2.19}$$

Under a given basis (ϕ_k), the equation (5.2.1) can be transformed to an algebraic system

$$\mathcal{A}\mu = \eta. \tag{5.2.20}$$

Similar to (5.2.2), a linear iterative method for (5.2.20) can be written as

$$\mu^{k+1} = \mu^k + \mathcal{B}(\eta - \mathcal{A}\mu^k), \quad k = 0, 1, 2, \cdots, \tag{5.2.21}$$

where $\mathcal{B} \in \mathbb{R}^{n\times n}$ is an iterator of the matrix $\mathcal{A}$.

Proposition 1 *Assume that* $\tilde{u} = \mu, \tilde{f} = \beta$ *and* $\eta = \mathcal{M}\beta$. *Then* u *is the solution of* (5.2.1) *if and only if* μ *is the solution of* (5.2.20). *The linear iterations* (5.2.2) *and* (5.2.21) *are equivalent if and only if* $\widetilde{B} = \mathcal{B}\mathcal{M}$. *In this case* $\kappa(\mathcal{B}\mathcal{A}) = \kappa(BA)$.

In the following, we shall call $\mathcal{B}$ the algebraic representation of B.

Using the property of the operator defined by (5.2.19), we can show the following simple result.

Proposition 2 *The scheme* (5.2.2) *represents the Richardson iteration for the equation* (5.2.20) *if* B *is given by*

$$Bv = \omega\rho(\mathcal{A})^{-1} \sum_{i=1}^{n} (v, \phi_i)\phi_i, \quad \forall\, v \in \mathcal{V},$$

and it represents the damped Jacobi iteration if B *is given by*

$$Bv = \omega \sum_{i=1}^{n} (A\phi_i, \phi_i)^{-1} (v, \phi_i)\phi_i, \quad \forall\, v \in \mathcal{V}.$$

5.2.4 Preconditioned conjugate gradient method

The well-known conjugate gradient method is the basis of all the preconditioning techniques to be studied in this paper. The preconditioned conjugate gradient (PCG) method can be viewed as a conjugate gradient method applied to the preconditioned system:

$$BAu = Bf. \tag{5.2.22}$$

Here $B : V \mapsto V$ is another SPD operator and known as a preconditioner for A. Note that BA is symmetric with respect to the inner product $(B^{-1}\cdot, \cdot)$. One version of this algorithm is as follows: *Given* u_0; $r_0 = f - Au_0$; $p_0 = Br_0$; *For* $k = 1, 2, \ldots,$

$$\begin{aligned} u_k &= u_{k-1} + \alpha_k p_{k-1},\ r_k = r_{k-1} - \alpha_k A p_{k-1},\ p_k = Br_k + \beta_k p_{k-1}, \\ \alpha_k &= (Br_{k-1}, r_{k-1})/(Ap_{k-1}, p_{k-1}),\ \beta_k = (Br_k, r_k)/(Br_{k-1}, r_{k-1}). \end{aligned}$$

It is well-known that

$$\|u - u_k\|_A \le 2\Big(\frac{\sqrt{\kappa(BA)} - 1}{\sqrt{\kappa(BA)} + 1}\Big)^k \|u - u_0\|_A, \tag{5.2.23}$$

which implies that PCG converges faster with smaller condition number $\kappa(BA)$.

Observing the formulae in the PCG method and the convergence estimate (5.2.23), one sees that the efficiency of a PCG method depends on two main factors: the action of B and the size of $\kappa(BA)$. Hence, a good preconditioner should have the properties that the action of B is relatively easy to compute and that $\kappa(BA)$ is relatively small (at least smaller than $\kappa(A)$).

5.3 Iterative methods by subspace correction

Following Xu (1992) (see also Bramble-Pasciak-Wang-Xu [8, 7]), a general framework of constructing linear iterative methods and/or preconditioners can be obtained by the concept of *space decomposition* and *subspace correction*. This framework will be presented here from a purely algebraic point of view. Some simple examples are given for illustration and more important applications are given in the later sections for multigrid methods. This framework can also be applied directly to domain decomposition methods.

The presentation here more or less follows Xu [32]. The main modification is that the subspace solvers here may not be symmetric. For related topics, we refer to Bramble [5].

5.3.1 Preliminaries

A decomposition of a vector space $\mathcal{V}$ consists of a number of subspaces $\mathcal{V}_i \subset \mathcal{V}$ (for $0 \le i \le J$) such that

$$\mathcal{V} = \sum_{i=0}^{J} \mathcal{V}_i. \tag{5.3.1}$$

This means that, for each $v \in \mathcal{V}$, there exist $v_i \in \mathcal{V}_i$ $(0 \le i \le J)$ such that $v = \sum_{i=0}^{J} v_i$. This representation of v may not be unique in general, namely (5.3.1) is not necessarily a direct sum.

For each i, we define $Q_i, P_i : \mathcal{V} \mapsto \mathcal{V}_i$ and $A_i : \mathcal{V}_i \mapsto \mathcal{V}_i$ by

$$(Q_i u, v_i) = (u, v_i), \quad (P_i u, v_i)_A = (u, v_i)_A, \quad u \in \mathcal{V}, v_i \in \mathcal{V}_i, \tag{5.3.2}$$

and

$$(A_i u_i, v_i) = (A u_i, v_i), \quad u_i, v_i \in \mathcal{V}_i. \tag{5.3.3}$$

Q_i and P_i are both orthogonal projections and A_i is the restriction of A on $\mathcal{V}_i$ and is SPD. It follows from the definition that

$$A_i P_i = Q_i A. \tag{5.3.4}$$

This identity is of fundamental importance and will be used frequently in this chapter. A consequence of it is that, if u is the solution of (5.2.1), then

$$A_i u_i = f_i \tag{5.3.5}$$

with $u_i = P_i u$ and $f_i = Q_i f$. This equation may be regarded as the restriction of (5.2.1) to $\mathcal{V}_i$.

We note that the solution u_i of (5.3.5) is the best approximation of the solution u (5.2.1) in the subspace $\mathcal{V}_i$ in the sense that

$$J(u_i) = \min_{v \in \mathcal{V}_i} J(v), \quad \text{with } J(v) = \frac{1}{2}(Av, v) - (f, v)$$

and

$$\|u - u_i\|_A = \min_{v \in \mathcal{V}_i} \|u - v\|_A.$$

The subspace equation (5.3.5) will be in general solved approximately. To describe this, we introduce, for each i, another non-singular operator $R_i : \mathcal{V}_i \mapsto \mathcal{V}_i$ that represents an approximate inverse of A_i in certain sense. Thus an approximate solution of (5.3.5) may be given by $\hat{u}_i = R_i f_i$.

(5.3.6) Example. Consider the space $\mathcal{V} = R^n$ and the simplest decomposition:

$$\mathbb{R}^n = \sum_{i=1}^{n} \operatorname{span}\{e^i\},$$

where e^i is the i-th column of the identity matrix. For a SPD matrix $A = (a_{ij}) \in \mathbb{R}^{n\times n}$

$$A_i = a_{ii}, \quad Q_i y = y_i e^i,$$

where y_i the i-th component of $y \in \mathbb{R}^n$.

5.3.2 *Basic algorithms*

From the viewpoint of subspace correction, most linear iterative methods can be classified into two major algorithms, namely the *parallel subspace correction* (PSC) method and the *successive subspace correction* method (SSC).

PSC: Parallel subspace correction

This type of algorithm is similar to Jacobi method. The idea is to correct the residue equation on each subspace in parallel.

Let $u^{\rm old}$ be a given approximation of the solution u of (5.2.1). The accuracy of this approximation can be measured by the residual: $r^{\rm old} = f - Au^{\rm old}$. If $r^{\rm old} = 0$ or very small, we are done. Otherwise, we consider the residual equation:

$$Ae = r^{\rm old}.$$

Obviously $u = u^{\rm old} + e$ is the solution of (5.2.1). Instead we solve the restricted equation to each subspace $\mathcal{V}_i$

$$A_i e_i = Q_i r^{\rm old}.$$

It should be helpful to note that the solution e_i is the best possible correction $u^{\rm old}$ in the subspace $\mathcal{V}_i$ in the sense that

$$J(u^{\rm old} + e_i) = \min_{e\in\mathcal{V}_i} J(u^{\rm old} + e), \quad \text{with } J(v) = \frac{1}{2}(Av, v) - (f, v)$$

and

$$\|u - (u^{\rm old} + e_i)\|_A = \min_{e\in\mathcal{V}_i} \|u - (u^{\rm old} + e)\|_A.$$

As we are only seeking for a correction, we only need to solve this equation approximately using the subspace solver R_i described earlier

$$\hat{e}_i = R_i Q_i r^{\rm old}.$$

An update of the approximation of u is obtained by

$$u^{new} = u^{\rm old} + \sum_{i=0}^{J} \hat{e}_i$$

which can be written as

$$u^{new} = u^{\rm old} + B(f - Au^{\rm old}),$$

where

$$B = \sum_{i=0}^{J} R_i Q_i. \tag{5.3.7}$$

We have therefore

Algorithm 5.3.8 *Given $u_0 \in \mathcal{V}$, apply the iterative scheme* (5.2.2) *with B given in* (5.3.7).

(5.3.9) Example. With $\mathcal{V} = \mathbb{R}^n$ and the decomposition given by 5.3.1, the corresponding (5.3.8) is just the Jacobi iterative method.

It is well-known that the Jacobi method is not convergent for all SPD problems (see Theorem 2, hence (5.3.8) is not always convergent. However the preconditioner obtained from this algorithm is of great importance. We note that The operator B given by (5.3.7) is SPD if each $R_i : \mathcal{V}_i \to \mathcal{V}_i$ is SPD.

Algorithm 5.3.10 *Apply the CG method to equation* (5.2.1), *with B defined by* (5.3.7) *as a preconditioner.*

(5.3.11) Example. The preconditioner B corresponding to 5.3.1 is

$$B = \mathrm{diag}(a_{11}^{-1}, a_{22}^{-1}, \cdots, a_{nn}^{-1})$$

which is the well-known diagonal preconditioner for the SPD matrix A.

SSC: Successive subspace correction

This type of algorithm is similar to the Gauss-Seidel method.

To improve the PSC method that makes simultaneous correction, we here make the correction in one subspace at a time by using the most updated approximation of u. More precisely, starting from $v^{-1} = u^{\text{old}}$ and correcting its residue in $\mathcal{V}_0$ gives

$$v^0 = v^{-1} + R_0 Q_0 (f - Av^{-1}).$$

By correcting the new approximation v^1 in the next space $\mathcal{V}_1$, we get

$$v^1 = v^0 + R_1 Q_1 (f - Av^0).$$

Proceeding this way successively for all $\mathcal{V}_i$ leads to

Algorithm 5.3.12 *Given $u^0 \in \mathcal{V}$.*

for $k = 0, 1, \ldots$ *till convergence*

$\quad v \leftarrow u^k$

$\quad$ for $i = 0 : J \quad v \leftarrow v + R_i Q_i (f - Av)$ endfor

$\quad u^{k+1} \leftarrow v.$

endfor

(5.3.13) Example. Corresponding to decomposition in Example 5.3.1, the Algorithm 5.3.12 is the Gauss-Seidel iteration.

(5.3.14) Example. More generally, decompose $\mathbb{R}^n$ as

$$\mathbb{R}^n = \sum_{i=0}^{J} \mathrm{span}\{e^{l_i}, e^{l_i+1}, \cdots, e^{l_{i+1}-1}\},$$

where $1 = l_0 < l_1 < \cdots < l_{J+1} = n+1$. Then (5.3.8), (5.3.10) and (5.3.12) are the block Jacobi method, block diagonal preconditioner and block Gauss-Seidel method respectively.

Let $T_i = R_i Q_i A$. By (5.3.4), $T_i = R_i A_i P_i$. Note that $T_i : \mathcal{V} \mapsto \mathcal{V}_i$ is symmetric with respect to $(\cdot,\cdot)_A$ and nonnegative and that $T_i = P_i$ if $R_i = A_i^{-1}$.

If u is the exact solution of (5.2.1), then $f = Au$. Let v^i be the $i-th$ iterate (with $v^0 = u^k$) from Algorithm 5.3.12, we have by definition

$$u - v^{i+1} = (I - T_i)(u - v^i), \quad i = 0, \cdots, J.$$

A successive application of this identity yields

$$u - u^{k+1} = E_J(u - u^k), \tag{5.3.15}$$

where

$$E_J = (I - T_J)(I - T_{J-1}) \cdots (I - T_1)(I - T_0). \tag{5.3.16}$$

(5.3.17) Remark. It is interesting to look at the operator E_J in the special case that $R_i = \omega A_i^{-1}$ for all i. The corresponding SSC iteration is a generalization of the classic SOR method. In this case, we have

$$E_J = (I - \omega P_J)(I - \omega P_{J-1}) \cdots (I - \omega P_1)(I - \omega P_0).$$

One trivial fact is that E_J is invertible when $\omega \neq 1$. Following an argument by Nicolaides [28] for the SOR method, let us take a look at the special case $\omega = 2$. Since, obviously, $(I - 2P_i)^{-1} = I - 2P_i$ for each i, we conclude that $E_J^{-1} = E_J^*$ where $*$ is the adjoint with respect to the inner product $(\cdot,\cdot)_A$. This means that E_J is an orthogonal operator and, in particular, $\|E_J\|_A = 1$. As a consequence, the SSC iteration can not converge when $\omega = 2$. In fact, as we shall see in Proposition 3 below, in this special case, that the SSC method converges if and only if $0 < \omega < 2$.

The symmetrization of Algorithm (5.3.12) can also be implemented as follows.

Algorithm 5.3.18 *Given* $u^0 \in \mathcal{V}$, $v \leftarrow u^0$
  for $k = 0, 1, \ldots$ *till convergence*
    for $i = 0 : J$ *and* $i = J : -1 : 0$ $\quad v \leftarrow v + R_i Q_i (f - Av)$ endfor
  endfor

The advantage of the symmetrized algorithm is that it can be used as a preconditioner. In fact, (5.3.18) can be formulated in the form of (5.2.2) with operator B defined as follows: For $f \in \mathcal{V}$, let $Bf = u^1$ with u^1 obtained by (5.3.18) applied to (5.2.1) with $u^0 = 0$.

(5.3.19) Colorization and parallelization of SSC iteration. Associated with a given partition (5.3.1), a coloring of the set $\mathcal{J} = \{0, 1, 2, \ldots, J\}$ is a disjoint decomposition:

$$\mathcal{J} = \bigcup_{t=1}^{J_c} \mathcal{J}(t)$$

such that

$$P_i P_j = 0 \quad \text{for any} \quad i, j \in \mathcal{N}(t), i \neq j (1 \leq t \leq J_c).$$

We say that i, j have the same color if they both belong to some $\mathcal{J}(t)$.

The important property of the coloring is that the SSC iteration can be carried out in parallel in each color.

Algorithm 5.3.20 (Colored SSC) *Given* $u^0 \in \mathcal{V}$, $v \leftarrow u^0$

for $k = 0, 1, \ldots$ *till convergence*

for $t = 1 : J_c \quad v \leftarrow v + \sum_{i \in \mathcal{J}(t)} R_i Q_i (f - Av)$ endfor

endfor

We note that the terms under the sum in the above algorithm can be evaluated in parallel (for each t, namely within the same color).

5.3.3 Convergence theory

The purpose of this section is to establish an abstract theory for algorithms described in previous sections.

In view of Theorem 1, it suffices to study (5.3.10) and (5.3.12). Two fundamental theorems will be presented.

For the preconditioner (5.3.10), we need to estimate the condition number of

$$T = BA = \sum_{i=0}^{J} T_i,$$

where B is defined by (5.3.7) and $T_i = R_i A_i P_i$.

It is interesting to note the following special case:

$$BA = \sum_{i=0}^{J} P_i, \text{ if } R_i = A^{-1}.$$

For (5.3.12), we need to establish the contraction property: there exists a constant $0 < \delta < 1$ such that

$$\|E_J\|_A \leq \delta \quad \text{with} \quad \|E_J\|_A = \sup_{v \in \mathcal{V}} \frac{\|E_J v\|_A}{\|v\|_A},$$

where E_J is given by (5.3.16). Applying this estimate to (5.3.15) yields $\|u - u^k\|_A \le \delta^k \|u - u^0\|_A$.

Important parameters

The convergence theory here is to be built upon several parameters associated with the space decomposition and subspace solvers.

Parameter ω_1

The first constant, named ω_1, is the smallest constant satisfying

$$(T_i v, T_i v)_A \le \omega_1 (T_i v, v)_A \quad \forall\, v \in \mathcal{V}, 0 \le i \le J. \tag{5.3.21}$$

or equivalently

$$(v_i, A_i v_i) \le \omega_1 (R_i^{-1} v_i, v_i) \quad \forall\, v \in \mathcal{V}, 0 \le i \le J. \tag{5.3.22}$$

We assume that R_i is chosen in such a way that ω_1 is well-defined. If all R_i are SPD, then ω_1 is obviously well defined and in fact

$$\omega_1 = \max_{0 \le i \le J} \rho(R_i A_i) = \max_{0 \le i \le J} \rho(T_i).$$

The constant ω_1 is, in most cases, very easy to estimate and its boundedness often comes as an assumption. For example, while all the subspace solvers are exact, namely $R_i = A_i^{-1}$, then $\omega_1 = 1$. As we shall see late, the convergence of an SSC method is assured if the following condition holds:

$$\omega_1 < 2.$$

This condition is equivalent to saying that the symmetrized schemes for all R_i are convergent schemes (see Theorem 1) and in particular the iterative schemes given by all R_i are convergent schemes.

Parameter K_0 *and* $\bar{K}_0$

The parameter K_0 to be introduced now plays the most crucial role in most applications and it is also most difficult to estimate in applications. It measures the correlation between space decomposition and the choice of subspace solvers. We define

$$K_0 = \sup_{\|v\|_A = 1} \inf_{v_i \in \mathcal{V}_i, \sum v_i = v} \sum_i (R_i^{-1} v_i, v_i).$$

and

$$\bar{K}_0 = \sup_{\|v\|_A = 1} \inf_{v_i \in \mathcal{V}_i, \sum v_i = v} \sum_i (\bar{R}_i^{-1} v_i, v_i).$$

In other words, for any $v \in \mathcal{V}$, there exists a decomposition $v = \sum_{i=0}^J v_i$ for $v_i \in \mathcal{V}_i$ such that

$$\sum_{i=0}^{J} (R_i^{-1} v_i, v_i) \le K_0 (Av, v). \tag{5.3.23}$$

Lemma 2 *Assume, for any $v \in \mathcal{V}$, there is a decomposition $v = \sum_{i=0}^{J} v_i$ with $v_i \in \mathcal{V}_i$ satisfying*

$$\sum_{i=0}^{J} (v_i, v_i)_A \le C_0 (v, v)_A, \tag{5.3.24}$$

then

$$\bar{K}_0 \le \frac{C_0}{\bar{\omega}_0}, \quad \text{with} \quad \bar{\omega}_0 = \min_{0 \le i \le J} \lambda_{\min}(\bar{R}_i A_i)$$

and, if all R_i are SPD,

$$K_0 \le \frac{C_0}{\omega_0} \quad \text{with} \quad \text{and} \quad \omega_0 = \min_{0 \le i \le J} \lambda_{\min}(R_i A_i).$$

The above lemma is most useful in domain decomposition applications. A good upper bound of K_0 relies on a good lower bound of ω_0, which means that each subspace solver R_i should resolve the whole range of the spectrum of A_i. In another word, the subspace problems should be very well solved or preconditioned.

The constant C_0 in (5.3.24) only depends on the partition (decomposition) of the space and it is sometimes called *partition constant.*

Lemma 3 *Assume, for any $v \in \mathcal{V}$, there is a decomposition $v = \sum_{i=0}^{J} v_i$ with $v_i \in \mathcal{V}_i$ satisfying*

$$\sum_{i=0}^{J} \rho(A_i)(v_i, v_i) \le \hat{C}_0 (v, v)_A,$$

then then

$$K_0 \le \frac{\hat{C}_0}{\breve{\omega}_0} \quad \text{with} \quad \breve{\omega}_0 = \min_{0 \le i \le J} (\lambda_{\min}(\bar{R}_i)\rho(A_i)),$$

and, if all R_i are SPD,

$$K_0 \le \frac{\hat{C}_0}{\hat{\omega}_0} \quad \text{with} \quad \text{and} \quad \hat{\omega}_0 = \min_{0 \le i \le J} (\lambda_{\min}(R_i)\rho(A_i)).$$

The above lemma is most useful in multigrid applications. A good upper bound of K_0 relies on a good lower bound of $\hat{\omega}_0$, which means that each subspace solver R_i only needs to resolve the "upper" range of the spectrum of A_i. In another word, each subspace solver R_i should be spectrally equivalent to $(\rho(A_i))^{-1}$.

Parameter K_1 and $\bar{K}_1$

This parameter measures the interaction among subspaces together with the subspaces solvers.

If each R_i is SPD, we define $\epsilon_{ij} \in (0,1]$, for $j < i$,

$$\epsilon_{ij}^2 = \rho(P_j T_i P_j)/\omega_1 \quad \text{and} \quad \epsilon_{ji} = \epsilon_{ij}, \quad \epsilon_{ii} = 1. \tag{5.3.25}$$

And we define $\bar{\epsilon}_{ij} \in (0,1]$, for $j < i$,

$$\bar{\epsilon}_{ij}^2 = \rho(P_j \bar{T}_i P_j) \quad \text{and} \quad \epsilon_{ji} = \epsilon_{ij}, \quad \epsilon_{ii} = 1, \tag{5.3.26}$$

where with $\bar{R}_i$ being the symmetrization of R_i (see § 5.2.1),

$$\bar{T}_i = \bar{R}_i A_i P_i. \tag{5.3.27}$$

Note that, for each $i \geq j$, ϵ_{ij} and $\bar{\epsilon}_{ij}$ are the smallest numbers satisfying

$$(T_i v_j, v_j)_A \leq \omega_1 \epsilon_{ij}^2 (v_j, v_j)_A, \quad (\bar{T}_i v_j, v_j)_A \leq \bar{\epsilon}_{ij}^2 (v_j, v_j)_A \quad \forall\, v_j \in \mathcal{V}_j.$$

Lemma 4 *If each R_i is SPD, then*

$$(T_i u, T_j v)_A \leq \omega_1 \epsilon_{ij} (T_i u, u)_A^{\frac{1}{2}} (T_j v, v)_A^{\frac{1}{2}} \quad \forall\, u, v \in \mathcal{V}; \tag{5.3.28}$$

Proof: Without loss of generality, we may assume that $i \geq j$. It follows from Cauchy-Schwarz inequality that

$$\begin{aligned} (T_i u, T_j v)_A &\leq (T_i u, u)_A^{\frac{1}{2}} (T_i T_j v, T_j v)_A^{\frac{1}{2}} \\ &\leq \sqrt{\omega_1} \epsilon_{ij} (T_i u, u)_A^{\frac{1}{2}} (T_j v, T_j v)_A^{\frac{1}{2}} \\ &\leq \omega_1 \epsilon_{ij} (T_i u, u)_A^{\frac{1}{2}} (T_j v, v)_A^{\frac{1}{2}}. \end{aligned}$$

(5.3.29) Remark. Clearly $\epsilon_{ij} \leq 1$ and, $\epsilon_{ij} = 0$ if $P_i P_j = 0$. If $\epsilon_{ij} < 1$, the inequality (5.3.28) is often known as the *strengthened Cauchy-Schwarz inequality*.

Definition 5.3.30

$$K_1 = \min_{\mathcal{J}_0 \subset \{0:J\}} \left(|\mathcal{J}_0| + \max_{i \in \mathcal{J}_0^c} \sum_{j \in \mathcal{J}_0^c} \epsilon_{ij} \right).$$

and

$$\bar{K}_1 = \min_{\mathcal{J}_0 \subset \{0:J\}} \left(|\mathcal{J}_0| + \max_{i \in \mathcal{J}_0^c} \sum_{j \in \mathcal{J}_0^c} \bar{\epsilon}_{ij} \right).$$

Roughly speaking, K_1 is bounded if the matrix (ϵ_{ij}) is sparse except for a few rows and columns.

Lemma 5 *The parameter $K1$ admits the following estimates:*

1. $K_1 \leq J+1$.

2. $K_1 \leq 1+\rho((\epsilon_{ij})_{i,j=1:J}) \leq 1+\max_{1\leq i\leq J}\sum_{j=1}^{n}\epsilon_{ij}$.

3. If $\epsilon_{ij} \lesssim \gamma^{|i-j|}$ *or* $\bar{\epsilon}_{ij} \lesssim \gamma^{|i-j|}$ *for some* $\gamma \in (0,1)$, *then* $\hat{K}_1 \lesssim \frac{1}{1-\gamma}$ *or* $\bar{K}_1 \lesssim \frac{1}{1-\gamma}$.

Lemma 6

$$\sum_{i>j}(\bar{T}_i u_i, T_j v_j)_A \leq (\bar{K}_1-1)\Big(\sum_{i=0}^{J}(\bar{T}_i u_i, u_i)_A\Big)^{1/2}\Big(\sum_{j=0}^{J}(T_j v_j, T_j v_j)_A\Big)^{1/2}$$

If each R_i is SPD, then

$$\sum_{i>j}(T_i u_i, T_j v_j)_A \leq \omega_1(K_1-1)\Big(\sum_{i=0}^{J}(T_i u_i, u_i)_A\Big)^{1/2}\Big(\sum_{j=0}^{J}(T_j v_j, v_j)_A\Big)^{1/2}$$

If each R_i is SPD, then for any $S \subset \{0:J\}\times\{0:J\}$,

$$\sum_{i,j\in S}(T_i u_i, T_j v_j)_A \leq \omega_1 K_1\Big(\sum_{i=0}^{J}(T_i u_i, u_i)_A\Big)^{1/2}\Big(\sum_{j=0}^{J}(T_j v_j, v_j)_A\Big)^{1/2}$$

Convergence theory

With the parameters ω_1, K_0 and K_1 introduced above, the convergence estimates for PSC and SSC methods can be neatly presented. The analysis for PSC preconditioner is relatively easy whereas the analysis for SSC iteration is less straightforward.

We first give a lower bound for the spectrum of the PSC preconditioner.

Lemma 7 *Assume that all R_i are SPD. The PSC preconditioner B given by (5.3.7) satisfies*

$$\lambda_{\min}(BA) = K_0^{-1}.$$

Proof: If $v=\sum_{i=0}^{J}v_i$ is a decomposition that satisfies (5.3.23), then

$$(v,v)_A = \sum_{i=0}^{J}(v_i, v)_A = \sum_{i=0}^{J}(v_i, P_i v)_A,$$

and by the Cauchy-Schwarz inequality

$$\begin{aligned}\sum_{i=0}^{J}(v_i, P_i v)_A &= \sum_{i=0}^{J}(v_i, A_i P_i v) \leq \sum_{i=0}^{J}(R_i^{-1}v_i, v_i)^{\frac{1}{2}}(R_i A_i P_i v, v)_A^{\frac{1}{2}} \\ &\leq \Big(\sum_{i=0}^{J}(R_i^{-1}v_i, v_i)\Big)^{\frac{1}{2}}\Big(\sum_{i=0}^{J}(T_i v, v)_A\Big)^{\frac{1}{2}} \leq \sqrt{K_0}\|v\|_A(Tv,v)_A^{\frac{1}{2}}.\end{aligned}$$

Consequently

$$\|v\|_A^2 \le K_0(Tv, v)_A$$

This implies that $\lambda_{\min}(BA) \ge K_o^{-1}$.

Now for $v = \sum_{i=0}^J v_i$ with $v_i = T_i T^{-1} v$, we have

$$\begin{aligned} K_0 &\le \max_{v\in\mathcal{V}} \frac{\sum_{i=0}^J (R_i^{-1} T_i T^{-1} v, T_i T^{-1} v)}{\|v\|_A^2} \\ &= \max_{v\in\mathcal{V}} \frac{(T^{-1}v, v)_A}{(v,v)_A} = (\lambda_{\min}(BA))^{-1}. \end{aligned}$$

The desired estimate then follows.

Theorem 3 *Assume all R_i are SPDE. The PSC preconditioner B given by (5.3.7) satisfies*

$$\lambda_{\min}(BA) = K_0^{-1} \quad \textit{and} \quad \lambda_{\max}(BA) \le \omega_1 K_1,$$

and

$$\kappa(BA) \le \omega_1 K_0 K_1.$$

And in view of Lemmas 2 and 3,

$$\kappa(BA) \le \frac{\omega_1}{\omega_0} C_0 K_1, \quad \kappa(BA) \le \frac{\omega_1}{\hat{\omega}_0} \hat{C}_0 K_1,$$

Proof: By Lemma 6,

$$\|Tv\|_A^2 = \sum_{i,j=0}^J (T_i v, T_j v)_A \le K_1 (Tv, v)_A \le K_1 \|Tv\|_A \|v\|_A,$$

which implies that $\lambda_{\max}(BA) \le K_1$.

To present our next theorem, let us first prove a very simple but important lemma.

Lemma 8 *Denote $E_{-1} = I$ and for $0 \le i \le J$,*

$$E_i = (I - T_i)(I - T_{i-1}) \cdots (I - T_1)(I - T_0).$$

Then

$$I - E_i = \sum_{j=0}^i T_j E_{j-1}, \tag{5.3.31}$$

and for any $v \in \mathcal{V}$,

$$\|v\|_A^2 - \|E_J v\|_A^2 = \sum_{i=0}^J (\bar{T}_i E_{i-1} v, E_{i-1} v)_A \tag{5.3.32}$$

where $\bar{T}_i$ *is given by* (5.3.27).

Furthermore if each R_i *is symmetric then*

$$\|v\|_A^2 - \|E_J v\|_A^2 \ge (2-\omega_1)\sum_{i=0}^{J}(T_i E_{i-1}v, E_{i-1}v)_A \tag{5.3.33}$$

Proof: The identity (5.3.31) follows immediately from the trivial identity $E_{i-1} - E_i = T_i E_{i-1}$. Similar to (5.2.8) and (5.2.7), we have

$$\|E_{i-1}v\|_A^2 - \|E_i v\|_A^2 = ((2I - T_i)E_{i-1}v, T_i E_{i-1}v)_A = (\bar{T}_i E_{i-1}v, E_{i-1}v)_A.$$

Summing up these inequalities with respect to i gives (5.3.32). The estimate (5.3.33) follows by combining (5.3.32) and (5.2.13).

Again let us take a look at the special case that $R_i = \omega A_i^{-1}$ for each i. In this case, we have

$$\|v\|_A^2 - \|E_J v\|_A^2 = \omega(2-\omega)\sum_{i=0}^{J}\|P_i E_{i-1}v\|_A^2.$$

This identity implies immediately that a necessary condition for the convergence for the related SSS method is that $0 < \omega < 2$. In fact, like in SOR method, it is not hard to see that this condition is also sufficient for the convergence (see Corollary 1 below). Thus, we have the following simple generalization of a classic result for the SOR method (see also Remark 5.3.2).

Proposition 3 *The SSC method with* $R_i = \omega A_i^{-1}$ *for each* i *converges if and only if* $0 < \omega_1 < 2$.

Lemma 9 *Assume that* $\omega_1 < 2$. *If each* R_i *is SPD, then*

$$\sum_{i=0}^{J}(T_i v, v)_A \le (1+K_1)^2 \sum_{i=0}^{J}(T_i E_{i-1}v, E_{i-1}v)_A \quad \forall\, v \in \mathcal{V}, \tag{5.3.34}$$

and in general

$$\sum_{i=0}^{J}(\bar{T}_i v, v)_A \le \left(1 + \sqrt{\frac{\omega_1}{2-\omega_1}}(\bar{K}_1 - 1)\right)^2 \sum_{j=0}^{J}(\bar{T}_j E_{j-1}v, E_{j-1}v)_A \tag{5.3.35}$$

Proof: By (5.3.31)

$$\begin{aligned}(T_i v, v)_A &= (T_i v, E_{i-1}v)_A + (T_i v, (I - E_{i-1})v)_A \\ &= (T_i v, E_{i-1}v)_A + \sum_{j=0}^{i-1}(T_i v, T_j E_{j-1}v)_A.\end{aligned}$$

Applying the Cauchy-Schwarz inequality gives,

$$\sum_{i=0}^{J}(T_i v, E_{i-1}v)_A \le \left(\sum_{i=0}^{J}(T_i v, v)_A\right)^{\frac{1}{2}} \left(\sum_{i=0}^{J}(T_i E_{i-1}v, E_{i-1}v)_A\right)^{\frac{1}{2}},$$

and, by Lemma 6,

$$\sum_{i=0}^{J}\sum_{j=0}^{i-1}(T_i v, T_j E_{j-1}v)_A$$

$$\le \omega_1(K_1 - 1)\left(\sum_{i=0}^{J}(T_i v, v)_A\right)^{\frac{1}{2}} \left(\sum_{j=0}^{J}(T_j E_{j-1}v, E_{j-1}v)_A\right)^{\frac{1}{2}}.$$

Combining these three formulae then leads to (5.3.34) and hence completes the proof for (5.3.36).

With arguments similar to the above (essentially by replacing T_i by $\bar{T}_i$ in the above proof), it is easy to obtain that

$$\sum_{i=0}^{J}(\bar{T}_i v, v)_A \le \left(\sum_{i=0}^{J}(\bar{T}_i v, v)_A\right)^{1/2} \left(\sum_{j=0}^{J}(\bar{T}_j E_{j-1}v, E_{j-1}v)_A\right)^{1/2}$$

$$+(\bar{K}_1 - 1)\left(\sum_{i=0}^{J}(\bar{T}_i v, v)_A\right)^{\frac{1}{2}} \left(\sum_{j=0}^{J}(T_j E_{j-1}v, T_j E_{j-1}v)_A\right)^{\frac{1}{2}}.$$

After canceling the common factor and using the following inequalities (see (5.2.10) and (5.2.9)):

$$(T_j w, w) \le (2-\omega_1)^{-1}(\bar{T}_j w, w), \quad (T_j w, T_j w) \le \omega_1(2-\omega_1)^{-1}(\bar{T}_j w, w),$$

The estimate (5.3.35) then follows easily.

Now we are in a position to present our second fundamental theorem.

Theorem 4 *Assume that* $\omega_1 < 2$. *If each* R_i *is SPD, then the iterator* E_J *(given by* (5.3.16)*) for the Algorithm 5.3.12 satisfies*

$$\|E_J\|_A^2 \le 1 - \frac{2-\omega_1}{K_0(1+\omega_1(K_1-1))^2}; \tag{5.3.36}$$

and, in general,

$$\|E_J\|_A^2 \le 1 - \frac{2-\omega_1}{\bar{K}_0(\sqrt{2-\omega_1} + \sqrt{\omega_1}(\bar{K}_1 - 1))^2} \tag{5.3.37}$$

Proof: The estimate in (5.3.36) is obviously equivalent to

$$\|v\|_A^2 \le \frac{K_0(1+K_1)^2}{2-\omega_1}(\|v\|_A^2 - \|E_J v\|_A^2) \quad \forall\, v \in \mathcal{V}.$$

Estimate (5.3.36) then follows by combining (5.3.34) with (5.3.32) and (5.2.10).

The second estimate (5.3.37) then follows by combining (5.3.35) with the fact that $\lambda_{\min}(\sum_i \bar{T}_i) = \bar{K}_0^{-1}$ (similar to Lemma 7).

As a direct consequence of the above theorem, we have the following simple result.

Corollary 1 *A sufficient condition for the convergence of the SSC method is that*

$$\omega_1 < 2 \tag{5.3.38}$$

The condition (5.3.38) is also necessary in some sense, see Proposition 3.

(5.3.39) Remark. Note that the convergence estimate in Theorem 4 is independent of the order of how (5.3.12) is proceeded. Namely, if we shuffle the order in the decomposition (5.3.1), the corresponding estimate in (4) remains unchanged.

Theorem 5 *Under the assumptions in Lemma 2,*

$$\kappa(BA) \le \frac{\omega_1}{\omega_0} C_0 K_1$$

and

$$\|E_J\|_A^2 \le \begin{cases} 1 - \dfrac{(2-\omega_1)\omega_0}{C_0(1+\omega_1(K_1-1))^2} & \textit{if each } R_i \textit{ is SPD} \\[2ex] 1 - \dfrac{\bar{\omega}_0}{C_0(\sqrt{2-\omega_1}+\sqrt{\omega_1}(K_1-1))^2} & \textit{otherwise.} \end{cases}$$

Theorem 6 *Under the assumptions in Lemma 3,*

$$\kappa(BA) \le \frac{\omega_1}{\hat{\omega}_0} \hat{C}_0\ K_1$$

and

$$\|E_J\|_A^2 \le \begin{cases} 1 - \dfrac{\hat{\omega}_0}{\hat{C}_0(1+\omega_1(K_1-1))^2} & \textit{if each } R_i \textit{ is SPD} \\[2ex] 1 - \dfrac{(2-\omega_1)\breve{\omega}_0}{\hat{C}_0(\sqrt{2-\omega_1}+\sqrt{\omega_1}(\bar{K}_1-1))^2} & \textit{otherwise.} \end{cases}$$

5.3.4 Matrix representations of PSC and SSC methods

The PSC and SSC have been presented above in terms of projections and operators in abstract vector spaces. We shall now translate all these algorithms into explicit algebraic forms by using the simple techniques in § 5.2.3.

For each k, let $\mathcal{I}_k \in \mathbb{R}^{n\times n_k}$ be the matrix representation of the natural inclusion $I_k : \mathcal{V}_k \mapsto \mathcal{V}$; To derive the algebraic representation of the preconditioner (5.3.7), we rewrite it in a slightly different form

$$B = \sum_{k=0}^{J} I_k R_k Q_k.$$

Applying (5.2.18) and the easily verifiable identity $\widetilde{Q}_k = \mathcal{M}_k^{-1}\mathcal{I}_k^t\mathcal{M}$ gives

$$\widetilde{B} = \sum_{k=0}^{J} \widetilde{I}_k \widetilde{R}_k \widetilde{Q}_k = \sum_{k=0}^{J} \mathcal{I}_k(\mathcal{R}_k\mathcal{M}_k)(\mathcal{M}_k^{-1}\mathcal{I}_k^t\mathcal{M}) = \mathcal{B}\mathcal{M}.$$

Here $\mathcal{R}_k$ is the algebraic representation of R_k and

$$\mathcal{B} = \sum_{k=0}^{J} \mathcal{I}_k\mathcal{R}_k\mathcal{I}_k^t. \tag{5.3.40}$$

Different choices of R_k yield the following three main different preconditioners:

$$\mathcal{B} = \begin{cases} \sum_{k=0}^{J} \rho(\mathcal{A}_k)^{-1}\mathcal{I}_k\mathcal{I}_k^t & \text{Richardson;} \\ \sum_{k=0}^{J} \mathcal{I}_k\mathcal{D}_k^{-1}\mathcal{I}_k^t & \text{Jacobi;} \\ \sum_{k=0}^{J} \mathcal{I}_k\mathcal{G}_k\mathcal{I}_k^t & \text{Gauss-Seidel.} \end{cases}$$

Here $\mathcal{G}_k = (\mathcal{D}_k - \mathcal{U}_k)^{-1}\mathcal{D}_k(\mathcal{D}_k - \mathcal{L}_k)^{-1}$, $\mathcal{A}_k = \mathcal{D}_k - \mathcal{L}_k - \mathcal{U}_k$, $\mathcal{D}_k$ is the diagonal of $\mathcal{A}_k$, $-\mathcal{L}_k$ and $-\mathcal{U}_k$ are, respectively, the lower and upper triangular parts of $\mathcal{A}_k$.

Following (1), we get

Proposition 4 *The PSC preconditioner for the stiffness matrix $\mathcal{A}$ is given by* (5.3.40) *and* $\kappa(\mathcal{B}\mathcal{A}) = \kappa(BA)$.

Similarly, we can derive the algebraic representation of (5.3.12) for solving (5.2.20).

Algorithm 5.3.41 $\mu^0 \in \mathbb{R}^n$ *is given. Assume that* $\mu^k \in \mathbb{R}^n$ *is obtained. Then* μ^{k+1} *is defined by*

$$\mu^{k+i/J} = \mu^{k+(i-1)/J} + \mathcal{I}_i\mathcal{R}_i\mathcal{I}_i^t(\eta - \mathcal{A}\mu^{k+(i-1)/J})$$

for $i = 0 : J$.

5.4 Finite element approximations

In the following sections, we shall introduce the multigrid methods. Our presentations will be confined on a second order elliptic model problem with the linear finite element discretization.

This section is devoted to some basic properties of finite element spaces that will be used for the analysis of multigrid algorithms.

5.4.1 A model problem and finite element discretization

We consider the boundary-value problem:

$$\begin{aligned} -\nabla \cdot a\nabla U &= F \quad \text{in } \Omega, \\ U &= 0 \quad \text{on } \partial\Omega, \end{aligned} \tag{5.4.1}$$

where $\Omega \subset \mathbb{R}^d$ is a polyhedral domain and a is a smooth function (or piecewise smooth) on $\bar{\Omega}$ with a positive lower bound.

Let $H^1(\Omega)$ be the standard Sobolev space consisting of square integrable functions with square integrable (weak) derivatives of first order, and $H_0^1(\Omega)$ the subspace of $H^1(\Omega)$ consisting of functions that vanish on $\partial\Omega$. Then $U \in H_0^1(\Omega)$ is the solution of (5.4.1) if and only if

$$a(U, \chi) = (F, \chi) \quad \forall \chi \in H_0^1(\Omega), \tag{5.4.2}$$

where

$$a(U, \chi) = \int_\Omega a\nabla U \cdot \nabla \chi dx, \quad (F, \chi) = \int_\Omega F\chi dx.$$

Introduce the fractional order Sobolev spaces

$$H^{m+\sigma}(\Omega)(m \geq 0, 0 < \sigma < 1)$$

defined by the completion of smooth functions in the following norm:

$$\|v\|_{m+\sigma,\Omega} = \left(\|v\|_{H^m(\Omega)}^2 + |v|_{H^{m+\sigma}(\Omega)}^2\right)^{\frac{1}{2}},$$

where

$$|v|_{m+\sigma,\Omega}^2 = \sum_{|\alpha|=m} \int_\Omega \int_\Omega \frac{|D^\alpha v(x) - D^\alpha v(y)|^2}{|x-y|^{d+2\sigma}} dx\, dy.$$

It is well-known that there exists a constant $\alpha \in (0, 1]$ such that

$$\|U\|_{1+\alpha} \leq C\|F\|_{\alpha-1}, \tag{5.4.3}$$

for the solution U of (5.4.2), where C is a constant depending on the domain Ω and the coefficient $a(x)$.

Assume that Ω is triangulated with $\Omega = \cup_i \tau_i$, where τ_i's are nonoverlapping simplexes of size h, with $h \in (0,1]$ and quasi-uniform. i.e. there exist constants C_0 and C_1 not depending on h such that each simplex τ_i is contained in (contains) a ball of radius $C_1 h$ (respectively $C_0 h$). Define

$$\mathcal{V} = \{v \in H_0^1(\Omega) : v|_\tau \in \mathcal{P}_1(\tau_i), \quad \forall\, \tau_i\},$$

where $\mathcal{P}_1$ is the space of linear polynomials.

We shall now mention some properties of the finite element space. For any $v \in \mathcal{V}$, we have

$$\|v\|_{0,\infty,\Omega} \lesssim h^{-d/p}\|v\|_{L^p(\Omega)}, p \geq 1, \tag{5.4.4}$$

$$\|v\|_{1,\Omega} \lesssim h^{-1}\|v\|, \tag{5.4.5}$$

$$\|v\|_{1+\sigma,\Omega} \lesssim h^{-\sigma}\|v\|_{H^1(\Omega)} \quad \sigma \in (0, \frac{1}{2}), \tag{5.4.6}$$

$$\|v\|_{s,\Omega} \lesssim h^{-s}\|v\|_{0,\Omega} \quad s, t \in [0,1],\ t \leq s, \tag{5.4.7}$$

$$\|v\|_{L^\infty(\Omega)} \lesssim c_d(h)\|v\|_{H^1(\Omega)}, \tag{5.4.8}$$

where $c_1(h) = 1, c_2(h) = |\log h|^{\frac{1}{2}}$ and $c_d(h) = h^{\frac{2-d}{2}}$ for $d \geq 3$. The *inverse* inequalities (5.4.4) and (5.4.5) can be found, for example, in Ciarlet [19] and a proof of the discrete Sobolev inequality (5.4.8) can be found in Bramble and Xu [11]. A proof of (5.4.6) and (5.4.7) may be found in Bramble, Pasciak and Xu [10] and Xu [31].

Theorem 7 *Assume that $P_h : H_0^1(\Omega) \mapsto \mathcal{V}$ is the Galerkin projection with respect to $a(\cdot,\cdot)$, then*

$$\|(I - P_h)u\|_{1-\alpha} \lesssim h^\alpha \|u\|_1 \quad \forall\, u \in H_0^1(\Omega), \tag{5.4.9}$$

and

$$\|(I - P_h)u\|_1 \lesssim h^s \|u\|_{1+s}, \quad \forall\, u \in H_0^1(\Omega) \cap H^{1+s}(\Omega),\ 0 \leq s \leq \alpha \tag{5.4.10}$$

where α is as in (5.4.3).

Defining the L^2 projection $Q_h : L^2(\Omega) \mapsto \mathcal{V}$ by

$$(Q_h v, \chi) = (v, \chi), \quad \forall\, v \in L^2(\Omega), \chi \in \mathcal{V},$$

we have

$$\|v - Q_h v\| + h\|Q_h v\|_{H^1(\Omega)} \lesssim Ch\|v\|_{H^1(\Omega)}. \tag{5.4.11}$$

This estimate is well-know, we refer to [31, 11] for a rigorous proof and related results.

By interpolation, we have (for $\sigma \in (0, \frac{1}{2})$)

$$\|Q_h v\|_{H^\sigma(\Omega)} \lesssim \|v\|_{H^\sigma(\Omega)} \quad \forall\, v \in H_0^1(\Omega). \tag{5.4.12}$$

and

$$\|v - Q_h v\|_{1-\alpha} \lesssim h^\alpha \|v\|_1 \quad \forall\, v \in H_0^1(\Omega). \tag{5.4.13}$$

The finite element approximation to the solution of (4.1) is the function $u \in \mathcal{V}$ satisfying

$$a(u, v) = (F, v) \quad \forall v \in \mathcal{V}. \tag{5.4.14}$$

Define a linear operator $A : \mathcal{V} \mapsto \mathcal{V}$ by

$$(Au, v) = a(u, v), \quad u, v \in \mathcal{V}. \tag{5.4.15}$$

The equation (5.4.14) is then equivalent to (5.2.1) with $f = Q_h F$. The space $\mathcal{V}$ has a natural (nodal) basis $\{\phi_i\}_{i=1}^n$ ($n = \dim\mathcal{V}$) satisfying

$$\phi_i(x_l) = \delta_{il} \quad \forall\, i, l = 1, \ldots, n,$$

where $\{x_l : l = 1, \ldots, n\}$ is the set of all interior nodal points of $\mathcal{V}$. By means of these nodal basis functions, the solution of (5.4.14) is reduced to solving an algebraic system (5.2.20) with $\mathcal{A} = ((a\nabla\phi_i, \nabla\phi_l))_{n\times n}$ and $\eta = ((f, \phi_i)_{n\times 1})$.

It is well-known that

$$h^d|\nu|^2 \lesssim \nu^t \mathcal{A}\nu \lesssim h^{d-2}|\nu|^2 \quad \text{and} \quad h^d|\nu|^2 \lesssim \nu^t \mathcal{M}\nu \lesssim h^d|\nu|^2 \quad \forall\, \nu \in \mathbb{R}^n. \tag{5.4.16}$$

Hence $\kappa(\mathcal{A}) \lesssim h^{-2}$ and $\kappa(\mathcal{M}) \lesssim 1$.

5.4.2 Finite element spaces on multiple levels

This section is to study the interaction between finite element spaces with different scales. We assume that Ω has been triangulated with a nested sequence of quasi-uniform triangulations $\mathcal{T}_k = \{\tau_k^i\}$ of size h for $k = 0, \ldots, j$ where the quasi-uniformity constants are independent of k. These triangulations should be nested in the sense that any triangle τ_{k-1}^l can be written as a union of triangles of $\{\tau_k^i\}$ (see Figure 5.1. We further assume that there is a constant $\eta > 1$, independent of k, such that

$$h_k \eqsim \eta^{-k}.$$

Associated with each $\mathcal{T}_k$, a finite element space $\mathcal{M}_k \subset H_0^1(\Omega)$ can be defined. One has

$$\mathcal{M}_1 \subset \mathcal{M}_2 \subset \ldots \subset \mathcal{M}_k \subset \ldots \subset \mathcal{M}_J. \tag{5.4.17}$$

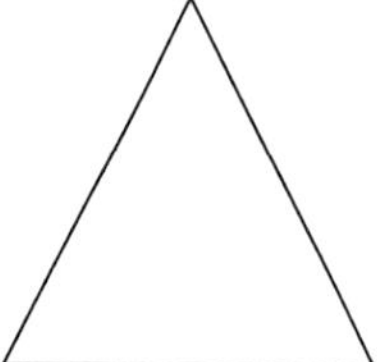 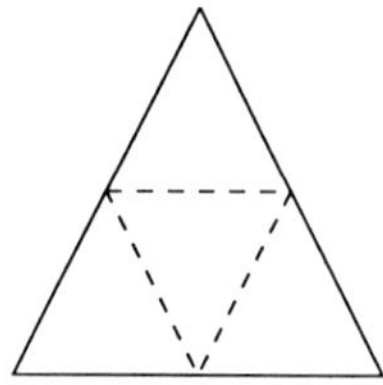 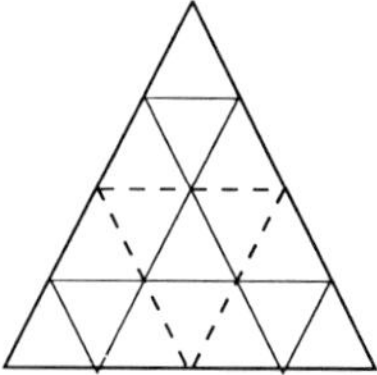

FIGURE 5.1. A typical multilevel grids

For each k, we define the interpolant $I_k : C(\bar{\Omega}) \mapsto \mathcal{M}_k$ by

$$(I_k u)(x) = u(x) \quad \forall\, x \in \mathcal{N}_k.$$

Here $\mathcal{N}_k$ is the set of all nodes in $\mathcal{T}_k$.

Let $Q_k, P_k : H_0^1(\Omega) \mapsto \mathcal{M}_k$ be the L^2 and H^1 projection defined by

$$(Q_k u, v_k) = (u, v_k), \quad (\nabla P_k u, \nabla v_k) = (\nabla u, \nabla v_k) \quad \forall\, u \in H_0^1(\Omega), v_k \in \mathcal{M}_k. \tag{5.4.18}$$

Lemma 10 *Let R_k be any one of I_k, Q_k or P_k. Then*

1. $R_i R_j = R_{i \wedge j}$.
2. $(R_i - R_{i-1})(R_j - R_{j-1}) = 0$ *if* $i \neq j$.
3. $(R_k - R_{k-1})^2 = R_k - R_{k-1} = (I - R_{k-1})R_k$.

Lemma 11 *lm:Ik*

$$\|(I_k - I_{k-1})v\|^2 + h_k^2 \|I_k v\|_A^2 \lesssim c_d(k) h_k^2 \|v\|_A^2, \quad v \in \mathcal{V}, \tag{5.4.19}$$

where $c_d(k) = 1, J - k$ *and* $2^{(d-2)(J-k)}$ *for* $d = 1, 2$ *and* $d \geq 3$, *respectively.*

5.4.3 Regularity and approximation property

Associated with each $\mathcal{M}_k$, we define, as in (5.3.3), $A_k : \mathcal{M}_k \mapsto \mathcal{M}_k$. The following result is instrumental in multigrid analysis.

Theorem 8 *Assume α is as in* (5.4.3). *Then*

$$A((I - P_{k-1})u, u) \lesssim (\lambda_k^{-1} \|A_k u\|^2)^\alpha A(u, u)^{1-\alpha} \quad \forall\, u \in \mathcal{M}_k. \tag{5.4.20}$$

Proof: Let $u \in \mathcal{M}_k$. Applying Cauchy-Schwarz's inequality and the following norm equivalence (See Bank and Dupont [2])

$$\|A_k^{s/2} v\| \eqsim \|v\|_s \quad \forall\, H^s(\Omega) \cap H_0^1(\Omega) \quad s \in [0, 1].$$

we deduce that

$$\begin{aligned} A((I-P_{k-1})u,u) &\le (A_k^{\frac{1+\alpha}{2}}u, A_k^{\frac{1-\alpha}{2}}(I-P_{k-1})u) \\ &\le \|A_k^{\frac{1+\alpha}{2}}u\|\|A_k^{\frac{1-\alpha}{2}}(I-P_{k-1})u\| \\ &\le \|A_k^{\frac{1+\alpha}{2}}u\|\|(I-P_{k-1})u\|_{1-\alpha}. \end{aligned}$$

By Hölder's inequality,

$$\|A_h^{\frac{1+\alpha}{2}}u\| \le \left(A(u,u)^{1-\alpha}\|A_h u\|^{2\alpha}\right)^{1/2}. \tag{5.4.21}$$

Note that $\lambda_k \lesssim h_k^{-2}$, the theorem the follows by combining above inequalities with Theorem 7.

5.4.4 Strengthened Cauchy-Schwarz inequalities

These types of inequalities were used as assumptions in Section 4 (see (5.3.28)). Here we shall establish them for multilevel spaces.

Lemma 12 *Let $i \ge j$; then*

$$a(u,v) \lesssim \gamma^{i-g} h_i^{-1}\|u\|_A\|v\| \quad \forall\, u \in \mathcal{M}_i, v \in \mathcal{M}_i.$$

Here, we recall, that $\gamma \in (0,1)$ is a constant such that $h_j \eqsim \gamma^{2j}$.

Proof: Given $K \in \mathcal{T}_j$, it follows from Green's identity that

$$\begin{aligned} \int_K a\nabla u\cdot\nabla v &= \int_K \nabla a\cdot\nabla u v + \int_{\partial K} a\frac{\partial u}{\partial n}v \\ &\le \|u\|_{1,K}\|v\| + \|\nabla u\|_{0,\partial K}\|v\|_{0,\partial K} \\ &\lesssim \|u\|_{1,K}\|v\| + (h_j^{-1/2}\|\nabla u\|_{0,K})(h_i^{-1/2}\|v\|_{0,K}) \\ &\lesssim (h_j h_i)^{-1/2}\|\nabla u\|_{0,K}\|v\|_{0,K} \\ &\lesssim \gamma^{i-j}h_i^{-1}\|\nabla u\|_{0,K}\|v\|_{0,K}. \end{aligned}$$

A repeated applications of Cauchy-Schwarz inequality yield

$$a(u,v) = \sum_{K\in\mathcal{T}_j}\int_K a\nabla u\cdot\nabla v \lesssim \gamma^{i-j}h_j^{-1}\sum_{K\in\mathcal{T}_j}\|u\|_{H^1(K)}\|v\|_{L^2(K)}$$

$$\lesssim \gamma^{i-j}h_j^{-1}\left(\sum_{K\in\mathcal{T}_j}\|u\|_{H^1(K)}^2\right)^{\frac12}\left(\sum_{K\in\mathcal{T}_j}\|v\|_{L^2(K)}^2\right)^{\frac12} = \gamma^{i-j}h_j^{-1}\|u\|_A\|v\|.$$

The inequality in the previous lemma is a generalization of the strengthen

Cauchy inequality for hierarchical basis functions in Yserentant [36]. Our proof is similar in nature to that in [36], but appears to be a little shorter and more straightforward.

Lemma 13 *Let $\mathcal{V}_i = (I_i - I_{i-1})\mathcal{V}$ or $\mathcal{V}_i = (Q_i - Q_{i-1})\mathcal{V}$; then*

$$a(u,v) \lesssim \gamma^{|i-j|}\|u\|_A\|v\|_A \quad \forall\, u \in \mathcal{V}_i, v \in \mathcal{V}_j. \tag{5.4.22}$$

Proof: By (5.4.11), we have

$$\|v\| \lesssim h_i\|v\|_A \quad \forall\, v \in \mathcal{V}_i.$$

The result then follows directly from Lemma 12.

Lemma 14 *Assume that $T_k = R_k A_k P_k$ and that $R_k : \mathcal{M}_k \mapsto \mathcal{M}_k$ satisfies*

$$\|R_k A_k v\|^2 \lesssim \lambda_k^{-1}(A_k v, v) \quad \forall\, v \in \mathcal{M}_k,$$

where $\lambda_k = \rho(A_k)$. Then, for $0 \le i,j \le J$

$$(T_i u, T_j v)_A \lesssim \gamma^{\frac{|i-j|}{2}}(T_i u,u)_A^{\frac{1}{2}}(T_j v,v)_A^{\frac{1}{2}} \quad \forall\, u,v \in \mathcal{V}.$$

Proof: If $i \le j$, an application of Lemma 12 yields

$$(u_i, T_j v)_A \lesssim \gamma^{j-i}h_j^{-1}\|u_i\|_A\|T_j v\|.$$

By the assumption on R_k,

$$\|T_j v\| = \|R_j A_j P_j v\| \lesssim h_j\|A_j^{\frac{1}{2}}P_j v\| \lesssim h_j\|v\|_A.$$

Consequently

$$(u_i, T_j v)_A \lesssim \gamma^{j-i}\|u_i\|_A\|v\|_A \quad \forall\, u_i \in \mathcal{V}_i, v \in \mathcal{V}.$$

The second inequality follows from the Cauchy-Schwarz inequality and the inequality just proved:

$$\begin{aligned}(T_i u, T_j v)_A &\le (T_j v, v)_A^{\frac{1}{2}}(T_j T_i u, T_i u)_A^{\frac{1}{2}}\\ &\lesssim \gamma^{\frac{j-i}{2}}(T_j v,v)_A^{\frac{1}{2}}\|T_i u\|_A \lesssim \gamma^{\frac{j-i}{2}}(T_i u,u)_A^{\frac{1}{2}}(T_j v,v)_A^{\frac{1}{2}}.\end{aligned}$$

5.4.5 An equivalent norm using multigrid splitting

If nested multilevel finite element spaces $\mathcal{M}_k$ are allowed to get refined in an infinite way, namely $k \to \infty$, then the Sobolev space H_0^1 can be characterized by these finite element spaces in a very elegant way. We shall give such a characterization.

Theorem 9 *For all* $v \in H_0^1(\Omega)$,

$$\|v\|_1^2 \eqsim \sum_{k=0}^{\infty} \|(Q_k - Q_{k-1})v\|_1^2 \eqsim \sum_{k=0}^{\infty} h_k^{-2} \|(Q_k - Q_{k-1})v\|.$$

Proof: Let $\tilde{Q}_k = Q_k - Q_{k-1}$ and $v_i = (P_i - P_{i-1})v$. It follows that

$$\begin{aligned}
\|\tilde{Q}_k v_i\|_1^2 &\lesssim h_k^{-2\alpha} \|\tilde{Q}_k v_i\|_{1-\alpha}^2 \quad \text{(by inverse inequality (5.4.7))} \\
&\lesssim h_k^{-2\alpha} \|v_i\|_{1-\alpha}^2 \quad \text{(by (5.4.12))} \\
&\lesssim h_k^{-2\alpha} h_i^{2\alpha} \|v_i\|_1^2 \quad \text{(by (5.4.13))}.
\end{aligned}$$

Note that $v = \sum_i v_i$ Let $i \wedge j = \min(i,j)$, we have

$$\begin{aligned}
&\sum_{k=0}^{\infty} \|(Q_k - Q_{k-1})v\|_1^2 \\
&= \sum_{k=0}^{\infty} \sum_{i,j=k}^{\infty} (\nabla \tilde{Q}_k v_i, \nabla \tilde{Q}_k v_j) \quad (\text{since } \tilde{Q}_k v_i = 0 \text{ if } i < k) \\
&= \sum_{i,j=1}^{\infty} \sum_{k=0}^{i \wedge j} (\nabla \tilde{Q}_k v_i, \nabla \tilde{Q}_k v_j) \quad \text{(change the order of sum: Fubini theorem)} \\
&\lesssim \sum_{i,j=1}^{\infty} \sum_{k=0}^{i \wedge j} h_k^{-2\alpha} h_i^{\alpha} h_j^{\alpha} \|v_i\|_1 \|v_j\|_1 \lesssim \sum_{i,j=1}^{\infty} h_{i \wedge j}^{-2\alpha} h_i^{\alpha} h_j^{\alpha} \|v_i\|_1 \|v_j\|_1 \\
&\lesssim \sum_{i,j=1}^{\infty} \eta^{\alpha|i-j|} \|v_i\|_1 \|v_j\|_1 \lesssim \sum_{i=1}^{\infty} \|v_i\|_1^2 = \|v\|_1^2.
\end{aligned}$$

To prove the other inequality, we use the strengthened Cauchy-Schwarz inequality and obtain (Lemma 12)

$$\|v\|_1^2 = \sum_{i,j=1}^{\infty} (\nabla \tilde{Q}_i v, \nabla \tilde{Q}_j v) \lesssim \sum_{i,j=1}^{\infty} \gamma^{|i-j|} \|\tilde{Q}_i v\|_1 \|\tilde{Q}_j v\|_1 \lesssim \sum_{i=1}^{\infty} \|\tilde{Q}_i v\|_1^2.$$

Theorem 10 *For all* $v \in H_0^1(\Omega)$,

$$\|v\|_1^2 \eqsim \sum_{k=0}^{\infty} h_k^{-2} \|(I - Q_{k-1})v\|^2.$$

Proof: By previous theorem, we obviously have

$$\sum_{k=0}^{\infty} h_k^{-2}\|(I-Q_{k-1})v\| \geq \sum_{k=0}^{\infty} h_k^{-2}\|(Q_k-Q_{k-1})v\| \gtrsim \|v\|_1^2.$$

The proof for the other direction of inequality is identical to that of the previous theorem except using $\tilde{Q}_k = I - Q_{k-1}$ instead of $Q_k - Q_{k-1}$.

Theorem 11 *For all* $v \in \tilde{H}_0^s(\Omega)(-1 \leq s \leq 1)$,

$$\|v\|_s^2 \eqsim \sum_{k=0}^{\infty} \|(Q_k-Q_{k-1})v\|_s^2 \eqsim \sum_{k=0}^{\infty} h_k^{-2s}\|(Q_k-Q_{k-1})v\|.$$

Proof: Set $B = \sum_{k=0}^{\infty} h_k^{-2}(Q_k - Q_{k-1})$. We then have $\|v\|^2 = (B^0 v, v)$ and, by previous theorem, $\|v\|_1^2 \eqsim (Bv, v)$. An application of operator interpolation then gives that $\|v\|_s^2 = (B^s v, v)$ which implies the desired result.

(5.4.23) Remark. The above theorem is also valid for $-3/2 \leq s \leq 3/2$.

(5.4.24) Remark. A relevant interesting identity is as follows:

$$\|v\|_1^2 \eqsim \sum_{k=0}^{\infty} h_k^2\|A_k P_k\|^2 \quad \forall\, v \in H_0^1(\Omega).$$

5.5 Multigrid methods

This section is devoted to multigrid methods and their convergence properties. The following topics will be studied: classic multigrid iterative methods, BPX preconditioners, hierarchical basis methods, methods for locally refined meshes and full multigrid principle.

5.5.1 Analysis for smoothers

The most crucial step in developing a multigrid solver is the design of a relaxation scheme. A relaxation scheme is also the most problem-dependent part of a multigrid solver as most other parts (such as prolongation and restriction operators) are usually quite standard. The role of relaxation is not to reduce the overall error, but to smooth it out (namely damp out the non-smooth or high frequency components) so that it can be well approximated by functions on a coarser grid.

The smoother will be analyzed by three approaches in this section. The first approach is through numerical experiments, which would give an intuitive idea on the numerical behavior of a smoother. The second approach is Brandt's local mode analysis. This approach, using local Fourier analysis, can give a good insight on the role of a smoother. The third approach is to build technical machineries for the convergence analysis of multigrid methods.

A model problem and some numerical examples

Consider the Poisson equation with homogeneous Dirichlet condition on unit square discretized with uniform triangulation, the discretized equation can be expressed as

$$4u_{ij} - (u_{i+1,j} + u_{i-1,j} + u_{i,j+1} + u_{i,j-1}) = b_{i,j}, \quad 1 \le i, j \le n. \qquad (5.5.1)$$

The damped Jacobi and Gauss-Seidel methods are among the most popular relaxation schemes for this problem. The damped Jacobi (or Richardson) iteration can be written as

$$4\tilde{u}_{ij} = \omega(\bar{u}_{i+1,j} + \bar{u}_{i-1,j} + \bar{u}_{i,j+1} + \bar{u}_{i,j-1}) + b_{i,j} \qquad (5.5.2)$$

where $\tilde{u}_{ij}$ denote the new value of u while $\bar{u}_{ij}$ denote the old value of u, and the (point) Gauss-Seidel iteration (with lexigraphical order on nodal points, from left to right and bottom to top):

$$4\tilde{u}_{ij} = (\bar{u}_{i+1,j} + \tilde{u}_{i-1,j} + \bar{u}_{i,j+1} + \tilde{u}_{i,j-1}) + b_{i,j}. \qquad (5.5.3)$$

The Gauss-Seidel method has a good smoothing property. Let us illustrate this by a simple numerical example. Consider the equation (5.5.1) with an initial residual $u - u^0$ shown on the left plot of Figure 5.2. The initial residual apparently contains a lot of oscillations. The middle plot in Figure 5.2 is the residual after 2 Gauss-Seidel iterations. As we see the error components are smoothed out very quickly with only two Gauss-Seidel iterations although the global errors are still very large. The right plot in Figure 5.2 is the residual after one hundred iterations and as we see the error is still quite big.

Basic ideas in a multigrid strategy

The above numerical examples show that, high frequency errors, which involve local variations in the solution, are well annihilated by simple relaxation methods such as Gauss-Seidel iterations. Low-frequency or more global errors are much more insensitive to the application of simple relaxation methods. In fact, as shown in Figure 5.3, the convergence rate of the Gauss-Seidel iteration consists of a rather rapid initial residual reduction phase, which gradually develops into a much slower residual reduction phase, corresponding to a situation where all high-frequency errors have been damped down and low-frequency errors dominate. A multigrid

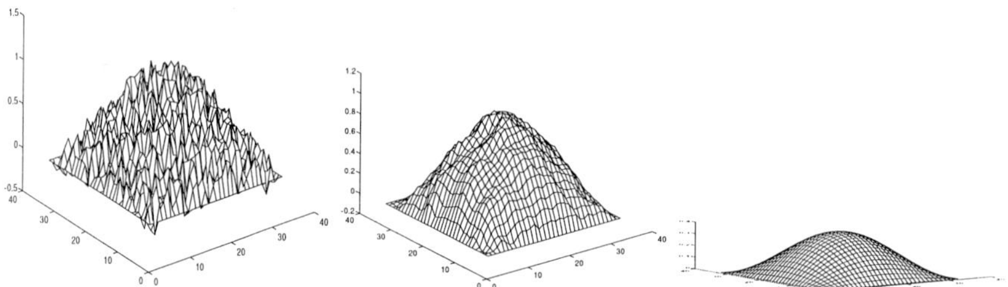

FIGURE 5.2. Residual after 0, 2 and 100 iterations, respectively, with 961 unknowns.

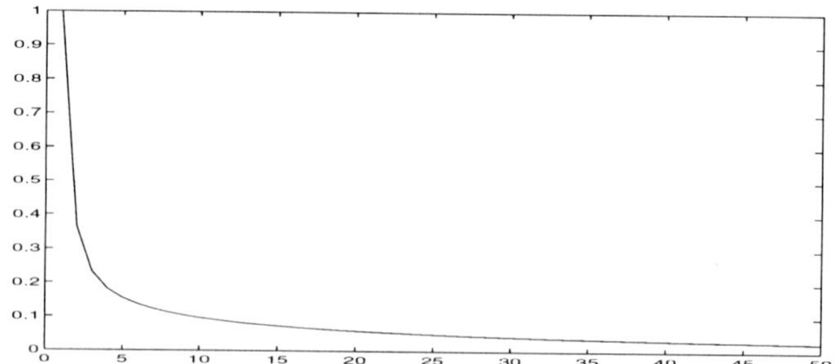

FIGURE 5.3. Convergence history of Gauss-Seidel method within 50 iterations

methodology capitalizes on the this rapid initial high-frequency errors associated with an initial solution on the fine grid, using a simple relaxation scheme such as Gauss-Seidel iteration. Therefore, the solution is transfered to a coarse grid. On this grid, the low-frequency errors of the fine grid manifest themselves as high frequency errors, and are thus damped out efficiently using the same relaxation scheme. The coarse grid corrections computed in this manner are interpolated back to the fine grid in order to update the solution. This procedure can be applied recursively on a sequence of coarser and coarser grids, where each grid-level is responsible for eliminating a particular frequency bandwidth of errors.

Multigrid strategies may be applied to any existing relaxation technique. The success of the overall solution strategy depends on a close matching between the bandwidth of errors which can be efficiently smoothed on a given grid using the particular chosen relaxation strategy, with a careful construction of a sequence of coarse grids, in order to represent the entire error frequency range.

Brandt's local mode analysis

The local mode analysis of Brandt [13] is a general effective tool to analyze and predict the performance of a multigrid solver and in particular the performance of a smoother. This method is based on the fact that a relaxation process is often a local process in which the information propagates just few mesh-sizes per sweep. Therefore, one can assume the problem to be in an unbounded domain, with constant (frozen) coefficients, in which case the algebraic error can be expanded in terms of Fourier series.

The local mode analysis for smoother can in fact be applied in a rigorous fashion to the model problem discussed in previous section. Let us first recall the discrete Fourier theory. For clarity, we confine our discussion in two dimensional case. The discrete Fourier transform theory says that every discrete function $u : I_n \mapsto \mathbb{R}$, with $I_n = \{(i,j) : 0 \le i,j \le n\}$, can be written as

$$u_{i,j} = \sum_{\theta\in\Theta_n} c_\theta \psi_{i,j}(\theta), \quad \psi_{i,j}(\theta) = e^{\mathbf{i}(i\theta_1 + j\theta_2)}, \quad \mathbf{i} = \sqrt{-1}, \quad \theta = (\theta_1, \theta_2).$$

where

$$c_\theta = \frac{1}{(n+1)^2} \sum_{(k,l)\in I_n} u_{k,l}\psi_{k,l}(-\theta).$$

and

$$\Theta_n = \{\frac{2\pi}{n+1}(k,l) \quad -m \le k,l \le m+p\},$$

where $p = 1, m = (n+1)/2$ for odd n and $p = 0, m = n/2+1$ for even n.

We now first use the discrete Fourier transform to analyze the damped Jacobi method. Let $\tilde{\epsilon}_{i,j} = u_{i,j} - \tilde{u}_{i,j}$ and $\bar{\epsilon}_{i,j} = u_{i,j} - \bar{u}_{i,j}$. It is easy to see that

$$\tilde{\epsilon}_{ij} = \bar{\epsilon}_{i,j} - \frac{\omega}{4}(4\bar{\epsilon}_{ij} - (\bar{u}_{i+1,j} + \bar{u}_{i-1,j} + \bar{u}_{i,j+1} + \bar{u}_{i,j-1})). \tag{5.5.4}$$

We write

$$\tilde{\epsilon}_{i,j} = \sum_{\theta\in\Theta_n} \tilde{c}_\theta \psi_{i,j}(\theta) \tag{5.5.5}$$

and

$$\bar{\epsilon}_{i,j} = \sum_{\theta\in\Theta_n} \bar{c}_\theta \psi_{i,j}(\theta). \tag{5.5.6}$$

Substituting the above expressions into (5.5.4) and comparing the coefficients of each $\psi_{ij}(\theta)$, we obtain that

$$\lambda(\theta) = 1 - \omega(1 - \frac{\cos\theta_1 + \cos\theta_2}{2}). \tag{5.5.7}$$

where

$$\lambda(\theta) \equiv \frac{\tilde{c}(\theta)}{\bar{c}(\theta)} \tag{5.5.8}$$

is called the amplification factor of the local mode $\psi_{i,j}(\theta)$.

The smoothing factor introduced by Brandt is the following quantity

$$\bar{\rho} = \sup\{|\lambda(\theta)|, \pi/2 \le |\theta_k| \le \pi, k = 1, 2\}. \tag{5.5.9}$$

Roughly speaking, the smoothing factor $\bar{\rho}$ is the maximal amplification factor corresponding to those high frequency local modes that oscillate within $2h$ range (and hence can not be resolved by coarse grid of size $2h$).

For the damped Jacobi method, it is easy to see that

$$\bar{\rho} = \max\{|1 - 2\omega|, |1 - \omega/2|, |1 - 3\omega/2|\}.$$

The optimal ω that minimizes the smoothing factor is

$$\omega = 4/5, \quad \bar{\rho} = 3/5.$$

For $\omega = 1$ we have $\bar{\rho} = 1$. This means that the undamped Jacobi method for this model problem, although convergent as an iterative method by itself, should not be used as a smoother.

We next exam the smoothing property of the Gauss-Seidel iteration. Unlike the Jacobi method, Gauss-Seidel method depends on the ordering of the unknown. The most natural ordering is perhaps the lexicographic order which was used in the numerical examples given earlier and the corresponding Gauss-Seidel method reads

$$\tilde{\epsilon}_{ij} = \frac{1}{4}(\bar{\epsilon}_{i+1,j} + \tilde{\epsilon}_{i-1,j} + \bar{\epsilon}_{i,j+1} + \tilde{\epsilon}_{i,j-1}). \tag{5.5.10}$$

Again using the Fourier transform (5.5.5) and (5.5.6), we obtain the local amplification factor as follows:

$$\lambda(\theta) = \frac{e^{\mathbf{i}\theta_1} + e^{\mathbf{i}\theta_1}}{4 - e^{-\mathbf{i}\theta_1} - e^{-\mathbf{i}\theta_2}}.$$

It is elementary to see that

$$\bar{\rho} = |\lambda(\pi/2, \cos^{-1}(4/5))| = 1/2.$$

This means that Gauss-Seidel method is a better smoother than the damped Jacobi method.

A more interesting ordering for the Gauss-Seidel method is the so-called red-black ordering. In this particular example, we say two grid points belong to the same color (see (5.3.2)) if and only then they are not neighbors (in either horizontal or vertical direction). It is easy to see that the uniform grid in our example can be grouped into two colors, often called red color and black color. The red-black ordering is to first order all the nodes in one color and then order the other points in another color. (The actual ordering within the same color is not crucial).

The smoothing factor for the Gauss-Seidel method with red-black ordering can not be obtained as easily as the lexigraphical ordering, but it can indeed be proved that

$$\bar{\rho} = 1/4.$$

This means that the Gauss-Seidel method with red-black ordering is a better smoother than the one with the lexigraphical ordering. Furthermore red-black Gauss-Seidel has much better parallel feature (see (5.3.2)).

General smoother analysis

We shall now develop some technical results concerning the smoothing property of the Gauss-Seidel method. We choose to study Gauss-Seidel method since it is one of the better smoothers for our model problems and also it is less obvious to analyse. The analysis for other smoother is relatively simple (see the analysis for Richardson in (5.2.15)).

Lemma 15 *For the stiffness matrix* $\mathcal{A} = \mathcal{D} - \mathcal{L} - \mathcal{U}$

$$\|(\mathcal{D} - \mathcal{L})\xi\|_2 \eqsim h^{d-2}\|\xi\|_2 \quad \forall\, \xi \in \mathbb{R}^N.$$

Proof: Because of the sparsity, it is trivial to prove that

$$\|(\mathcal{D} - \mathcal{L})\xi\|_2 \lesssim h^{d-2}\|\xi\|_2.$$

Now it follows that

$$\begin{aligned} h^{2-d}(\xi, \xi) &\lesssim \frac{1}{2}(\mathcal{D}\xi, \xi) \leq \frac{1}{2}((\mathcal{A} + \mathcal{D})\xi, \xi) \\ &= ((\mathcal{D} - \mathcal{L})\xi, \xi) \leq \|(\mathcal{D} - \mathcal{L})\xi\|_2\|\xi\|_2. \end{aligned}$$

This completes the proof.

The following result is an operator interpretation of the algebraic result given in Lemma 15, which means that the Gauss-Seidel is basically like Richardson iteration.

Lemma 16 *Assume that $R : \mathcal{V} \mapsto \mathcal{V}$ represents the iterator for symmetric Gauss-Seidel iteration. Then*

$$R \eqsim h^2 \eqsim \lambda_h^{-1}.$$

Proof: By definition, the matrix representation of R is

$$\tilde{R} = (\mathcal{D} - \mathcal{U})^{-1}\mathcal{D}(\mathcal{D} - \mathcal{L})^{-1}\mathcal{M}.$$

Given $v \in \mathcal{V}$, let $\nu = \tilde{v}$, then it is easy to see that

$$(Rv, v) = \|\mathcal{D}^{\frac{1}{2}}(\mathcal{D} - \mathcal{L})^{-1}\mathcal{M}\nu\|_2^2$$

Thus it is equivalent to showing that

$$\|\mathcal{D}^{\frac{1}{2}}(\mathcal{D} - \mathcal{L})^{-1}\mathcal{M}\nu\|_2^2 \eqsim h^2(\mathcal{M}\nu, \nu).$$

Making a change of variable $\xi = (\mathcal{D} - \mathcal{L})^{-1}\mathcal{M}\nu$ and using the fact that $\mathcal{M}^{-1} \eqsim h^{-d}$, the above relation can be reduced to

$$\|(\mathcal{D} - \mathcal{L})\xi\|_2^2 \eqsim h^{d-2}(\mathcal{D}\xi, \xi) \eqsim h^{2(d-2)}\|\xi\|_2^2,$$

which was given by Lemma 15

Lemma 17 *For the stiffness matrix $\mathcal{A} = \mathcal{D} - \mathcal{L} - \mathcal{U}$*

$$(\mathcal{A}\xi, \xi) \le \frac{2}{1 + k_0}((\mathcal{D} - \mathcal{L})\xi, \xi) \quad \forall\, \xi \in \mathbb{R}^N,$$

where k_0 is the maximal number of nonzero row entries of A.

If $R : \mathcal{V} \mapsto \mathcal{V}$ represents the iterator for Gauss-Seidel iteration. Then

$$(Av, v) \le \frac{2}{1 + k_0}(R^{-1}v, v) \quad \forall\, v \in V_h.$$

Proof: It is easy to see that the desired estimate is equivalent to the following:

$$(\mathcal{A}\xi, \xi) \le k_0(\mathcal{D}\xi, \xi) \quad \forall\, \xi \in \mathbb{R}^N$$

which can be obtained by a simple of application of Cauchy-Schwarz inequality.

5.5.2 *A basic multigrid cycle: backslash (\) cycle*

Although, as we shall see, multigrid methods have many variants, there is one particular multigrid algorithm which can be viewed a basic multigrid cycle. This algorithm is sometimes called the backslash (\) cycle (we shall explain below why this algorithm is given this name).

We shall first present this method from a more classic point of view. This more classic approach makes it easier to introduce many different kinds of classic multigrid methods and also make it possible to use more classic approach to analyze the convergence of multigrid methods.

A multigrid process can be viewed as defining a sequence of operators $B_k : \mathcal{M}_k \mapsto \mathcal{M}_k$ which are approximate inverse of A_k in the sense that $\|I - B_k A_k\|_A$ is bounded away from 1. A typical way of defining such a sequence of operators is the following *backslash* cycle multigrid procedure.

Algorithm 5.5.11 *For $k = 0$, define $B_0 = A_0^{-1}$. Assume that $B_{k-1} : \mathcal{M}_{k-1} \mapsto \mathcal{M}_{k-1}$ is defined. We shall now define $B_k : \mathcal{M}_k \mapsto \mathcal{M}_k$ which is an iterator for the equation of the form*

$$A_k v = g.$$

1. Fine grid smoothing: *For $v^0 = 0$ and $l = 1, 2, \cdots, m$*

$$v^l = v^{l-1} + R_k(g - A_k v^{l-1})$$

2. Coarse grid correction: $e_{k-1} \in \mathcal{M}_{k-1}$ *is the approximate solution of the residual equation $A_{k-1}e = Q_{k-1}(g - Av^m)$ by the iterator B_{k-1}:*

$$e_{k-1} = B_{k-1} Q_{k-1}(g - Av^m).$$

Dcfinc

$$B_k g = v^m + e_{k-1}.$$

After the first step, the residual $v - v^m$ is small on high frequencies. In another word, $v - v^m$ is smoother (see the middle plot in Figure 5.2) and hence it can be very well approximated by a coarse space $\mathcal{M}_{k-1}$. The second step in the above algorithm plays role of correcting the low frequencies by the coarser space $\mathcal{M}_{k-1}$ and the coarse grid solver B_{k-1} given by induction.

With the above defined B_k, we may consider the following simple iteration

$$u^{k+1} = u^k + B_J(f - Au^k) \tag{5.5.12}$$

There are many different ways to make use of B_k, which will be discussed late.

Before we now study its convergence, we now discuss briefly the algebraic version of the above algorithm.

Let $\Phi^k = (\phi_1^k, \cdots, \phi_{n_k}^k)$ be the nodal basis vector for the space $\mathcal{M}_k$, we define the so-called prolongation matrix $\mathcal{I}_k^{k+1} \in \mathbb{R}^{n_{k+1}\times n_k}$ as follows

$$\Phi^k = \Phi^{k+1}\mathcal{I}_k^{k+1}, \tag{5.5.13}$$

Algorithm 5.5.14 (Matrix version) *Let* $\mathcal{B}_0 = \mathcal{A}_0^{-1}$. *Assume that* $\mathcal{B}_{k-1} \in \mathbb{R}^{n_{k-1}\times n_{k-1}}$ *is defined; then for* $\eta \in \mathbb{R}^{n_k}$, $\mathcal{B}_k \in \mathbb{R}^{n_k\times n_k}$ *is defined as follows:*

1. Fine grid smoothing: *For* $\nu^0 = 0$ *and* $l = 1, 2, \cdots, m$

$$\nu^l = \nu^{l-1} + \mathcal{R}_k(\eta - \mathcal{A}_k\nu^{l-1})$$

2. Coarse grid correction: $\varepsilon_{k-1} \in \mathbb{R}^{n_{k-1}}$ *is the approximate solution of the residual equation* $\mathcal{A}_{k-1}\varepsilon = (\mathcal{I}_{k-1}^k)^t(\eta - \mathcal{A}_k\nu^m)$ *by using* $\mathcal{B}_{k-1}$

$$\varepsilon_{k-1} = \mathcal{B}_{k-1}(\mathcal{I}_{k-1}^k)^t(\eta - \mathcal{A}_k\nu^m).$$

Define

$$\mathcal{B}_k\eta = \nu^m + \mathcal{I}_{k-1}^k\varepsilon_{k-1}.$$

The above algorithm is given in recurrence. But it can also be easily implemented in a non-recursive fashion. For such type of implementation, we refer to Algorithm 5.5.19

5.5.3 A convergence analysis using full elliptic regularity

With the assumption of full elliptic regularity (namely $\alpha = 1$ in (5.4.3), see also Theorem 8), a very sharp convergence estimate can be obtained in a very simple and elegant fashion.

We shall assume that the smoothers R_k are SPD and satisfies

$$\frac{c_0}{\lambda_k}(v, v) \le (R_k v, v) \le (A_k^{-1}v, v) \quad \forall\, v \in \mathcal{M}_k. \tag{5.5.15}$$

We would like to remark that the above assumptions on R_k can be much weakened and, for example, R_k do not need to by symmetric (in this case, assumptions need to be made on the symmetrization of R_k, see, e.g., § 5.5.5).

If the regularity estimate (5.4.3) holds with $\alpha = 1$, then there exists a positive constant c_1 independent of mesh parameters such that (see Theorem 8)

$$\|(I - P_{k-1})v\|_A^2 \le c_1\lambda_k^{-1}\|A_k v\|^2 \quad \forall\, v \in \mathcal{M}_k. \tag{5.5.16}$$

The next technical result shows that any function smoothened by local relaxation can be well approximated by a coarser grid.

Lemma 18

$$\|(I-P_{k-1})K_k^m v\|_A^2 \le \frac{c_1}{2mc_0}(\|v\|_A^2 - \|K_k^m v\|_A^2).$$

Proof:

$$\begin{aligned}\|(I-P_{k-1})K_k^m v\|_A^2 &\le c_1\lambda_k^{-1}\|A_kK_k^m v\|^2\\ &= \frac{c_1}{c_0}(R_kA_kK_k^m v, A_kK_k^m v) = \frac{c_1}{c_0}((I-K_k)K_k^{2m}v, v)_A\\ &\le \frac{c_1}{2mc_0}(\|v\|_A^2 - \|K_k^m v\|_A^2).\end{aligned}$$

The proof is completed by using the following elementary inequality:

$$((I-K_k)K_k^{2m}v, v)_A \le \frac{1}{2m}\sum_{j=0}^{2m-1}((I-K_k)K_k^j v, v)_A \quad (5.5.17)$$

$$\le \frac{1}{2m}(\|v\|_A^2 - \|K_k^m v\|_A^2).$$

Theorem 12 *For the Algorithm 5.5.11, we have*

$$\|I - B_kA_k\|_A^2 \le \frac{c_1}{2mc_0 + c_1}, \qquad 1 \le k \le J.$$

Proof: By definition of Algorithm 5.5.11, we have

$$I - B_kA_k = (I - P_{k-1}B_{k-1}A_{k-1})(I - R_kA_k)^m$$

and, thus, for all $v \in \mathcal{M}_k$

$$\begin{aligned}&\|(I-B_kA_k)v\|_A^2\\ &= \|(I-P_{k-1})K_k^m v\|_A^2 + \|(I-B_{k-1}A_{k-1})P_{k-1}K_k^m v)\|_A.\end{aligned}$$

Let $\delta = c_1/(2mc_0 + c_1)$. We shall prove the above estimate by induction. First of all it is obviously true for $k = 0$. Assume it holds for $k-1$. In the case of k, we have from the above identity that

$$\begin{aligned}\|(I-B_kA_k)v\|_A^2 &\le \|(I-P_{k-1})K_k^m v\|_A^2 + \delta\|P_{k-1}K_k^m v\|_A^2\\ &\le (1-\delta)\|(I-P_{k-1})K_k^m v\|_A^2 + \delta\|K_k^m v\|_A^2\\ &\le (1-\delta)\frac{c_1}{2mc_0}(\|v\|_A^2 - \|K_k^m v\|_A^2) + \delta\|K_k^m v\|_A^2\\ &\le \delta\|v\|_A^2.\end{aligned}$$

5.5.4 V-cycle and W-cycle

Two important variants of the above backslash cycle are the so-called V-cycle and W-cycle.

A V-cycle algorithm is obtained from the backslash cycle by performing more smoothings after the coarse grid corrections. Such an algorithm, roughly speaking, is like a backslash (\) cycle plus a slash (/) (a reversed backslash) cycle. The detailed algorithm is given as follows.

Algorithm 5.5.18 *For $k = 0$, define $B_0 = A_0^{-1}$. Assume that $B_{k-1} : \mathcal{M}_{k-1} \mapsto \mathcal{M}_{k-1}$ is defined. We shall now define $B_k : \mathcal{M}_k \mapsto \mathcal{M}_k$ which is an iterator for the equation of the form*

$$A_k v = g.$$

1. pre-smoothing: *For $v^0 = 0$ and $l = 1, 2, \cdots, m$*

$$v^l = v^{l-1} + R_k(g - A_k v^{l-1})$$

2. Coarse grid correction: *$e_{k-1} \in \mathcal{M}_{k-1}$ is the approximate solution of the residual equation $A_{k-1}e = Q_{k-1}(g - Av^m)$ by the iterator B_{k-1}:*

$$e_{k-1} = B_{k-1}Q_{k-1}(g - Av^m).$$

3. post-smoothing: *For $v^{m+1} = v^m + e_{k-1}$ and $l = m+2, 2, \cdots, 2m$*

$$v^l = v^{l-1} + R_k(g - A_k v^{l-1})$$

A non-recursive implementation

The recursive formulation of the above algorithm makes it a little less straightforward to code sometimes. A non-recursive version of the algorithm is given below in terms of matrices and vectors.

Algorithm 5.5.19 (V-cycle computation of $\mathcal{B}\beta$)

for $l = J : 1$,
 $\alpha_l = \mathcal{R}_l\beta_l, \beta_{l-1} = (\mathcal{I}_{l-1}^l)^t(\beta_l - \mathcal{A}_l\alpha_l)$;
endfor

for $l = 2 : J$,
 $\alpha_l = \mathcal{R}_l^t(\alpha_l + \mathcal{I}_{l-1}^l\alpha_{l-1})$;
endfor

$\mathcal{B}\beta = \alpha_J$.

The reason why this algorithm is called V-cycle is quite clear with the above implementation. The algorithm starts on the finest level and traverses all the grids, one at a time, until it reaches the coarsest grid. Then it traverses all the grids until it reaches the finest level.

It is easy to see that the operators B_k defined by the above V-cycle algorithm satisfy

$$I - B_k A_k = (I - B_k^{(\backslash)} A_k)^*(I - B_k^{(\backslash)} A_k)$$

where $B_k^{(\backslash)}$ correspond the operator defined by the backslash cycle Algorithm 5.5.11 and $*$ is the adjoint operator with respect to the $A-$inner product. Consequently,

$$\|I - B_k A_k\|_A = \|I - B_k^{(\backslash)} A_k\|_A^2.$$

This means that the convergence of the V-cycle is a consequence of the convergence of backslash cycle.

The W-cycle, roughly speaking, is like a V-cycle plus another V-cycle. In a V-cycle iteration, the coarse grid correction is only performed once, while in a W-cycle, the coarse grid correction is performed twice.

Algorithm 5.5.20 *For $k = 0$, define $B_0 = A_0^{-1}$. Assume that $B_{k-1} : \mathcal{M}_{k-1} \mapsto \mathcal{M}_{k-1}$ is defined. We shall now define $B_k : \mathcal{M}_k \mapsto \mathcal{M}_k$ which is an iterator for the equation of the form*

$$A_k v = g.$$

1. pre-smoothing: *For $v^0 = 0$ and $l = 1, 2, \cdots, m$*

$$v^l = v^{l-1} + R_k(g - A_k v^{l-1})$$

2. Coarse grid correction: *$e_{k-1} \in \mathcal{M}_{k-1}$ is the approximate solution of the residual equation $A_{k-1}e = Q_{k-1}(g - Av^m)$ by applying iterator B_{k-1} twice, $e_{k-1} = w^2$, with*

$$w^j = w^{j-1} + B_{k-1}(Q_{k-1}(g - Av^m) - A_{k-1}w^{j-1}, \quad j = 0, 1, 2, \quad w^0 = 0.$$

3. post-smoothing: *For $v^{m+1} = v^m + e_{k-1}$ and $l = m+2, 2, \cdots, 2m$*

$$v^l = v^{l-1} + R_k(g - A_k v^{l-1}).$$

Again it is also easy to see that the convergence of the W-cycle is an easy consequence of the convergence of V-cycle. But it is often the case that W-cycle is easier to analyze. We shall now give an optimal estimate for the convergence of W-cycle based on the elliptic regularity assumption (5.4.3) (which implies that (5.4.20) holds).

Theorem 13 *Under the elliptic regularity assumption* (5.4.3) *for some $\alpha \in (0, 1]$, the W-cycle iteration admits the following estimate*

$$\|I - B_k A_k\|_A^2 \le \frac{M}{m^\alpha + M} \tag{5.5.21}$$

for some constant M that is independent of mesh parameters.

Proof: Let $\delta = \frac{M}{m^\alpha + M}$. The estimate (5.5.21) will be proved by induction. As there is nothing to prove for $k = 0$, we assume that (5.5.21) is valid for $k-1$. By definition, the following recurrence relation holds for any $v \in \mathcal{M}_k$.

$$A((I - B_k A_k)v, v) = A((I - P_{k-1})\tilde{v}, \tilde{v}) + A((I - B_{k-1}A_{k-1})P_{k-1}\tilde{v}, P_{k-1}\tilde{v})$$

where $\tilde{v} = K_k^m v$.

$$\begin{aligned}
&\|(I - B_k A_k)v\|_A^2 \\
&\quad \le \|(I - P_{k-1})\tilde{v}\|_A^2 + \delta\|P_{k-1}\tilde{v}\|_A^2 \quad \text{(by induction)} \\
&\quad \le (1-\delta^2)\|(I - P_{k-1})\tilde{v}\|_A^2 + \delta^2\|\tilde{v}\|_A^2 \\
&\quad \le (1-\delta^2)(c_1\lambda_k^{-1}\|A_k\tilde{v}\|^2)^\alpha \|\tilde{v}\|_A^{2(1-\alpha)} + \delta^2\|\tilde{v}\|_A^2 \quad \text{(by (5.4.20))} \\
&\quad \le (1-\delta^2)\left[\frac{c_1}{2mc_0}(\|v\|_A^2 - \|\tilde{v}\|_A^2)\right]^\alpha \|\tilde{v}\|_A^{2(1-\alpha)} + \delta^2\|\tilde{v}\|_A^2. \quad \text{(by (5.5.17))}
\end{aligned}$$

Note that $\|\tilde{v}\|_A^2/\|v\|_A^2 \in [0,1]$, the desired estimate then easily follows if we can prove the following elementary inequality

$$(1-\delta^2)\left[\frac{c_1}{2mc_0}(1-t)\right]^\alpha t^{1-\alpha} + \delta^2 \le \delta \quad t \in [0,1]. \tag{5.5.22}$$

By Hölder inequality, for any $\eta > 0$, we have

$$(1-t)^\alpha t^{1-\alpha} \le \alpha\eta(1-t) + (1-\alpha)\eta^{\frac{\alpha}{\alpha-1}} t.$$

Thus (5.5.22) is a consequence of the following inequality

$$(1-\delta^2)\left[\frac{c_1}{2mc_0}\right]^\alpha \alpha\eta(1-t) + ((1-\alpha)\eta^{\frac{\alpha}{\alpha-1}} + \delta^2)t \le \delta \quad t \in [0,1].$$

The choice of η is made to minimize the above left hand side, name to equalize the coefficients of $1-t$ and t. The proof may be completed with some more elementary manipulations by choosing sufficient large M in the expression of δ.

5.5.5 Subspace correction interpretation

We shall now discuss the multigrid method from the subspace correction point of view. Let $\mathcal{M}_k$ $(k = 0, 1, \cdots, J)$ be the multilevel finite element spaces defined as in the preceding section. Again let $\mathcal{V} = \mathcal{M}_J$, but set $\mathcal{V}_k = \mathcal{M}_{J-k}$. In this case, the decomposition (5.3.1) is trivial.

We observe that, with the above choice of subspaces $\mathcal{V}_k$, there are redundant overlappings in the decomposition (5.3.1). The point is that these overlappings can be used advantageously in the choice of the subspace solvers in a simple fashion. Roughly speaking, the subspace solvers need only to take care of those "non-overlapped parts" (which correspond to the so-called *high frequencies*). As we know, the methods like Gauss-Seidel method discussed earlier just satisfies such requirements.

With the above ingredients, the successive subspace correction method Algorithm (5.3.12) can be stated as follows.

Algorithm 5.5.23 *Given* $u^0 \in \mathcal{V}$.

```
for k = 0, 1, ... till convergence
    v ← u^k
    for i = J : −1 : 0    v ← v + R_i Q_i (f − Av)   endfor
    u^{k+1} ← v.
endfor
```

Lemma 19 *For the Algorithm 5.5.11 with* $m = 1$, *we have*

$$I - B_J A_J = (I - T_0)(I - T_1) \cdots (I - T_J) \tag{5.5.24}$$

where

$$T_0 = P_0, T_k = R_k A_k P_k, \quad 1 \leq k \leq J.$$

Hence, with such defined operators B_k, *the iteration* (5.5.12) *is mathematically equivalent to Algorithm 5.5.23.*

Furthermore if B_J^m *is obtained from Algorithm 5.5.11 for general* $m \geq 1$, *then*

$$\|I - B_J^m A_J\|_A \leq \|I - B_J A_J\|_A.$$

Based on the above lemma, different proofs may be obtained on the convergence of the backslash cycle multigrid method. In particular, the framework given in § 5.3 can be applied. This new analysis does not depend crucially on the elliptic regularity and hence can be applied more easily to more complicated situations such as the problems with large discontinuous jump coefficients and locally refined meshes (see § (5.5.9)) .

We now consider the more general case in which we do not assume any elliptic regularity for the underlying partial differential equations. (5.4.1).

We assume that all the smoothers R_k satisfy

$$(R_k v, v) \leq \omega_1 (A_k^{-1} v, v) \quad \forall\, v \in \mathcal{M}_k. \tag{5.5.25}$$

and, if R_k are all symmetric

$$\frac{c_0}{\lambda_k}(v, v) \leq (R_k v, v) \leq \frac{c_1}{\lambda_k}(v, v) \quad \forall\, v \in \mathcal{M}_k. \tag{5.5.26}$$

or, in general

$$\frac{c_0}{\lambda_k}(v, v) \leq (\bar{R}_k v, v) \leq \frac{c_1}{\lambda_k}(v, v) \quad \forall\, v \in \mathcal{M}_k. \tag{5.5.27}$$

where $\bar{R}_k$ is the symmetrization of R_k (see § 5.2.1).

By Lemma 17, the Gauss-Seidel methods satisfy the above assumptions.

The idea is to use the general framework of § 5.3. According to the theory there, we need to estimate two basic parameters, namely $\bar{K}_0$ and $\bar{K}_1$.

By Theorem 9 and (5.5.26), there exists a constant C_0 independent of mesh parameters such that

$$K_0 \le C_0. \tag{5.5.28}$$

Lemma 20 *The $\bar{\epsilon}_{ij}$ defined in* (5.3.25) *satisfy, for some $\gamma \in (0,1)$ independent of mesh parameters,*

$$\bar{\epsilon}_{ij} \lesssim \gamma^{|i-j|/2}. \tag{5.5.29}$$

Proof: Let $i \ge j$. It follows from Lemma 12 and (5.5.26) that

$$(\bar{T}_i v_j, v_j)_A \lesssim \gamma^{i-j} h_i^{-1} \|v_j\|_A \|\bar{T}_i v_j\| \lesssim \gamma^{i-j}(v_j, v_j)_A \quad \forall\, v_j \in \mathcal{V}_j.$$

The desired estimate then follows from the definition of $\bar{\epsilon}_{ij}$.

By definition, we conclude from above lemma that there exists a constant C_1 independent of mesh parameters such that

$$\bar{K}_1 \le C_1. \tag{5.5.30}$$

With the above results, the following result follows directly from Theorem 4.

Theorem 14 *Assume that the smoothers R_k satisfy* (5.5.26) *and* (5.5.25) *with $\omega_1 < 2$, then the backslash cycle Algorithm 5.5.11 or Algorithm* (5.5.23) *satisfy*

$$\|I - B_J A_J\|_A^2 \le 1 - \frac{2-\omega_1}{C}$$

for some positive constant C independent of mesh parameters.

Another convergence analysis using full elliptic regularity

If the full elliptic regularity is valid, however, a more straightforward proof can also be obtained in this new framework. We shall first present such a proof.

Theorem 15 *For the iteration* (5.5.12) *with B_J given by Algorithm 5.5.11 with one smoothing on each level, then*

$$\|I - B_J A_J\|_A^2 \le 1 - \frac{c_0}{c_1}.$$

Proof: Denote $E_{-1} = I$ and for $0 \le i \le J$,

$$E_i = (I - T_i)(I - T_{i-1}) \cdots (I - T_1)(I - T_0).$$

Note that $E_J = (I - B_J A_J)^*$. It follows that, denoting $\tilde{P}_i = P_i - P_{i-1}$,

$$
\begin{aligned}
\|v\|_A^2 &= \sum_{i=0}^{J} (\tilde{P}_i v, v)_A \\
&= \sum_{i=0}^{J} (\tilde{P}_i v, E_{i-1} v)_A \quad (\text{since } (I - E_{i-1})v \in \mathcal{M}_{i-1}) \\
&= \sum_{i=0}^{J} (\tilde{P}_i v, A_i P_i E_{i-1} v) \\
&\le \sqrt{c_1} \sum_{i=0}^{J} \lambda_i^{-1/2} \|\tilde{P}_i v\|_A \|A_i P_i E_{i-1} v\| \quad (\text{by } (5.5.16)) \\
&\le \sqrt{\frac{c_1}{c_0}} \sum_{i=0}^{J} \|\tilde{P}_i v\|_A (R_i A_i P_i E_{i-1} v, A_i P_i E_{i-1} v)^{1/2} \quad (\text{by } (5.5.15)) \\
&\le \sqrt{\frac{c_1}{c_0}} \sum_{i=0}^{J} \|\tilde{P}_i v\|_A (T_i E_{i-1} v, E_{i-1} v)_A^{1/2} \\
&\le \sqrt{\frac{c_1}{c_0}} \left(\sum_{i=0}^{J} \|\tilde{P}_i v\|_A^2 \right)^{1/2} \left(\sum_{i=0}^{J} (T_i E_{i-1} v, E_{i-1} v)_A \right)^{1/2} \\
&\le \sqrt{\frac{c_1}{c_0}} \left(\sum_{i=0}^{J} \|\tilde{P}_i v\|_A^2 \right)^{1/2} \left(\sum_{i=0}^{J} (T_i E_{i-1} v, E_{i-1} v)_A \right)^{1/2} \\
&\le \sqrt{\frac{c_1}{c_0}} \|v\|_A \left(\|v\|_A^2 - \|E_J v\|_A^2 \right)^{1/2}.
\end{aligned}
$$

Consequently,

$$\|E_J v\|_A^2 \le (1 - \frac{c_0}{c_1}) \|v\|_A^2 \quad \forall\, v \in \mathcal{V}.$$

The desired estimate then follows easily.

5.5.6 Full multigrid cycle

We shall now describe a more efficient multigrid technique, called full multigrid cycle, originally proposed by Brandt.

On each level of finite element space $\mathcal{V}_k$, there corresponds to a finite element approximation $u_k \in \mathcal{V}_k$ such that

$$a(u^{(k)}, v) = (F, v) \quad \forall v \in \mathcal{V}. \tag{5.5.31}$$

Similar to (5.4.10), the best error estimate in H^1 norm is

$$\|U - u^{(k)}\|_1 = O(h_k^{\alpha}). \tag{5.5.32}$$

If $\mu^{(k)} \in \mathbb{R}^{n_k}$ is the nodal value vector of $u^{(k)}$, then

$$A_k \mu^{(k)} = b^{(k)}. \tag{5.5.33}$$

where $b^{(k)} = ((F, \phi_i^k))$. It can be proved that, with $\mathcal{I}_k^{k+1}$ given by (5.5.13)

$$b^{(k)} = (\mathcal{I}_k^{k+1})^t b^{(k+1)} \quad \text{with } b^{(J)} = b.$$

The full multigrid method is based on the following two observations:

1. $u^{(k-1)} \in \mathcal{V}_{k-1} \subset \mathcal{V}_k$ is close to $u^{(k)} \in \mathcal{V}_k$ and hence can be used as an initial guess for an iterative scheme for solving $u^{(k)}$.

2. Each $u^{(k)}$ can be solved within its truncation error shown in (5.5.32) by a multigrid iterative scheme.

Algorithm 5.5.34 $\hat{\mu}^{(1)} \leftarrow A_1^{-1} b^{(1)}$. *For* $k = 2 : J$

1. $\hat{\mu}^{(k)} \leftarrow \mathcal{I}_{k-1}^k \hat{\mu}^{(k-1)}$,

2. Iterate $\hat{\mu}^{(k)} \leftarrow \hat{\mu}^{(k)} + \mathcal{B}_k(b^{(k)} - \mathcal{A}_k \hat{\mu}^{(k)})$ *for* m *times.*

The most important fact on the full multigrid method is that it has an optimal computational complexity $O(N)$ to compute the solution within truncation error.

Proposition 5 *Assume that that* C_0 *is a positive constant satisfying (for all k)*

$$\|\mu^{(k)} - \mu^{(k-1)}\|_{\mathcal{A}} \le C_0 h_k^\alpha.$$

Then

$$\|\mu^{(k)} - \hat{\mu}^k\|_{\mathcal{A}} \le h_k^\alpha$$

if

$$m \ge \frac{\log(2^\alpha + C_0)}{|\log \delta|}.$$

Proof: By definition, $\|\mu^{(1)} - \hat{\mu}^1\|_{\mathcal{A}} = 0$. Now assume that

$$\|\mu^{(k-1)} - \hat{\mu}^{k-1}\|_{\mathcal{A}} \le h_{k-1}^\alpha.$$

Then

$$\begin{aligned}
\|\mu^{(k)} - \hat{\mu}^k\|_{\mathcal{A}} &\le \delta^m \|\mu^{(k)} - \hat{\mu}^{k-1}\|_{\mathcal{A}} \\
&\le \delta^m \|\mu^{(k)} - \mu^{k-1}\|_{\mathcal{A}} + \delta^m \|\mu^{(k-1)} - \hat{\mu}^{k-1}\|_{\mathcal{A}} \\
&\le \delta^m (C_0 + 2^\alpha) h_k^\alpha \\
&\le h_k^\alpha
\end{aligned}$$

5.5.7 BPX multigrid preconditioners

We shall now describe a parallelized version of the multigrid method studied earlier. This method was first proposed in Bramble-Pasciak-Xu [9] and Xu [31], and is now often known as the BPX preconditioner in the literature.

Basic algorithm and theory

There are different ways of deriving the BPX preconditioners. The method was originally resulted from an attempt to parallelizing the classic multigrid method. With the current multigrid theoretical technology, the derivation of this method is not so difficult. We shall first derive this preconditioner based on Theorem 9 and then, in the next subsection, study the method using the framework of subspace correction.

By Theorem 9, we have

$$(Av, v) \eqsim \sum_{k=0}^{J} h_k^{-2} \|(Q_k - Q_{k-1})v\|^2 = (\hat{A}v, v) \quad v \in \mathcal{V}.$$

where

$$\hat{A} = \sum_k h_k^{-2}(Q_k - Q_{k-1}).$$

Using Lemma (10), we can show that

$$\hat{A}^{-1} = \sum_k h_k^2(Q_k - Q_{k-1}).$$

Using the fact that $h_k = 2h_{k+1}$, we deduce that

$$\begin{aligned}(\hat{A}^{-1}v, v) &= \sum_{k=0}^{J} h_k^2((Q_k - Q_{k-1})v, v) = \sum_{k=0}^{J} h_k^2(Q_k v, v) - \sum_{k=0}^{J-1} h_{k+1}^2(Q_k v, v) \\ &\eqsim h_J^2(v, v) + \sum_{k=0}^{J-1} h_k^2(Q_k v, v) \eqsim \sum_{k=0}^{J} h_k^2(Q_k v, v) = (\tilde{B}v, v)\end{aligned}$$

where

$$\tilde{B} = \sum_{k=0}^{J} h_k^2 Q_k.$$

If $R_k : \mathcal{M}_k \mapsto \mathcal{M}_k$ is an SPD operator satisfying

$$(R_k v_k, v_k) \eqsim h_k^2(v_k, v_k) \quad \forall\, v_k \in \mathcal{M}_k \tag{5.5.35}$$

then, for

$$B = \sum_{k=0}^{J} R_k Q_k \tag{5.5.36}$$

we have

$$(Bv,v) \eqsim (\tilde{B}v,v) \eqsim (A^{-1}v,v)$$

namely

$$\kappa(BA) \eqsim 1.$$

Theorem 16 *Assume that R_k's satisfy* (5.5.35)*; then the preconditioner* (5.3.7) *satisfies*

$$\kappa(BA) \eqsim 1.$$

We note that all the relaxation methods mentioned earlier, such as Richardson, Jacobi and symmetric Gauss-Seidel satisfy (5.5.35).

Subspace correction approach

In § 5.5.5, the slash cycle multigrid method is interpreted as a successive subspace correction method. Correspondingly, the BPX preconditioner can be interpreted as the relevant PSC (parallel subspace correction) preconditioner. It is possible to use the abstract theory in § 5.3 to derive some estimates for the BPX preconditioner somewhat more refined than that in Theorem 16, we leave the details to the interested readers. In § 5.5.9, we shall use this approach to analyze the BPX preconditioner for locally refined meshes.

Implementation

Again, in view of (5.3.40), the algebraic representation of the preconditioner given by (5.3.40) is

$$\mathcal{B} = \sum_{k=0}^{J} \mathcal{I}_k \mathcal{R}_k \mathcal{I}_k^t, \tag{5.5.37}$$

where $\mathcal{I}_k \in \mathbb{R}^{n\times n_k}$ is the representation matrix of the nodal basis $\{\phi_i^k\}$ in $\mathcal{M}_k$ in terms of the nodal basis $\{\phi_i\}$ of $\mathcal{M}$, i.e. $(\phi_1^k,\cdots,\phi_{n_k}^k) = (\phi_1,\cdots,\phi_n)\Phi\mathcal{I}_k$.

Let $\mathcal{I}_k^{k+1} \in \mathbb{R}^{n_{k+1}\times n_k}$ be as defined in (5.5.13), then

$$\mathcal{I}_k = \mathcal{I}_{J-1}^J \cdots \mathcal{I}_{k+1}^{k+2}\mathcal{I}_k^{k+1}.$$

This identity is very useful for the efficient implementation of (5.5.37) on both serial and parallel fashions.

If $\mathcal{R}_k$ are given by the Richardson iteration, we have

$$\mathcal{B} = \sum_{k=0}^{J} h_k^{2-d} \mathcal{I}_k \mathcal{I}_k^t. \tag{5.5.38}$$

From (5.5.37) or (5.5.38), we see that the preconditioner depends entirely on the transformation between the nodal bases on multilevel spaces.

Let, for $1 \le l \le J$,

$$\mathcal{B}_l = \sum_{k=0}^{l} \mathcal{I}_k^l \mathcal{R}_k (\mathcal{I}_k^l)^t.$$

By definition $\mathcal{B} = \mathcal{B}_J$ and

$$\mathcal{B}_l = \mathcal{R}_l + \mathcal{I}_{l-1}^l \mathcal{B}_{l-1} (\mathcal{I}_{l-1}^l)^t.$$

We shall use the above recurrence relation to compute the action of $\mathcal{B}$. Assume that m_l is the number of operations that are needed to compute the action $\mathcal{B}_l \alpha_l$ for $\alpha_l \in \mathbb{R}^{n_l}$. By the identity

$$\mathcal{B}_l \alpha_l = \mathcal{R}_l \alpha_l + \mathcal{I}_{l-1}^l [B_{l-1} (\mathcal{I}_{l-1}^l)^t \alpha_l]$$

we get, for some constant $c_0 > 0$,

$$m_l \le m_{l-1} + c_0 n_l$$

from which we conclude that

$$m_J \le m_1 + c_0 \sum_{l=2}^{J} n_l \le c_1 n$$

for some positive constant c_1. This means that the action of B_J can be carried out within $O(n)$ operations.

Algorithm 5.5.39 (Computation of $\mathcal{B}\alpha$)

for $l = J : 1$,
  $\alpha_{l-1} = (\mathcal{I}_{l-1}^l)^t \alpha_l$;
end

$\beta_0 = \mathcal{R}_0 \alpha_0$;
for $l = 1 : J$,
  $\beta_l = \mathcal{R}_l \alpha_l + \mathcal{I}_{l-1}^l \beta_{l-1}$;
end
$\mathcal{B}\alpha = \beta_J$.

As it is discussed above, the number of operations needed in the above algorithm is $O(n)$. We also note that all the vectors α_l for $1 \le k \le J$ need to be stored, but the whole storage space for these vectors is also only $O(n)$.

5.5.8 Hierarchical basis methods

Assume that we are given a nested sequence of multigrid subspace of $H_0^1(\Omega)$

$$\mathcal{M}_1 \subset \mathcal{M}_2 \subset \ldots \subset \mathcal{M}_k \subset \ldots \subset \mathcal{M}_J.$$

as described in §sc:nestedMG. The so-called *hierarchical basis* refers to the special set of nodal basis functions

$$\{\phi_i^k : x_i^k \in \mathcal{N}_k \setminus \mathcal{N}_{k-1}, k = 0, \cdots, J\}. \tag{5.5.40}$$

It is easy to see that this set of functions does form a basis of $\mathcal{M}$. For $d \neq 2$, it is often more convenient to use the scaled HB as follows

$$\{h_k^{2-d}\phi_i^k : x_i^k \in \mathcal{N}_k \setminus \mathcal{N}_{k-1}, k = 0, \cdots, J\}. \tag{5.5.41}$$

With a proper ordering, we shall denote the scaled HB by $\{\psi_i, i = 1 : N\}$.

The HB in multiple dimensions is formally a direct generalization of the HB in one dimensional case. But the property for the corresponding stiffness matrix in multiple dimensions is not all as clear as in one dimension where the stiffness matrix is an identity matrix in some special case. In this section, we shall show that at least in two dimensions, hierarchical basis is still very useful.

The hierarchical basis in two dimensions was first analyzed by Yserentant in his pioneering paper [36]. The work of Yserentant was apparently motivated by the famous unpublished technical report of Bank and Dupont [1]. Incidently these three authors got together wrote another important paper Bank-Dupont-Yserentant [3] on a Gauss-Seidel (or multiplicative) variant of the hierarchical basis method. The presentation of the materials in this section is of course mostly based on the aforementioned papers, and moreover it also adopts the view of subspace correction from Xu [32] (see also Xu [31].

Preliminaries

We shall now discuss multigrid subspaces that are directly related to the HB. Consider the part of the HB functions on level k as follows

$$\{\phi_i^k : x_i^k \in \mathcal{N}_k \setminus \mathcal{N}_{k-1}\}. \tag{5.5.42}$$

It is easy to see that the above set of functions spans to the following subspace

$$\mathcal{V}_k = (I_k - I_{k-1})\mathcal{M} = (I - I_{k-1})\mathcal{M}_k, \quad \text{for } k = 0 : J. \tag{5.5.43}$$

Here, we recall, $I_{-1} = 0$ and $I_k : \mathcal{M} \mapsto \mathcal{M}_k$ is the nodal value interpolant. The above subspaces obviously give rise to a direct sum decomposition of the space $\mathcal{M}$ as follows

$$\mathcal{V} = \oplus_{k=0}^J \mathcal{V}_k.$$

In fact, for any $v \in \mathcal{V}$, we have the following unique decomposition

$$v = \sum_{k=0}^J v_k \quad \text{with} \quad v_k = (I_k - I_{k-1})v.$$

With the subspaces $\mathcal{V}_k$ given by (5.5.43), the operators A_k are all well conditioned. In fact, by (5.4.19) and (5.4.5), we can see that

$$A(v,v) \eqsim h_k^{-2}\|v\| \quad \forall\, v \in \mathcal{V}_k.$$

As a result, the subspace equations can be effectively solved by elementary iterative methods such as Richardson, Jacobi and Gauss-Seidel methods.

Stiffness matrix in terms of hierarchical basis

The easiest way of understanding the HB is perhaps, like in one dimension, through the study of the property of the corresponding stiffness matrix. As one may expect, the condition number of the HB stiffness matrix should be smaller than the NB stiffness matrix. This is indeed the case in two and three dimensions.

Theorem 17 *Assume that $\hat{A}$ is the stiffness matrix under the scaled hierarchical basis, then*

$$\kappa(\hat{A}) \lesssim \kappa_d(h) \tag{5.5.44}$$

where

$$\kappa_d(h) \lesssim \begin{cases} 1 & \text{if } d=1; \\ |\log h|^2 & \text{if } d=2; \\ h^{2-d} & \text{if } d \geq 3. \end{cases} \tag{5.5.45}$$

In fact, the estimates given in the above theorem can be proven to be sharp. The most interesting case is obviously $d = 2$ for which $\kappa(\hat{A}) \lesssim |\log h|^2$. Compared with the conditioning of the stiffness matrix under the NB, this is a great improvement. It is also in the case $d = 2$ that the HB is most useful. Indeed for $d = 3$, the $\kappa(\hat{A}) = O(h^{-1})$ is also one magnitude smaller than the condition number of the NB stiffness matrix, but such an improvement is not attractive as we shall see that a much better approach (such as BPX preconditioner) is available. There is no doubt that, as far as $\kappa(\hat{A})$ is concerned, the HB is of no use for $d \geq 4$.

Proof: Given $\alpha \in \mathbb{R}^N$, set $v = \sum_{i=1}^N \alpha_i \psi_i$. We can write

$$v = \sum_k v_k \quad \text{with} \quad v_k = (I_k - I_{k-1})v = \sum_{x_i^k \in \mathcal{N}_k \setminus \mathcal{N}_{k-1}} \alpha_i^k \phi_i^k.$$

It follows that

$$|v_k|_1^2 \eqsim \sum_{x_i^k \in \mathcal{N}_k \setminus \mathcal{N}_{k-1}} v_k^2(x_i^k) = \sum_{x_i^k \in \mathcal{N}_k \setminus \mathcal{N}_{k-1}} (\alpha_i^k)^2 = |\alpha|^2.$$

Thus

$$\alpha^t \hat{A} \alpha \;=\; a(v,v) = \sum_{kl} a(v_k, v_l)$$

$$
\begin{aligned}
&\lesssim \sum_{kl} \gamma^{|k-l|} |v_k|_1 |v_l|_1 \quad \text{(by Lemma 13)} \\
&\lesssim \sum_k |v_k|_1^2 \eqsim |\alpha|^2.
\end{aligned}
$$

This implies that $\lambda_{\max}(\hat{\mathcal{A}}) \lesssim 1$. On the other hand

$$
\begin{aligned}
|\alpha|^2 &\eqsim \sum_k |v_k|_1^2 = \sum_k |(I_k - I_{k-1})v|_1^2 \\
&\lesssim \sum_k (J-k+1)|v|_1^2 \quad \text{(by (5.4.19))} \\
&\lesssim \kappa_d(h)|v|_1^2 \lesssim \kappa_d(h)\alpha^t \hat{\mathcal{A}}\alpha
\end{aligned}
$$

This proves that $\lambda_{\min}(\hat{\mathcal{A}}) \gtrsim \kappa_d(h)^{-1}$.

The above proof is essentially the same as in Yserentant [36] (and see also Ong [29] for $d=3$).

Subspace correction approach and general case

Following Xu [32], we shall now study the HB method from the viewpoint of space decomposition and subspace correction.

In view of (5.3.40), the algebraic representation of the PSC preconditioner is

$$
\mathcal{H} = \sum_{k=0}^{J} \mathcal{S}_k \mathcal{R}_k \mathcal{S}_k^t, \tag{5.5.46}
$$

where $\mathcal{S}_k \in \mathbb{R}^{n\times(n_k - n_{k-1})}$ is the representation matrix of the nodal basis $\{\phi_i^k\}$ in $\mathcal{M}_k$, with $x_i^k \in \mathcal{N}_k \setminus \mathcal{N}_{k-1}$, in terms of the nodal basis $\{\phi^i\}$ of $\mathcal{M}$.

A special case

If $\mathcal{R}_k$ is given by the Richardson iteration: $\mathcal{R}_k = h_k^{2-d}\mathcal{I}$, we have

$$
\mathcal{H} = \sum_{k=0}^{J} h_k^{2-d} \mathcal{S}_k \mathcal{S}_k^t = \hat{\mathcal{S}}\hat{\mathcal{S}}^t.
$$

where

$$
\mathcal{S} = (h_1^{1-d/2}\mathcal{S}_1, h_2^{1-d/2}\mathcal{S}_2, \cdots, h_J^{1-d/2}\mathcal{S}_J)
$$

is the representation matrix of the HB in terms of NB. Obviously the HB stiffness matrix and NB stiffness matrix are related by $\hat{\mathcal{A}} = \mathcal{S}^t\mathcal{A}\mathcal{S}$. Therefore,

$$
\kappa(\mathcal{H}\mathcal{A}) = \kappa(\hat{\mathcal{A}}),
$$

and, as a result of Theorem 17,

$$
\kappa(\mathcal{H}\mathcal{A}) \eqsim \kappa_d(h). \tag{5.5.47}
$$

The above estimate apparently also holds for the more general $\mathcal{H}$ when $\mathcal{R}_k$ is given by either Jacobi or symmetric Gauss-Seidel since either of this iteration satisfies the following spectrally equivalence

$$\mathcal{R}_k \eqsim h_k^{2-d}. \tag{5.5.48}$$

In fact, the estimate (5.5.47) also follows easily from the general theory for the PSC preconditioner. To see this,

For the SSC iterative method, it is more convenient to choose $\mathcal{R}_k$ to be the symmetric Gauss-Seidel as the other two methods need to be properly scaled to assume that $\omega_1 \in (0, 2)$. The resulting algorithm is

Algorithm 5.5.49 *$\mu^0 \in \mathbb{R}^n$ is given. Assume that $\mu^k \in \mathbb{R}^n$ is obtained. Then μ^{l+1} is defined by*

$$\mu^{l+(k+1)/(J+1)} = \mu^{l+k/(J+1)} + \mathcal{S}_k\mathcal{R}_k\mathcal{S}_k^t(\eta - \mathcal{A}\mu^{l+(i-1)/J})$$

for $k = 0 : J$.

It can be proved late that

$$\|\mu - \mu_l\|_{\mathcal{A}} \le \left(1 - \frac{c}{\kappa_d(h)}\right)^l \|\mu - \mu_0\|_{\mathcal{A}}. \tag{5.5.50}$$

Convergence analysis

Lemma 21 *We assume that R_k is either Richardson, or Jacobi or symmetric Gauss-Seidel iteration, then*

$$\lambda_k^{-2}\|v\|_A^2 \lesssim (R_kA_kv, v)_A \le \omega_1(v, v)_A, \quad \forall\, v \in \mathcal{V}_k.$$

where ω_1 is a constant and for symmetric Gauss-Seidel, $\omega_1 = 1$.

Proof: The proof for Richardson or Jacobi method is straightforward. The proof for the symmetric Gauss-Seidel method is almost identical to that of Lemma 16 and the detail is left to the reader.

Lemma 22

$$K_0 \lesssim c_d \quad \textit{and} \quad K_1 \lesssim 1,$$

where $c_1 = 1, c_2 = J^2$ and $c_d = 2^{(d-2)J}$ for $d \ge 3$.

Proof: For $v \in \mathcal{V}$, it follows from (5.4.19) that

$$\sum_{k=0}^{J} h_k^{-2}\|v_k\|^2 \lesssim c_d\|v\|_A^2.$$

This gives the estimate of K_0. The estimate of K_1 follows from Lemma 13.

By using Theorem 3 and Lemma 3, we obtain

Theorem 18 *Assume that R_k satisfies* (21); *then the PSC preconditioner* (5.3.7), *with $\mathcal{V}_k$ given by* (5.5.43), *satisfies*

For the SSC iterative method, we apply Theorem 4 with Lemma 22 and get

Theorem 19 *The Algorithm 5.3.12 with the subspaces $\mathcal{V}_k$ given by* (5.5.43) *satisfies*

$$\|E_J\|_A^2 \le 1 - \frac{2-\omega_1}{Cc_d}$$

provided that R_k's satisfy (21) *with $\omega_1 < 2$.*

Compared with the usual multigrid method, the smoothing in the SSC hierarchical basis method is carried out only on the set of new nodes $\mathcal{N}_k \setminus \mathcal{N}_{k-1}$ on each subspace $\mathcal{M}_k$. The method proposed by Bank, Dupont and Yserentant [3] can be viewed as such an algorithm with R_k given by an appropriate Gauss-Seidel iterations. Numerical examples in [3] show that the SSC algorithm converges much faster than the corresponding SSC algorithm.

Relation with BPX preconditioners

Observing that S_k in (5.5.46) is a sub-matrix of $\mathcal{I}_k$ given in (5.5.37), we then have

$$(\mathcal{H}\alpha, \alpha) \le (\mathcal{B}\alpha, \alpha), \quad \forall\, \alpha \in \mathbb{R}^n.$$

In view of the above inequality, if we take

$$\hat{\mathcal{H}} = \sum_{k=0}^{J-1} h_k^{2-d} \mathcal{S}_k \mathcal{S}_k^t + I,$$

we obtain

$$(\mathcal{H}\alpha, \alpha) \le (\hat{\mathcal{H}}\alpha, \alpha) \le (\mathcal{B}\alpha, \alpha), \quad \forall\, \alpha \in \mathbb{R}^n.$$

Even though $\hat{H}$ appears to be a very slight variation of H, numerical experiments have shown a great improvement over H for $d = 2$. We refer to Xu and Qin [35] for the numerical results.

5.5.9 Locally refined grids

In practical computations, finite element grids are often locally refined (by using some error estimators or other adaptive strategies). In this subsection, we shall describe optimal multigrid procedures for adaptive grids. Our presentation here is based on [9], [7] and [6].

With appropriate rearrangement and grouping, we may assume that the mesh refinement can be done in the following fashion. We first start with the original domain Ω which is also denoted by Ω_0. We introduce a relatively coarse and quasiuniform triangulation of Ω_0 with a mesh size h_0 and denote the corresponding finite element space by $\mathcal{V}_0 \subset H_0^1(\Omega_0)$. Let Ω_1 be a subregion where we wish to increase the resolution and we do so by subdividing the elements of the first triangulation and get a new triangulation of Ω_1 with mesh size h_1 in Ω_1 and introduce an additional finite element space $\mathcal{V}_1 \subset H_0^1(\Omega_1)$. We repeat this process and finally get a collection of subdomains Ω_i together with the corresponding finite element spaces $\mathcal{V}_i$ defining on a triangulation of mesh size h_i for $i = 1, 2, \cdots, J$ for some integer $J > 1$. Throughout, we have

$$\Omega_i \subset \Omega_{i-1}, \qquad \mathcal{V}_{i-1} \cap H_0^1(\Omega_i) \subset \mathcal{V}_i \subset H_0^1(\Omega_i), \quad i = 1, 2, \cdots, J.$$

The finite element space on the repeatedly refined mesh can be written as

$$\mathcal{V} = \sum_{i=0}^{J} \mathcal{V}_i.$$

The only restrictions on the mesh domains $\{\Omega_k\}$ are that $\partial\Omega_k$ for $k \geq 1$ consists of edges of mesh triangles in the mesh $\mathcal{T}_{k-1}$ and that there is at least one edge from $\mathcal{T}_{k-1}$ contained in Ω_k.

Let $A : \mathcal{V} \mapsto \mathcal{V}$ be as defined by (5.4.15). Operators $A_i : \mathcal{V}_i \mapsto \mathcal{V}_i$ and $Q_i, P_i : \mathcal{V} \mapsto \mathcal{V}_i$ can be defined similarly as before. If we choose $R_i : \mathcal{V}_i \mapsto \mathcal{V}_i$ to be some appropriate approximate solvers of A_i's, we then have all the ingredients to define our PSC and SSC algorithms. In this setting, the coarse space $\mathcal{M}_0$ may not be very coarse, therefore we assume that, for the PSC type of algorithm, the first subspace solver R_0 is SPD and satisfies

$$(R_0^{-1}v, v) \eqsim (A_0 v, v) \quad \forall\, v \in \mathcal{M}_0, \tag{5.5.51}$$

and for the SSC type of algorithm, we assume that R_0 satisfies

$$\|(I - R_0 A_0)\|_A \lesssim \delta_0 \tag{5.5.52}$$

for some $\delta_1 \in (0, 1)$ independent of mesh parameters.

As for the other subspace solvers R_k for $k > 1$, for clarity, we assume that R_k is given by a Gauss-Seidel iteration or properly damped Jacobi iteration. But apparently other reasonable solvers can also be adopted.

We would like to remark that the corresponding PSC and SSC methods in this setting can be viewed as a "nested" multigrid method associated with the multilevel spaces given by

$$\mathcal{M}_k = \sum_{i=0}^{k} \mathcal{V}_i, \quad 0 \leq k \leq J$$

but with special coarse space solvers R_k only defined on the subspace $\mathcal{V}_k$, namely the smoothings are only carried in the refined regions.

To analyze the corresponding PSC and SSC methods, we introduce, for each k, an auxiliary finite element space $\hat{\mathcal{M}}_k$ which is defined on a quasi-uniform triangulation with mesh size h_k and satisfies $\mathcal{M}_k \subset \hat{\mathcal{M}}_k$ and $\hat{\mathcal{M}}_0 \subset \cdots \subset \hat{\mathcal{M}}_J$. It is easy to see that $\hat{\mathcal{M}}_k$ can be well defined. Corresponding to the space $\hat{\mathcal{M}}_k$, let $\hat{Q}_k : \mathcal{V} \mapsto \hat{\mathcal{M}}_k$ be the L^2 projection.

The following result (from [7]) plays a crucial role in our analysis for the algorithms discussed in this section.

Lemma 23 *Assume that* $h_{k-1}/h_k \leq C$. *There exists a sequence of linear operators* $\Pi_k : \mathcal{V} \mapsto \mathcal{M}_k$ *for* $k = 0, 1, 2, \cdots, J$ *with* $\Pi_J = I$ *such that, for any* $v \in \mathcal{V}$, $(\Pi_k - \Pi_{k-1})v \in \mathcal{V}_k$ *and*

$$\|(I - \Pi_k)v\| \lesssim \|(I - \hat{Q}_k)v\| \tag{5.5.53}$$

and

$$\|\Pi_k v\|_A \lesssim \|v\|_A. \tag{5.5.54}$$

Proof: The linear operator Π_k is then defined, for $v \in \mathcal{V}$, by $\Pi_k v = w$, where w is the unique function in $\mathcal{M}_k$ satisfying

$$w = \begin{cases} \hat{Q}_k v & \text{at the nodes of } \mathcal{M}_k \text{ in the interior of } \Omega_{k+1}, \\ v & \text{at the remaining nodes of } \mathcal{M}_k. \end{cases}$$

By this definition, it is clear that $(\Pi_k - \Pi_{k-1})v \in \mathcal{V}_k$. To establish (5.5.54), we first note that

$$\|\hat{Q}_k v - w\|^2 \leq C h_k^2 \Sigma' (\hat{Q}_k v(x_i^k) - v(x_i^k))^2 \leq C \|(I - \hat{Q}_k)v\|^2,$$

where the sum $\sum'$ is taken over the nodes x_i^k of $\mathcal{M}_k$ on $\partial\Omega_{k+1}$. Combining the above estimate with (5.4.11) yields

$$\|(I - \Pi_k)v\| \leq \|(I - \hat{Q}_k)v\| + \|\hat{Q}_k v - w\| \leq \|(I - \hat{Q}_k)v\|$$

This proves part of the estimate (5.5.53). The rest of (5.5.54) can be estimated similarly by using $\|\Pi_k v\|_A \leq \|(I - \hat{Q}_k)v\|_A + \|v - w\|_A$. This completes the proof.

Lemma 24 *Let* $\Pi_k : \mathcal{V} \mapsto \mathcal{M}_k$ *be the operator from in the previous lemma and* $\Pi_{-1} = 0$, *then, there exists a constant* c_0 *independent of mesh parameters such that*

$$\sum_{k=0}^{J} \lambda_k \|(\Pi_k - \Pi_{k-1})v\|_A^2 \leq c_0 \|v\|_A^2 \quad \forall\, v \in \mathcal{V}. \tag{5.5.55}$$

Proof: By Lemma 23, we have, for $k \geq 1$,

$$\begin{aligned} \|(\Pi_k - \Pi_{k-1})v\|_A^2 &\leq 2(\|(I - \Pi_k)v\|_A^2 + \|(I - \Pi_{k-1})v\|_A^2) \\ &\leq 2(\|(I - \hat{Q}_{k-1})v\|_A^2 + \|(I - \hat{Q}_{k-1})v\|_A^2) \end{aligned}$$

Thus, combining (5.5.54),

$$\sum_{k=0}^{J} \lambda_k \|(\Pi_k - \Pi_{k-1})v\|_A^2 \lesssim \|v\|_A^2 + \sum_{k=0}^{J} \lambda_k \|(I - \hat{Q}_k)v\|_A^2.$$

The desired estimate then follows by using Theorem 10.

PSC version

Let us first consider the preconditioner corresponding to the PSC algorithm:

$$B = \sum_{k=0}^{J} R_k Q_k. \tag{5.5.56}$$

Thanks to Lemma 24, similar to the quasiuniform case, we can use our general framework in § 5.3 to obtain the following theorem.

Theorem 20 *If R_0 satisfies* (5.5.51) *and R_k are given by Jacobi or symmetric Gauss-Seidel, then the PSC preconditioner* (5.5.56) *yields a uniformly bounded condition number*

$$\kappa(BA) \eqsim 1.$$

As a special example, we may choose R_k as follows:

$$R_k v = h_k^{2-d} \sum_{x_i^k \in \mathcal{N}_k} (v, \phi_i^k)\phi_i^k \tag{5.5.57}$$

as we know this corresponds to Richardson iteration which is equivalent to Jacobi iteration. For such a choice, we obtain that

$$Bv = R_0 v + \sum_{k=1}^{J} h_k^{2-d} \sum_{x_i^k \in \mathcal{N}_k} (v, \phi_i^k)\phi_i^k. \tag{5.5.58}$$

We notice that (5.5.58) is exactly the preconditioner given in [9] for locally refined meshes.

The hierarchical basis type algorithms for these composite grids can be obtained by the decomposition with $\mathcal{V}_i = (I_i - I_{i-1})\mathcal{V}$ (here $I_i : \mathcal{V} \mapsto \mathcal{M}_i$ is the nodal value interpolation operator). It is easy to see that the SSC type preconditioner is

$$Hv = R_0 v + \sum_{k=1}^{J} h_k^{2-d} \sum_{x_i^k \in \mathcal{N}_k \setminus \mathcal{N}_{k-1}} (v, \phi_i^k)\phi_i^k. \tag{5.5.59}$$

This preconditioner is equivalent to what is given in [36] for refined meshes. We further point out the corresponding algorithm in [3] for the refined meshes is the SSC algorithm by choosing R_k to be some appropriate Gauss-Seidel iteration.

SSC version

The SSC version correspond to multigrid algorithms with smoothings done in the refined regions. Similarly we have the following convergence theorem.

Theorem 21 *If R_0 satisfies* (5.5.52) *and R_k are given by properly damped Jacobi or Gauss-Seidel, then the corresponding SSC iteration yields a uniform contraction:*

$$\|I - BA\|_A \leq \delta$$

for some $\delta \in [\delta_0, 1)$ independent of mesh parameters.

Additional bibliographic comments

The multilevel algorithm for finite element or finite difference equations was first developed in early sixties by the Russian mathematician Fedorenko [20]. In the early seventies, Brandt [12] brought this method to the attention of western countries and extensive research have been done on this method since then. It has become one of the most popular and also most powerful iterative methods in nowadays.

Multilevel method for composite grids can be traced back to Brandt [12] or composite grid in McCormick [25] (see also the references therein). The finite element space on so refined mesh is $\mathcal{V} = \sum_{i=0}^{J} \mathcal{V}_i$. PSC and SSC methods can be naturally obtained with Gauss-Seidel iteration as subspace solvers. The SSC iteration corresponds to a multigrid method with smoothings carried out only on the refined region discussed in Brandt [12]. The PSC preconditioner was first considered in Bramble-Pasciak-Xu [9].

The aforementioned refined grids may not give the minimal degree of freedoms from approximation point of view, but it is a computationally efficient approach that has the desirable structure for multigrid applications. If a more traditional graded meshes are used on each level, proper nested subspaces are then hard to come by and the corresponding nonnested multigrid methods are more complicated (cf. Zhang [38]).

5.6 Multigrid Methods for nonnested subspaces and varying bilinear forms

Theories presented in previous sections are based on the assumptions that the multilevel subspaces are nested in the sense the the coarse spaces are subspaces of the finer spaces and that the bilinear forms on each level are all the same. In this section, we shall present a more general theory that do not depend on these assumptions. Such a theory has been successfully applied to many situations and some example of applications will be briefly mentioned near the end of this section.

Assume we are given a Hilbert space H and a hierarchy of real finite dimensional subspaces of H

$$\mathcal{M}_0, \mathcal{M}_1, \mathcal{M}_2, \ldots, \mathcal{M}_J$$

which are related by the so-called prolongation operators $I_k : \mathcal{M}_{k-1} \mapsto \mathcal{M}_k$.

In addition, let $A_k(\cdot,\cdot)$ and $(\cdot,\cdot)_k$ be symmetric positive definite bilinear forms on $\mathcal{M}_k$. We shall develop multigrid algorithms for the solution of the following problem: Given $f \in \mathcal{M}_J$, find $u \in \mathcal{M}_J$ satisfying

$$A_J(u, \phi) = (f, \phi)_J \quad \forall \phi \in \mathcal{M}_J.$$

To define the multigrid algorithms, we need to define some auxiliary operators. For $k = 0, \ldots, J$, the operator $A_k : \mathcal{M}_k \mapsto \mathcal{M}_k$ is defined by

$$(A_k w, \phi)_k = A_k(w, \phi) \quad \forall w, \phi \in \mathcal{M}_k.$$

Clearly the operator A_k is symmetric positive definite (in both the $A_k(\cdot,\cdot)$ and $(\cdot,\cdot)_k$ inner products). In terms of the prolongation operator I_k, we have operators $I_k^t : \mathcal{M}_k \mapsto \mathcal{M}_{k-1}$ and $I_k^* : \mathcal{M}_k \mapsto \mathcal{M}_{k-1}$ defined by

$$(I_k^t w, \phi)_{k-1} = (w, I_k\phi)_k A_{k-1}(I_k^* w, \phi) = A_k(w, I_k\phi) \quad \forall w \in \mathcal{M}_k, \phi \in \mathcal{M}_{k-1}. \tag{5.6.1}$$

In other words, I_k^t are *iks* are the adjoint of I_k with the inner products $(\cdot,\cdot)_k$ and $A_k(\cdot,\cdot)$ respectively. I_k^t is often called *restriction* operator, which is another main ingredient of any multigrid algorithm.

Another important component of the multigrid algorithm is the *smoothing*, which will be represented by a sequence of linear operators $R_k : \mathcal{M}_k \mapsto \mathcal{M}_k$ for $1 \leq k \leq j$ to define the smoothing process. These operators may be symmetric or nonsymmetric with respect to the inner product $(\cdot,\cdot)_k$. If R_k is not symmetric, then we denote by R_k^t its adjoint and set

$$R_k^{(l)} = \begin{cases} R_k & \text{if } l \text{ is odd;} \\ R_k^t & \text{if } l \text{ is even.} \end{cases}$$

With the framework and notation given above, we are now in a position to define a multigrid algorithm, which will be characterized in terms of a sequence of recursively defined operators $B_k : \mathcal{M}_k \mapsto \mathcal{M}_k$. In the following, p, m_k are given positive integers and λ_k is either equal to $\rho(A_k)$ or an upper bound of $\rho(A_k)$ such that $\lambda_k \eqsim \rho(A_k)$.

Algorithm S

Step 1 $B_0 = A_0^{-1}$.

Step 2 *Assume B_{k-1} is defined. Then B_k is defined, for $g \in \mathcal{M}_k(A_k w = g)$, as follows:*

1. Pre-smoothing *on $\mathcal{M}_k$:*

$$\begin{aligned} w^0 &= 0 \\ w^l &= w^{l-1} + R_k^{(l+m_k)}(g - A_k w^{l-1}) \\ l &= 1, 2, \cdots, m_k. \end{aligned}$$

2. Correction *on $\mathcal{M}_{k-1}$: $w^{m_k+1} = w^{m_k} + I_k q^p$ where $q^p \in \mathcal{M}_{k-1}$ is defined as follows, with $r_k = g - A_k w^{m_k}$,*

$$\begin{aligned} q^0 &= 0 \\ q^l &= q^{l-1} + B_{k-1}(I_k^t r_k - A_{k-1} q^{l-1}), \\ l &= 1, 2, \cdots, p. \end{aligned}$$

3. Post-smoothing *on $\mathcal{M}_k$:*

$$\begin{aligned} w^l &= w^{l-1} + R_k^{(l+m_k+1)}(g - A_k w^{l-1}) \\ l &= m_k + 2, \cdots, 2m_k + 1. \end{aligned}$$

Define: $B_k g = w^{2m_k+1}$.

(5.6.2) Remark. Ordinarily, the above multigrid methods can be made more general. For example, In Step 1, B_1 may be defined by an iterative method which solves the equation approximately on $\mathcal{M}_1$. Another generalization is that the number of pre- and post-smoothings are not necessarily the same. Nevertheless we are not going to consider these more general cases here. However in some circumstances it seems crucial to our theory that the number of pre- and post smoothings should be the same, which will guarantee that the multigrid operators B_k are also symmetric(see Lemma 26). This is reasonable and important from many viewpoints. The most natural reason would be because the original problems are symmetric themselves.

A great advantage in setting out the algorithm by means of the operators B_k is that we have a very simple recurrence relation for the "residue" operator $E_k \stackrel{\text{def}}{=} I - B_k A_k$ as given in the following lemma.

Lemma 25 *Let $E_k = I - B_k A_k$ and $K_k = I - R_k A_k$. Then*

$$E_k = (\tilde{K}_k^{m_k})^* \left((I - I_k I_k^*) + I_k E_{k-1}^p I_k^*\right) \tilde{K}_k^{m_k}. \tag{5.6.3}$$

Furthermore, for any $u, v \in \mathcal{M}_k$

$$A_k(E_k u, v) = A_k((I - I_k I_k^*)\tilde{u}, \tilde{v}) + A_{k-1}(E_{k-1}^p I_k^* \tilde{u}, I_k^* \tilde{v}), \tag{5.6.4}$$

where $\tilde{u} = \tilde{K}_k^{m_k} u$ and

$$\tilde{K}_k^{m_k} = \begin{cases} (K_k^* K_k)^{\frac{m-1}{2}} K_k^*, & \text{if } m \text{ is odd;} \\ (K_k^* K_k)^{\frac{m}{2}}, & \text{if } m \text{ is even.} \end{cases}$$

The verification of the above lemma is straightforward by the definition of the algorithm. The next thing we want to address is that the **Algorithm S** defines a symmetric operator. More specifically, we have

Lemma 26 *B_k is symmetric with respect to $(\cdot,\cdot)_k$ and E_k is symmetric with respect to $A_k(\cdot,\cdot)$.*

We observe that in the algorithm stated above, p and m_k are free parameters. With different parameters, we will take account of the three types of the algorithms named in the following:

Definition 5.6.5 *The* **Algorithm S** *is known as the*

1. *V-cycle if $p = 1$ and $m_k = m \geq 1$ for $k = 0, \cdots, J$,*
2. *W-cycle if $p = 2$ and $m_k = m \geq 1$ for $k = 0, \cdots, J$,*
3. *Variable V-cycle if $p = 1$ and $\gamma_0 m_k \leq m_{k-1} \leq \gamma_1 m_k$ for $k = 0, \cdots, J$, where γ_0 and γ_1 are constants greater than 1.*

The subsequent multigrid convergence theory is largely based on two basic assumptions, one is concerned with the smoothing operators R_k and the other is relevant to the "regularity" of the underlying problem and the approximation property of the multilevel spaces. More refined convergence estimates depend on more assumptions on prolongation operators and such assumptions will be stated during the presentation of the convergence theory.

The assumption on the smoothing operator R_k which concern only one level of space is the same as in the nested case.

(A0) $$\|v\|^2 \leq C_0 \lambda_k (\bar{R}_k v, v)$$

where $\bar{R}_k$ is the symmetrization of R_k (see § 5.2.1).

A direct consequence of (**A0**) is

(A0′) $$\rho(K_k) < 1,$$

where $\rho(\cdot)$ denotes the spectral radius. This assumption will be used in place of (**A0**) to get some more general (but weaker) results.

The second assumption, usually called "regularity and approximation assumption", is that there exists a constant $\beta \in (0,1]$ and a constant C_1 such that

(A1) $$|A_k((I - I_k I_k^*)v, v)| \leq C_1 ({\lambda_k}^{-1} \|A_k v\|_k{}^2)^\beta A_k(v,v)^{1-\beta} \quad \forall v \in \mathcal{M}_k.$$

This is the most crucial assumption in the multigrid theory to be presented in this section. It relates the bilinear forms, different levels of spaces and prolongation operators. In the case of elliptic boundary value problems, its verification is strongly tied to the regularity property of the underlying partial differential equation.

It is not hard to see that (**A1**) implies that

$$A_k(I_k v, I_k v) \le \tilde{C}_1 A_{k-1}(v,v), \quad \forall\, v \in \mathcal{M}_{k-1}. \tag{5.6.6}$$

Theorem 22 *Under the assumptions* (**A0**) *and* (**A1**), *the Algorithm S has the following convergence properties (with a positive constant M depending only on C_0, C_1 and β):*

1. W-cycle converges uniformly for sufficiently many smoothings:

$$\|E_k\|_k \le \frac{M}{m^{\beta/2}} \text{ if } m^{\beta/2} \ge 2M.$$

2. Variable V-cycle converges uniformly for sufficiently large m_J:

$$\|E_k\|_k \le \frac{c_0}{m^{\beta/2}} \text{ if } m_J^{\beta/2} \ge c_0.$$

3. Variable V-cycle gives a uniform preconditioner with any fixed number of smoothings on the finest grid:

$$\kappa(B_J A_J) \le \frac{(M+m^\beta)^2}{m^{2\beta}}.$$

4. If the following condition holds

$$\textbf{(A2)} \qquad A_k(I_k v, I_k v) \le A_{k-1}(v,v) \quad \forall v \in \mathcal{M}_{k-1}.$$

$$A_k(I_k v, I_k v) \le A_{k-1}(v,v)$$

Then, the W-cycle and VV-cycle converge uniformly and V-cycle converges nearly uniformly with any given number of smoothings:

$$\|E_k\|_k \le \begin{cases} \frac{M}{m_k^\beta + M} & \text{for variable V-cycle} \\ \frac{M^\beta}{(m+M)^\beta} & \text{for W-cycle} \\ \frac{k^{1/\beta-1}M}{m^\beta + k^{1/\beta-1}M} & \text{for V-cycle} \end{cases}$$

5. If the following condition holds

$$A_k(I_k v, I_k v) \le 2A_{k-1}(v,v),$$

then the W-cycle converges uniformly with any given number of smoothings.

The proof of the above theorem is not very complicated and may be found in [9] and [31]

Applications

The above theory has found many applications. First of all, the development of the theory was motivated to applications to multigrid methods for unstructured grids. Unstructured grids refer to grids that do not have a natural multilevel structures. Most of the grids generated by traditional grid generators may fall into such a categories. In this application, one has to coarsen the given grid to obtain a sequence of often nonnested multilevel coarse grids. There have been many coarsening techniques available in two dimensions, see Chan and Smith [16], Bank and Xu [4] and others.

Another important application is to multigrid methods for nonconforming finite elements (especially to fourth order problems). Nonconforming elements often give rise to nonnested multigrid subspaces (even on nested multilevel grids). A major concern in this application, is the choice of prolongation operators which are often obtained by using proper averaging techniques. In most applications, estimates like (A2) or (A3) are hard to be satisfied and hence only W-cycle or variable V-cycle can be proved to be convergent with sufficently many smoothings, or variable V-cyle gives rise to an optimal preconditioner. One interesting exception of the above phenomenon is the work by Chen and Oswald [17] and Chen [18] where they have proved that (A2) or (A3) can be satisfied for some nonconforming P_1 elements in some special situations. For this application and related subjects, we refer to Brenner [14] (and the references cited there),

5.7 The auxiliary space method and an optimal preconiditioner for unstructured grids

In this section, an abstract framework of *auxiliary space method* is proposed and, as an application, an optimal multigrid technique is developed for general unstructured grids. The auxiliary space method is a (nonnested) two level preconditioning technique based on a simple relaxation scheme (smoother) and an auxiliary space (that may be roughly understood as a nonnested coarser space). An optimal multigrid preconditioner is then obtained for a discretized partial differential operator defined on an unstructured grid by using an auxiliary space defined on a more structured grid in which a further *nested* multigrid method can be naturally applied. This new technique make it possible to apply multigrid methods to general unstructured grids without too much more programming effort than traditional solution methods

The materials in this section are taken from Xu [33].

5.7.1 *The auxiliary space method*

The auxiliary space method is a general preconditioning approach based on a relaxation scheme and an auxiliary space. As mentioned in the introduction, this method can be interpreted in various ways but it may be best understood as a two level nonnested multigrid preconditioner. A detailed description of this approach will be described in details below and two theorems will be given for estimating the condition number of the preconditioned system.

Assume that a linear inner product space $\mathcal{V}$ is given together with a linear operator $A : \mathcal{V} \to \mathcal{V}$ that is SPD (symmetric, positive and definite) with respect to an inner product $(\cdot,\cdot)$. Consider the linear equation (5.2.1). The main ingredient in the new preconditioning technique is another auxiliary linear inner product space $\mathcal{V}_0$ together with an operator $A_0 : \mathcal{V}_0 \mapsto \mathcal{V}_0$ that is SPD with respect to an inner product $[\cdot,\cdot]$ on $\mathcal{V}_0$. This space, in most applications, may be viewed as certain approximation for $\mathcal{V}$. The space $\mathcal{V}_0$ needs not be a subspace of $\mathcal{V}$ in general, but it should be, in some sense, simpler than $\mathcal{V}$. The operator A_0 may be viewed as certain representation or approximation of A in the space $\mathcal{V}_0$ and A_0 is assumed to be preconditioned by another SPD operator $B_0 : \mathcal{V}_0 \mapsto \mathcal{V}_0$. In another word, the auxiliary space $\mathcal{V}_0$ is chosen in such a way that the equation given by A_0 can be more easily solved than (5.2.1).

The auxiliary space $\mathcal{V}_0$ is linked with the original space $\mathcal{V}$ by an operator $\Pi : \mathcal{V}_0 \mapsto \mathcal{V}$. If $\mathcal{V}_0$ is viewed as a "coarse" space, Π plays a role of prolongation in multigrid method. The "restriction" operator is given by its adjoint $\Pi^t : \mathcal{V} \mapsto \mathcal{V}_0$ defined by

$$[\Pi^t v, w] = (v, \Pi w) \quad v \in \mathcal{V}, w \in \mathcal{V}_0.$$

Another ingredient is an SPD operator $R : \mathcal{V} \mapsto \mathcal{V}$. The role of R is to resolve what can not be resolved, in preconditioning A, by the aforementioned space $\mathcal{V}_0$ and the operators defined on $\mathcal{V}_0$. By the multigrid terminology, R is like a smoother. In most applications, R is given by a simple relaxation scheme such as Jacobi and Gauss-Seidel method.

With the ingredients described above, the proposed preconditioner is as follows.

$$B = R + \Pi B_0 \Pi^t. \tag{5.7.1}$$

In the special case that $\mathcal{V}_0 \subset \mathcal{V}$ and $[\cdot,\cdot] = (\cdot,\cdot)$, Π can obviously be given by the natural inclusion operator and as a result Π^t is nothing but the orthogonal projection $Q_0 : \mathcal{V} \mapsto \mathcal{V}_0$. In this case, the preconditioner (5.7.1) is reduced to

$$B = R + B_0 Q_0$$

which is the two-level special case of the general nested multilevel preconditioner in Bramble-Pasciak-Xu [9].

By definition, for any $u, v \in \mathcal{V}_h$,

$$(BAu, v)_A = (RAu, v)_A + [B_0 A_0 \Pi^* u, \Pi^* v]_{A_0} \tag{5.7.2}$$

where $(\cdot,\cdot)_A = (A\cdot,\cdot)$, $[\cdot,\cdot]_{A_0} = [A_0\cdot,\cdot]$ and $\Pi^* = A_0^{-1}\Pi^t A$ satisfying

$$[\Pi^* v, w]_{A_0} = (v, \Pi w)_A \quad v \in \mathcal{V}, w \in \mathcal{V}_0.$$

Denote $\rho_A = \rho(A)$, the spectral radius of A. Also denote by $\|\cdot\|$ the norm induced by either $(\cdot,\cdot)$ or $[\cdot,\cdot]$. The main abstract result in this section is stated below.

Theorem 23 *Assume that there are some nonnegative constants α_0, α_1, λ_0, λ_1, and β_1 such that, for all $v \in \mathcal{V}$ and $w \in \mathcal{V}_0$,*

$$\alpha_0 \rho_A^{-1}(v,v) \le (Rv, v) \le \alpha_1 \rho_A^{-1}(v,v), \tag{5.7.3}$$

$$\lambda_0[w,w]_{A_0} \le [B_0 A_0 w, w]_{A_0} \le \lambda_1 [w,w]_{A_0}, \tag{5.7.4}$$

$$\|\Pi w\|_A^2 \le \beta_1 \|w\|_{A_0}^2, \tag{5.7.5}$$

and furthermore, assume that there exists a linear operator $P : \mathcal{V} \mapsto \mathcal{V}_0$ and positive constants β_0 and γ_0 such that,

$$\|Pv\|_{A_0}^2 \le \beta_0^{-1} \|v\|_A^2 \tag{5.7.6}$$

and

$$\|v - \Pi P v\|^2 \le \gamma_0^{-1} \rho_A^{-1} \|v\|_A^2. \tag{5.7.7}$$

Then the preconditioner given by (5.7.1) *satisfies*

$$\kappa(BA) \le (\alpha_1 + \beta_1\lambda_1)((\alpha_0\gamma_0)^{-1} + (\beta_0\lambda_0)^{-1}). \tag{5.7.8}$$

In particular, if P is a right inverse of Π namely $\Pi P v =$ for $v \in \mathcal{V}$, then

$$\kappa((\Pi B_0 \Pi^t)A) \le \frac{\beta_1}{\beta_0}\frac{\lambda_1}{\lambda_0}. \tag{5.7.9}$$

Proof: One first notes that (5.7.5) is equivalent to

$$\|\Pi^* v\|_{A_0}^2 \le \beta_1 \|v\|_A^2 \quad \forall\, v \in \mathcal{V}.$$

By (5.7.2) and the assumptions, one has

$$(BAv, v)_A \le \alpha_1 \rho_A^{-1}\|Av\|^2 + \lambda_1 \|\Pi^* v\|_{A_0}^2 \le (\alpha_1 + \lambda_1\beta_1)\|v\|_A^2.$$

This means that $\lambda_{\max}(BA) \le \alpha_1 + \beta_1\lambda_1$.

It follows that

$$\begin{aligned}(v,v)_A &= (v-\Pi P v, v)_A + [Pv, \Pi^* v]_{A_0}\\ &\le \|v-\Pi P v\|\|Av\| + \|Pv\|_{A_0}\|\Pi^* v\|_{A_0}\\ &\le (\gamma_0\rho_A)^{-1/2}\|v\|_A\alpha_0^{-1/2}\rho_A^{1/2}(RAv,Av)^{1/2} \quad \text{(by (5.7.7) and (5.7.3))}\\ &\quad +\beta_0^{-1/2}\|v\|_A\lambda_0^{-1/2}[B_0A_0\Pi^* v, \Pi^* v]_{A_0}^{1/2} \quad \text{(by (5.7.6) and (5.7.4))}\\ &\le ((\alpha_0\gamma_0)^{-1}+(\beta_0\lambda_0)^{-1})^{1/2}\|v\|_A(BAv,v)_A^{1/2} \quad \text{(by (5.7.2))}.\end{aligned}$$

This implies that $\lambda_{\min}(BA) \ge ((\alpha_0\gamma_0)^{-1} + (\beta_0\lambda_0)^{-1})^{-1}$. The estimate (5.7.8) then follows.

In the particular case that P is a right inverse of Π, one may take $\gamma_0 = \infty$ and $R = 0$, the proof of the corresponding estimate is then transparent.

The last estimate for a special case in the above theorem corresponds to the *fictitious space lemma* of Nepomnyaschikh [26, 27]. A necessary condition for Π to have a right inverse is that $\Pi : \mathcal{V}_0 \mapsto \mathcal{V}$ is surjective. As a consequence, $\dim \mathcal{V}_0 \ge \dim \mathcal{V}$, namely that $\mathcal{V}_0$ has to be at least as rich as $\mathcal{V}$. Furthermore the construction of Π also needs more caution. The introduction of an additional smoother (or relaxation operator) R greatly relaxes the constraints on the choice $\mathcal{V}_0$ and Π in the fictitious space approach and, hence the resulting preconditioner is potentially more flexible and more robust.

5.7.2 A special technique for unstructured grids

This subsection is to construct optimal multigrid preconditioners for finite element matrix from unstructured grids. The term "unstructured grids" here loosely mean those grids that do not possess natural or convenient multilevel structures. The main idea is to choose an auxiliary space from a rather structured grid in which a natural nested multilevel structure is available. This idea was briefly discussed in Xu [34].

In this section, the model problem in §3.1 will be used and the conforming piecewise linear finite element spaces over quasi-uniform triangulations $\mathcal{T}_h$ will be considered. Let Ω_h be the mesh domain determined by $\mathcal{T}_h$, namely

$$\bar{\Omega}_h = \cup_{\tau\in\mathcal{T}_h}\bar{\tau}.$$

To avoid unimportant technical difficulty, it is assumed, for all feasible h, that

$$\Omega_h \subset \Omega.$$

The finite element space $\mathcal{V}_h$ consists of continuous piecewise (with respect to $\mathcal{T}_h$) linear functions that vanish on $\bar{\Omega} \setminus \Omega_h$.

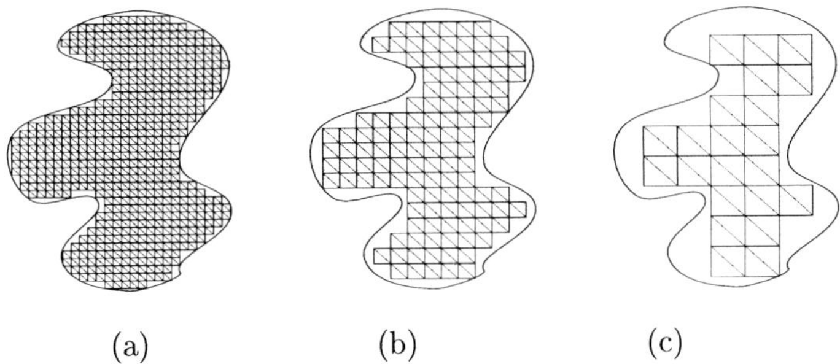

FIGURE 5.4. An example of an auxiliary grid together with two nested coarser grids.

A structured auxiliary finite element space

Given a uniform square partition of the whole space with mesh size

$$h_0 \equiv 2^{-J} \eqsim h$$

for some integer $J \eqsim |\log h|$, let $\mathcal{T}_0$ be the union of squares ($d = 2$) or cubes ($d = 3$) that are contained in Ω (see Figure 1(a)). Let Ω_0 denote the mesh domain determined by $\mathcal{T}_0$, namely

$$\bar{\Omega}_0 = \cup_{\tau \in \mathcal{T}_0} \bar{\tau}.$$

By construction, $\Omega_0 \subset \Omega$ and

$$\max_{x \in \bar{\Omega} \setminus \Omega_0} \operatorname{dist}(x, \partial\Omega) \lesssim h.$$

The set of interior nodes of $\mathcal{T}_0$ will be denoted by $\mathcal{N}_0 = \{x_j^0 : 1 \le j \le n_0\}$

An auxiliary finite element space $\mathcal{V}_0$ will be defined to be a space of continuous piecewise bilinear ($d = 2$) or trilinear ($d = 3$) functions that vanish on $\bar{\Omega} \setminus \Omega_0$. Alternatively, if $\mathcal{T}_0$ is a uniform triangulation (see Figure 1(a)) consisting of triangles ($d = 2$) or tetrahedrals ($d = 3$) from the aforementioned squares ($d = 2$) or cubes ($d = 3$), $\mathcal{V}_0$ may be defined to be a finite element space consisting of continuous piecewise linear functions that vanish on $\bar{\Omega} \setminus \Omega_0$.

The finite element space $\mathcal{V}_0$ will be used as the auxiliary space for the space $\mathcal{V}_h$.

Associated with the spaces $\mathcal{V}_h$ and $\mathcal{V}_0$, the operators: $A_h : \mathcal{V}_h \mapsto \mathcal{V}_h$ and $A_0 : \mathcal{V}_0 \mapsto \mathcal{V}_0$ are defined by

$$(A_h u, v) = a(u, v) \quad \forall\, u, v \in \mathcal{V}_h, \quad (A_0 u_0, v_0) = a(u_0, v_0) \quad \forall\, u_0, v_0 \in \mathcal{V}_0.$$

The operators between $\mathcal{V}_h$ and $\mathcal{V}_0$ will be the standard nodal value interpolants: $\Pi_h : \mathcal{V}_0 \mapsto \mathcal{V}_h$ and $\Pi_0 : \mathcal{V}_h \mapsto \mathcal{V}_0$. The approximation and stability properties of these two operators will be addressed in the following two lemmas.

Lemma 27 *For all* $v_0 \in \mathcal{V}_0$,

$$\|v_0 - \Pi_h v_0\| \lesssim h\|v_0\|_1, \quad \|\Pi_h v_0\|_1 \lesssim \|v_0\|_1. \tag{5.7.10}$$

Lemma 28 *For all* $v \in \mathcal{V}_h$,

$$\|v - \Pi_0 v\| \lesssim h\|v\|_1 \text{ and } \|\Pi_0 v\|_1 \lesssim \|v\|_1. \tag{5.7.11}$$

The proof of the above two lemmas can be found in Xu [33].

An optimal multigrid preconditioner

By the abstract approach in §2 and the auxiliary space $\mathcal{V}_0$, a preconditioner B_h for A_h can be obtained as follows

$$B_h = R_h + \Pi_h B_0 \Pi_h^t, \tag{5.7.12}$$

where $B_0 : \mathcal{V}_0 \mapsto \mathcal{V}_0$ is a given SPD preconditioner of A_0 and $R_h : \mathcal{V}_h \mapsto \mathcal{V}_h$ is an SPD smoother for A satisfying

$$(R_h v, v) \eqsim h^2(v, v).$$

By Theorem 23 and the estimates from the previous subsections, there exist some positive constants α, β, γ that are independent of h such that

$$\kappa(B_h A_h) \le (\alpha + \beta\lambda_{\max}(B_0 A_0))(\gamma + \lambda_{\min}^{-1}(B_0 A_0)).$$

The preconditioner B_0 may be obtained by a further multigrid method. Such type of multigrid method has been discussed by Kornhuber and Yserentant [23]. Setting $\hat{\mathcal{T}}_J = \mathcal{T}_0$ (with $h_J = h_0 = 2^{-J}$) and $\hat{\mathcal{V}}_J = \mathcal{V}_0$, triangulations $\hat{\mathcal{T}}_k$ $(1 \le k \le J)$ (with $h_k = 2^{-k}$) and their corresponding spaces $\hat{\mathcal{V}}_k$ can be defined similarly. As shown in Figure 1, if Figure 1(a) corresponds to $\hat{\mathcal{V}}_J = \mathcal{V}_0$, then Figure 1(b) and Figure 1 (c) correspond to $\hat{\mathcal{V}}_{J-1}$ and $\hat{\mathcal{V}}_{J-2}$ respectively. Evidently

$$\hat{\mathcal{V}}_1 \subset \hat{\mathcal{V}}_2 \subset \ldots \subset \hat{\mathcal{V}}_J.$$

Consequently, a BPX preconditioner B_0 may be obtained so that [1] $\kappa(B_0 A_0) \lesssim 1$ and hence $\kappa(B_h A_h) \lesssim 1$, or a hierarchical basis preconditioner B_0 may be obtained so that $\kappa(B_0 A_0) \lesssim |\log h|^2$ and hence $\kappa(BA) \lesssim |\log h|^2$.

Attention now is turned to the issues such as implementation and complexity of the aforementioned preconditioner. Let $\mathcal{A}_h$ and $\mathcal{A}_0$ be the stiffness matrices corresponding to the finite element spaces $\mathcal{V}_h$ and $\mathcal{V}_0$ respectively. Let $\mathcal{I}$ be the matrix representation of the interpolation Π_h, namely

$$(\Pi_h \psi_1, \ldots, \Pi_h \psi_{n_0}) = (\phi_1, \ldots, \phi_{n_h})\mathcal{I},$$

[1] Although the corresponding estimate was not optimal in [23] for their more general considerations, but it can be easily proved in the current context by the technique in [32].

where $\{\psi_i : i = 1 : n_0\}$ and $\{\phi_j : j = 1 : n_h\}$ are the nodal bases for $\mathcal{V}_h$ and $\mathcal{V}_0$ respectively.

The precondition matrix for the stiffness matrix $\mathcal{A}$ can be written as (cf. [32])

$$\mathcal{B}_h = \mathcal{R}_h + \mathcal{I}\mathcal{B}_0\mathcal{I}^t$$

where R_h represents Richardson, Jacobi or Gauss-Seidel iteration.

By definition, $\mathcal{I} = (\alpha_{ij}) \in \mathbb{R}^{n_h \times n_0}$ with $\alpha_{ij} = \psi_j(x_i^h)$. Obviously $\mathcal{I}$ is a sparse matrix with $O(n_h)$ nonzeros. The evaluation of α_{ij} depends on the location of x_i^h relative to the partition $\mathcal{T}_0$. Because of the regularity of $\mathcal{T}_0$, each x_i^h can be located in $\mathcal{T}_0$ with $O(1)$ operations by, for example, comparing the magnitude of the coordinates of x_i^h. Therefore $\mathcal{I}$ and $\mathcal{I}^t$ can be both obtained with $O(n_h)$ operations. For a more direct way of computing the action of $\mathcal{I}$, for example, if $\xi \in \mathbb{R}^{n_0}$ and $\eta = \mathcal{I}\xi$, then $\eta_i = w(x_i^h)$ with $w = \sum_{j=1}^{n_0} \xi_j\psi_j$. Again $w(x_i^h)$ can be easily obtained as long as the location of x_i^h is known in $\mathcal{T}_0$.

In summary, when B_0 is given by BPX preconditioner, the resulting preconditioner which may be called *BPX preconditioner for unstructured grids* has the following features: 1. one action of B requires only $O(n_h)$ operations; 2. the condition number of BA is uniformly bounded independent of h in both two and three dimensions; 3. it can be applied to unstructured grids.

REMARK. In practical computations, the auxiliary grid $\mathcal{T}_0$ can be more flexible than given above. For example, one does not have to take the elements that are exactly contained in Ω.

REMARK. For simplicity, the details for unstructured grids are only given for Dirichlet boundary value problems. Applications to more general cases are also possible. For Neumann boundary condition, for example, it is not sufficient that the auxiliary grid only consists of those regular elements that are contained in Ω. Instead, the auxiliary grid consists of all those regular elements that *intersect* with Ω. The application of the technique to locally refined meshes is a little more complicated. Again the idea is to use a structured refined mesh to define the auxiliary space. Locally refined meshes with multilevel structures were discussed in Brandt [12], McCormick [25] (see also the references therein), Bramble-Pasciak-Xu [9] and Bramble-Pasciak [6].

Some remarks

The main spirit of this section is that, with the help of an additional smoother, a quite rough auxiliary space can be used to construct an optimal preconditioner for a discretized partial differential operator. For an elliptic partial differential equation of order $2m$, for example, the auxiliary space $\mathcal{V}_0$ for a given finite element space $\mathcal{V}$ defined on a grid of size h needs only to satisfy the following approximation property:

$$\inf_{w\in\mathcal{V}_0} \|v-w\|_0 \lesssim h^m \|v\|_{m,h} \quad \forall\, v\in\mathcal{V} \tag{5.7.13}$$

where $\|\cdot\|_0$ is the L^2 norm and $\|\cdot\|_{m,h}$ is a (discretized) H^m norm.

The approximation property (5.7.13) is a very weak one and is certainly much weaker than the approximation property that $\mathcal{V}$ (as any reasonable finite element space) should have. One important point to address is that the role of $\mathcal{V}_0$ or the role of a coarse grid in a multigrid algorithm is to resolve the spectrum of the discretized differential operator and there is no reason that $\mathcal{V}_0$ should be comparable with $\mathcal{V}$ as far as their approximation properties are concerned. Roughly speaking, the spectral property of a discretized partial differential operator is mainly determined by the original partial differential operator rather than the underlying discretization space. Hence in order to capture the spectrum of a discretized operator, the auxiliary space only need to have an approximation property like (5.7.13) that is related to the order of the original differential operator but not, in certain sense, strongly related to the discretization space ($\mathcal{V}$).

The weak approximation property (5.7.13) makes it possible to use a simple and structured auxiliary space for preconditioning a complicated and unstructured problem. As the main application of this general idea, the main concrete conclusion of this section is that a finite element space defined on an unstructured grid can be well preconditioned by combining a simple relaxation scheme and a structured grid. As a consequence, it is possible to solve a finite element equation on a general unstructured grid by a multigrid approach with an optimal computational complexity.

5.8 References

[1] R. Bank and T. Dupont. Analysis of a two-level scheme for solving finite element equations. Technical Report Report CNA-159, Center for Numerical Analysis, University of Texas at Austin, 1980.

[2] R. E. Bank and T. Dupont. An optimal order process for solving elliptic finite element equations. *Math. Comp.*, 36:35–51, 1981.

[3] R. E. Bank, T. Dupont, and H. Yserentant. The hierarchical basis multigrid method. *Numer. Math.*, 52:427–458, 1988.

[4] R.E. Bank and J. Xu. An algorithm for coarsening unstructured meshes. *Numer. Math.*, 1995. (to appear).

[5] J. H. Bramble. *Multigrid Methods*, volume 294 of *Pitman Research Notes in Mathematical Sciences*. Longman Scientific & Technical, Essex, England, 1993.

[6] J. H. Bramble and J. E. Pasciak. New estimates for multigrid algorithms including the V–cycle. *Math. Comp.*, 60:447–471, 1993.

[7] J. H. Bramble, J. E. Pasciak, J. Wang, and J. Xu. Convergence estimates for multigrip algorithms without regularity assumptions. *Math. Comp.*, 57:23–45, 1991.

[8] J. H. Bramble, J. E. Pasciak, J. Wang, and J. Xu. Convergence estimates for product iterative methods with applications to domain decomposition. *Math. Comp.*, 57:1–21, 1991.

[9] J. H. Bramble, J. E. Pasciak, and J. Xu. Parallel multilevel preconditioners. *Math. Comp.*, 55:1–22, 1990.

[10] J. H. Bramble, J. E. Pasciak, and J. Xu. The analysis of multigrid algorithms with nonnested spaces or noninherited quadratic forms. *Math. Comp.*, 56:1–34, 1991.

[11] J. H. Bramble and J. Xu. Some estimates for a weighted l^2 projection. *Math. Comp.*, 56:463–476, 1991.

[12] A. Brandt. Multi–level adaptive solutions to boundary–value problems. *Math. Comp.*, 31:333–390, 1977.

[13] A. Brandt. *Multigrid techniques: 1984 guide with applications to fluid dynamics.* GMD–Studien Nr. 85. Gesellschaft für Mathematik und Datenverarbeitung, St. Augustin, 1984.

[14] S. C. Brenner. Multigrid methods for nonconforming finite elements. In J. Mandel and S. F. McCormick, editors, *Preliminary Proc. of the 4th Copper Mountain Conference on Multigrid Methods*, volume 1, pages 135–149, Denver, 1989. Computational Mathematics Group, Univ. of Colorado.

[15] W. L. Briggs. *A Multigrid Tutorial.* SIAM Books, Philadelphia, 1987.

[16] T. F. Chan and Barry Smith. Domain decomposition and multigrid methods for elliptic problems on unstructured meshes. In David Keyes and Jinchao Xu, editors, *Domain Decomposition Methods in Science and Engineering, Proceedings of the Seventh International Conference on Domain Decomposition, October 27-30, 1993, The Pennsylvania State University.* American Mathematical Society, Providence, 1994.

[17] Z. Chen. Multigrid and multilevel methods for nonconforming rotated q_1 elements. preprint, 1996.

[18] Z. Chen. On the convergence of nonconforming multigrid methods for second-order elliptic problems. preprint, 1996.

[19] P. G. Ciarlet. *The Finite Element Method for Elliptic Problems.* North–Holland, Amsterdam, New York, 1978.

[20] R. P. Fedorenko. A relaxation method for solving elliptic difference equations. *Z. Vycisl. Mat. i. Mat. Fiz.*, 1:922–927, 1961. Also in U.S.S.R. Comput. Math. and Math. Phys., 1 (1962), pp. 1092–1096.

[21] W. Hackbusch. *Multigrid Methods and Applications*, volume 4 of *Computational Mathematics.* Springer–Verlag, Berlin, 1985.

[22] W. Hackbusch. *Iterative Solution of Large Sparse Systems of Equations.* Springer–Verlag, Berlin, 1993.

[23] R. Kornhuber and H. Yserentant. Multilevel methods for elliptic problems on domains not resolved by the coarse grid. In *Domain Decomposition Methods in Scientific and Engineering Computing: Proceedings of the Seventh International Conference on Domain Decomposition*, volume 180 of *Contemporary Mathematics*, pages 49–60, Providence, Rhode Island, 1994. American Mathematical Society.

[24] S. F. McCormick. *Multigrid Methods*, volume 3 of *Frontiers in Applied Mathematics.* SIAM Books, Philadelphia, 1987.

[25] S. F. McCormick. *Multilevel Adaptive Methods for Partial Differential Equations*, volume 6 of *Frontiers in Applied Mathematics.* SIAM Books, Philadelphia, 1989.

[26] S. V. Nepomnyaschikh. Mesh theorems of traces, normalizations of function traces and their inversion. *Sov. J. Numer. Anal. Math. Modeling*, 6:151–168, 1991.

[27] S. V. Nepomnyaschikh. Decomposition and fictitious domains methods for elliptic boundary value problems. In D. E. Keyes, T. F. Chan, G. A. Meurant, J. S. Scroggs, and R. G. Voigt, editors, *Fifth International Symposium on Domain Decomposition Methods for Partial Differential Equations*, pages 62–72, Philadelphia, 1992. SIAM.

[28] R. A. Nicolaides. On a geometrical aspect of sor and the theory of consistent ordering for positive definite matrices. *Numer. Math.*, 23:99–104, 1974.

[29] M. E. G. Ong. The 3D linear hierarchical basis preconditioner and its shared memory parallel implementation. In *Vector and Parallel Computing: issues in applied research and development*, pages 273–283. Ellis Horwood Ltd. (J. Wiley & Sons), Chichester, 1989.

[30] P. Wesseling. *An Introduction to Multigrid Methods.* John Wiley & Sons, Chichester, 1992.

[31] J. Xu. *Theory of Multilevel Methods.* PhD thesis, Cornell University, 1989.

[32] J. Xu. Iterative methods by space decomposition and subspace correction: A unifying approach. *SIAM Review*, 34:581–613, 1992.

[33] J. Xu. An auxiliary space preconditioning technique with application to unstructured grids. *Numer. Math.*, 1995. (to appear).

[34] J. Xu. Multigrid and domain decomposition methods. In J. Xu W. Cai, J. Shu and Z. Shi, editors, *Numerical Analysis in Applied Sciences.* Science Press, 1995.

[35] J. Xu and J. Qin. Some remarks on a multigrid preconditioner. *SIAM J. Sci. Comput.*, 15:172–184, 1994.

[36] H. Yserentant. On the multi–level splitting of finite element spaces. *Numer. Math.*, 49:379–412, 1986.

[37] H. Yserentant. Old and new convergence proofs for multigrid methods. *Acta Numerica*, 1992.

[38] S. Zhang. Optimal order non–nested multigrid methods for solving finite element equations I-III. *Math. Comp.*, 1990,1991,1994.

6

Preconditioning Toeplitz Systems with Circulant Preconditioners

Raymond H. Chan*
Michael K. Ng†

In this chapter, we review the developments on using preconditioned conjugate gradient methods with circulant preconditioners for solving Toeplitz systems. One of the main results is that the complexity of solving a large class of n-by-n Toeplitz systems is reduced to $O(n \log n)$ operations as compared to $O(n \log^2 n)$ operations required by fast direct Toeplitz solvers. Different circulant preconditioners proposed for Toeplitz systems are surveyed.

6.1 Introduction

6.1.1 *Background*

An n-by-n matrix A_n is said to be *Toeplitz* if

$$A_n = \begin{bmatrix} a_0 & a_{-1} & \cdots & a_{2-n} & a_{1-n} \\ a_1 & a_0 & a_{-1} & & a_{2-n} \\ \vdots & a_1 & a_0 & \ddots & \vdots \\ a_{n-2} & & \ddots & \ddots & a_{-1} \\ a_{n-1} & a_{n-2} & \cdots & a_1 & a_0 \end{bmatrix}, \tag{6.1.1}$$

i.e., A_n is constant along its diagonals. The name Toeplitz originates from the work of Otto Toeplitz [75] in the early 1900's on bilinear forms related to Laurent series, see Grenander and Szegö [46] for details. We are interested in solving the Toeplitz system $A_n\mathbf{x} = \mathbf{b}$.

---

*Department of Mathematics, The Chinese University of Hong Kong, Shatin, Hong Kong. This research was supported by HKRGC Grant No. CUHK 316/94E

†Computer Sciences Laboratory, Research School of Information Sciences and Engineering, The Australian National University, Canberra ACT 0200, Australia. This research was supported by the Cooperative Research Centre for Advanced Computational Systems.

Toeplitz systems arise in a variety of applications in mathematics and engineering, see Bunch [10] and the references therein. The applications have motivated both mathematicians and engineers to develop specific algorithms catering to solving Toeplitz systems. We will call these algorithms *Toeplitz solvers.* Most of the early works on Toeplitz solvers were focused on direct methods. A straightforward application of the Gaussian elimination method will result in an algorithm of $O(n^3)$ complexity. However, since n-by-n Toeplitz matrices are determined by only $(2n-1)$ entries rather than n^2 entries, it is expected that the solution of Toeplitz systems can be obtained in less than $O(n^3)$ operations. There are a number of Toeplitz solvers that decrease the complexity to $O(n^2)$ operations, see, for instance, Levinson (1946) [62], Baxter (1961) [5], Trench (1964) [77], and Zohar (1974) [86]. These algorithms require the invertibility of the $(n-1)$-by-$(n-1)$ principal submatrix of A_n. Around 1980, fast direct Toeplitz solvers of complexity $O(n\log^2 n)$ were developed, see, for instance, Brent, Gustavson, and Yun (1980) [8], Bitmead and Anderson (1980) [7], Morf (1980) [66], de Hoog (1987) [51], and Ammar and Gragg (1988) [2]. These algorithms require the invertibility of the $\lfloor n/2\rfloor$-by-$\lfloor n/2\rfloor$ principal submatrix of A_n.

The stability properties of these direct methods for symmetric positive definite Toeplitz systems are discussed in Bunch [10]. It is noted that if A_n has a singular or ill-conditioned principal submatrix, then a *breakdown* or *near-breakdown* can occur in these algorithms. Such breakdowns will cause numerical instabilities in subsequent steps of the algorithms and result in inaccurately computed solutions. The question of how to avoid breakdowns or near-breakdowns by skipping over singular submatrices or ill-conditioned submatrices has been studied extensively, and various such algorithms have been proposed. For instance, T. Chan and Hansen (1992) [31] were the first to derive a *look-ahead* variant of the Levinson algorithm. The basic idea is to relax the inverse triangular decomposition slightly and to compute an inverse block factorization of the Toeplitz matrices with a block diagonal matrix instead of a scalar diagonal matrix. Other look-ahead extensions of fast Toeplitz solvers can be found in [41, 43, 47].

Recent research on using the preconditioned conjugate gradient method as an iterative method with circulant preconditioners for solving Toeplitz systems has brought much attention. One of the main important results of this methodology is that the complexity of solving a large class of Toeplitz systems can be reduced to $O(n\log n)$ operations as compared to the $O(n\log^2 n)$ operations required by fast direct Toeplitz solvers, provided that a suitable circulant preconditioner is chosen under certain conditions on the Toeplitz operator. Besides the reduction of the arithmetic complexity, there are large classes of important Toeplitz matrices where the fast direct Toeplitz solvers are notoriously unstable, e.g., indefinite and certain non-Hermitian Toeplitz matrices. Therefore, circulant preconditioned conjugate gradient methods provide alternatives to solving these Toeplitz systems. In this chapter, we will review results for these *iterative Toeplitz solvers* and give some insight in how to construct circulant preconditioners for them.

6.1.2 Toeplitz Matrices and Circulant Matrices

Let us begin by introducing the notation that will be used throughout the chapter. Let $\mathbf{C}_{2\pi}$ be the set of all 2π-periodic continuous real-valued functions defined on $[-\pi, \pi]$. For all $f \in \mathbf{C}_{2\pi}$, let

$$a_k = \frac{1}{2\pi}\int_{-\pi}^{\pi} f(\theta)e^{-ik\theta}d\theta, \quad k = 0, \pm 1, \pm 2, \cdots$$

be the Fourier coefficients of f. For all $n \geq 1$, let A_n be the n-by-n Toeplitz matrix with entries $a_{j,k} = a_{j-k}$, $0 \leq j, k < n$. The function f is called the *generating function* of the sequence of Toeplitz matrices A_n, see Grenander and Szegö [46]. Since f is a real-valued function, we have

$$a_{-k} = \bar{a}_k, \quad k = 0, \pm 1, \pm 2, \cdots.$$

It follows that A_n are Hermitian matrices. Note that when f is an even function, the matrices A_n are real symmetric. We emphasize that in practical applications, the functions f are readily available. Typical examples of generating functions are the kernels of Wiener-Hopf equations, see Gohberg and Fel'dman [44, p.82], the functions which give the amplitude characteristics of recursive digital filters, see Chui and A. Chan [34], the spectral density functions in stationary stochastic processes, see Grenander and Szegö [46, p.171], and the point-spread functions in image deblurring, see Jain [57, p.269].

We will solve the systems $A_n\mathbf{x} = \mathbf{b}$ by conjugate gradient methods. The convergence rate of the methods depends partly on how clustered the spectra of the sequence of matrices A_n are, see Axelsson and Barker [4, p.24]. The clustering of the spectra of a sequence of matrices is defined as follows:

Definition 1 *A sequence of matrices $\{A_n\}_{n=1}^{\infty}$ is said to have clustered spectra around 1 if for any given $\epsilon > 0$, there exist positive integers n_1 and n_2 such that for all $n > n_1$, at most n_2 eigenvalues of the matrix $A_n - I_n$ have absolute value larger than ϵ.*

For Toeplitz matrices, we note that there is a close relationship between the spectrum of A_n and its generating function f.

Theorem 1 (Grenander and Szegö [46, pp.63–65]) *Let $f \in \mathbf{C}_{2\pi}$. Then the spectrum $\lambda(A_n)$ of A_n satisfies*

$$\lambda(A_n) \subseteq [f_{\min}, f_{\max}], \quad \forall n \geq 1, \tag{6.1.2}$$

where $f_{\min}$ and $f_{\max}$ are the minimum and maximum values of f, respectively. Moreover, the eigenvalues $\lambda_j(A_n)$, $j = 0, 1, \ldots, n-1$, are equally distributed as $f(2\pi j/n)$, i.e.,

$$\lim_{n\to\infty} \frac{1}{n} \sum_{j=0}^{n-1} \left[g(\lambda_j(A_n)) - g(f(\frac{2\pi j}{n})) \right] = 0 \tag{6.1.3}$$

for any continuous function g defined on $[-\pi, \pi]$.

The equal distribution of eigenvalues of Toeplitz matrices indicates that the eigenvalues will not be clustered in general. To illustrate this, consider the 1-dimensional discrete Laplacian matrix

$$A_n = tridiag[-1, 2, -1].$$

Its eigenvalues are given by

$$\lambda_j(A_n) = 4\sin^2(\frac{\pi j}{n+1}), \quad 1 \le j \le n.$$

For $n = 32$, the eigenvalues of A_n are depicted in Figure 1.

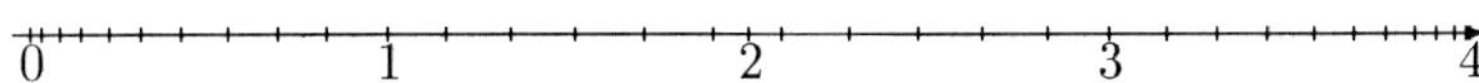

FIGURE 6.1. Spectrum of 1-D discrete Laplacian: $A_{32} = tridiag[-1, 2, -1]$.

An n-by-n matrix C_n is said to be *circulant* if

$$C_n = \begin{bmatrix} c_0 & c_{-1} & \cdots & c_{2-n} & c_{1-n} \\ c_1 & c_0 & c_{-1} & & c_{2-n} \\ \vdots & c_1 & c_0 & \ddots & \vdots \\ c_{n-2} & & \ddots & \ddots & c_{-1} \\ c_{n-1} & c_{n-2} & \cdots & c_1 & c_0 \end{bmatrix},$$

where $c_{-k} = c_{n-k}$ for $1 \le k \le n-1$. Circulant matrices are diagonalized by the Fourier matrix F_n, i.e.,

$$C_n = F_n^* \Lambda_n F_n \tag{6.1.4}$$

where the entries of F_n are given by

$$[F_n]_{j,k} = \frac{1}{\sqrt{n}} e^{2\pi i jk/n}, \quad 0 \le j, k \le n-1,$$

and Λ_n is a diagonal matrix holding the eigenvalues of C_n, see for instance Davis [37, p.73]. We note that Λ_n can be obtained in $O(n \log n)$ operations by taking the fast Fourier transform (FFT) of the first column of C_n. For the fast Fourier transform algorithm, we refer to Cooley and Tukey [36]. In fact, the diagonal entries λ_k of Λ_n are given by

$$\lambda_k = \sum_{j=0}^{n-1} c_j e^{2\pi ijk/n}, \quad k = 0, \ldots, n-1. \tag{6.1.5}$$

Once Λ_n is obtained, the products $C_n\mathbf{y}$ and $C_n^{-1}\mathbf{y}$ for any vector $\mathbf{y}$ can be computed by FFTs in $O(n \log n)$ operations using (6.1.4).

6.1.3 The Conjugate Gradient Method for Toeplitz Matrices

The conjugate gradient method is an iterative method for solving Hermitian positive definite matrix systems. The algorithm of the method can be found in Golub and Van Loan [45, pp. 516–527]. In each iteration, it requires two inner products of n-vectors and one multiplication of the coefficient matrix with an n-vector. Storage for four temporary n-vectors is needed but there is no explicit storage required for the coefficient matrix.

Let $f \in \mathbf{C}_{2\pi}$. For simplicity, let us assume for the moment that f is positive, i.e., $f_{\min} > 0$. Then by (6.1.2), A_n are positive definite for all n. Consider applying the conjugate gradient method to solve these symmetric positive definite Toeplitz systems $A_n\mathbf{x} = \mathbf{b}$. In each iteration, besides the two inner products required, one matrix-vector multiplication $A_n\mathbf{y}$ is also needed. That can be computed by FFTs by first embedding A_n into a $2n$-by-$2n$ circulant matrix, i.e.,

$$\begin{bmatrix} A_n & \times \\ \times & A_n \end{bmatrix} \begin{bmatrix} \mathbf{y} \\ \mathbf{0} \end{bmatrix} = \begin{bmatrix} A_n\mathbf{y} \\ \dagger \end{bmatrix}, \tag{6.1.6}$$

see Strang [71], and then carrying out the multiplication by using the decomposition as in (6.1.4). The matrix-vector multiplication thus requires $O(2n \log(2n))$ operations. It follows that the total number of operations per iteration is of $O(n \log n)$ operations. As for the storage required, besides the four temporary n-vectors, we need an extra $2n$-vector for storing the eigenvalues of the embedded circulant matrix given in (6.1.6).

The convergence rate of the conjugate gradient method is well studied, see Axelsson and Barker [4, p.24]. It depends on the condition number of the matrix A_n and how clustered the spectrum of A_n is. If the spectrum is not clustered, as is usually the case for Toeplitz matrices (cf. Theorem 1), a good estimate of the convergence rate is given in terms of the largest and smallest eigenvalues of A_n. Using (6.1.2), this estimate can be expressed as

$$\frac{\|\mathbf{e_q}\|_{A_n}}{\|\mathbf{e_0}\|_{A_n}} < 2\left(\frac{\sqrt{f_{\max}} - \sqrt{f_{\min}}}{\sqrt{f_{\max}} + \sqrt{f_{\min}}}\right)^q,$$

where $\mathbf{e_q}$ is the error vector at the qth iteration and $||\mathbf{x}||_{A_n}^2 \equiv \mathbf{x}^* A_n \mathbf{x}$. This indicates that the rate of convergence is linear. Thus, the method will converge in a constant number of iterations, and hence the complexity of solving the Toeplitz system is $O(n \log n)$. However, we remark that if $f_{\max}/f_{\min}$ is large, the constant in the operation count will be large and hence the convergence will be very slow.

One way to speed up the convergence rate of the method is to precondition the Toeplitz system. Thus, instead of solving $A_n \mathbf{x} = \mathbf{b}$, we solve the preconditioned system

$$P_n^{-1} A_n \mathbf{x} = P_n^{-1} \mathbf{b}. \tag{6.1.7}$$

The matrix P_n, called the preconditioner, should be chosen according to the following criteria:

- P_n should be constructed within $O(n \log n)$ operations.
- $P_n \mathbf{v} = \mathbf{y}$ should be solved in $O(n \log n)$ operations.
- The spectrum of $P_n^{-1} A_n$ should be clustered.

The first two criteria are to keep the operation count per iteration within $O(n \log n)$ as that is the count for the non-preconditioned system. The third criterion comes from the fact that the more clustered the eigenvalues are, the faster the convergence of the method will be, see for instance [65, pp. 249-251] and [4, pp. 27-28]. If $P_n^{-1} A_n$ has a clustered spectrum as defined in Definition 1, then the conjugate gradient method, when applied to solving the preconditioned system (6.1.7), converges superlinearly for large n, see [24]. More precisely, for any given $\epsilon > 0$, there exists a constant $c(\epsilon) > 0$ such that the error vector $\mathbf{e_q}$ of the preconditioned conjugate gradient method at the qth iteration satisfies

$$\frac{|||\mathbf{e_q}|||}{|||\mathbf{e_0}|||} \le c(\epsilon)\epsilon^q \tag{6.1.8}$$

where

$$|||\mathbf{v}|||^2 \equiv \mathbf{v}^* P_n^{-1/2} A_n P_n^{-1/2} \mathbf{v}.$$

The main aim of this chapter is to review different circulant preconditioners developed for Toeplitz systems that satisfy the three criteria mentioned above. For simplicity, we will drop the subscripts on matrices when their dimensions are apparent. In the next section, we study the use of circulant matrices as preconditioners for Toeplitz matrices. Finally, some remarks are given in §3.

6.2 Circulant Preconditioners for Toeplitz Systems

6.2.1 Circulant Preconditioners

In 1986, Strang [71] and Olkin [68] independently proposed the use of circulant matrices to precondition Toeplitz matrices in conjugate gradient iterations. Part of their motivation is to exploit the fast inversion of circulant matrices. Numerical results in [72, 68] suggest that the method converges very fast for a wide range of Toeplitz matrices. This has later been proved theoretically in [24] and in other papers for other circulant preconditioners. In this subsection, we will give a brief account of these developments.

With circulant matrices as preconditioners, in each iteration, we have to solve a circulant system. From (6.1.4), we see that circulant matrices can be diagonalized by discrete Fourier matrices, and hence the inversion of n-by-n circulant systems can be done in $O(n \log n)$ operations by using FFTs of size n. In contrast, by (6.1.6), we see that the cost of computing $A\mathbf{y}$, which is also required in each iteration whether the system is preconditioned or not, is done by using FFTs of size $2n$. Notice that if FFT is used to compute the discrete Fourier transform of a $2n$-vector for which the even discrete Fourier transform components are already known, then the cost is the same as carrying out a length n FFT, see Linzer [63] for details. Thus, the cost per iteration of the circulant preconditioned conjugate gradient method is roughly 1.25 times of that required by the method without using preconditioners. We remark that Huckle [55] has also discussed different ways to reduce the number of FFTs in the iterative scheme, even when n is not a power of 2. In particular, he has proposed a way to compute $A\mathbf{y}$ such that the computational cost per iteration of the preconditioned system is nearly the same as that required by the non-preconditioned system.

We emphasize that the use of circulant matrices as preconditioners for Toeplitz systems allows the use of FFT throughout the computations; and FFT is highly parallelizable and has been implemented on multiprocessors efficiently [1, p.238] and [73]. Since conjugate gradient methods are easily parallelizable too [6, p.165], the circulant preconditioned conjugate gradient method is well-adapted for parallel computing.

We remark that circulant approximations to Toeplitz matrices have been considered and used for some time in signal processing. However, in these applications, the circulant approximations thus obtained were used to replace the given Toeplitz matrices in subsequent computations. In contrast, circulant approximations are used here only as preconditioners for Toeplitz systems and the solutions to the Toeplitz systems are unchanged. In the following, we review some successful circulant preconditioners proposed for Toeplitz matrices.

Strang's Preconditioner

The first circulant preconditioner is proposed by Strang [71] in 1986 and is defined to be the matrix that copies the central diagonals of A and reflects them around to complete the circulant requirement. For A given by (6.1.1), the diagonals s_j of the Strang preconditioner $S = [s_{k-\ell}]_{0 \le k,\ell < n}$ are given by

$$s_j = \begin{cases} a_j, & 0 < j \le \lfloor n/2 \rfloor, \\ a_{j-n} & \lfloor n/2 \rfloor < j < n, \\ s_{n+j}, & 0 < -j < n. \end{cases}$$

One of the interesting properties of S is that S minimizes

$$||C - A||_1 \quad \text{and} \quad ||C - A||_\infty$$

over all Hermitian circulant matrices C, see [11].

Theorem 2 *Let f be an even positive function in the Wiener class, i.e., its Fourier coefficients are absolutely summable,*

$$\sum_{k=0}^{\infty} |a_k| < \infty.$$

Then for large n, the circulants S and S^{-1} are uniformly bounded in ℓ_2 norm.

Proof: By (6.1.5), the jth eigenvalue of S equal to the partial sum of from $k = 1 - m$ to m of the series $\sum_{k=0}^{\infty} a_k e^{-ik\theta}$ evaluated at the point $\theta = 2\pi j/n$. Since the infinite series is absolutely convergent and its sum satisfies $f \ge \delta > 0$, the partial sums are uniformly positive for large n. The result follows. $\square$

The spectra of these circulant preconditioned matrices have been analyzed by R. Chan and Strang [24].

Theorem 3 (R. Chan and Strang (1989) [24]) *Let f be an even positive function in the Wiener class, i.e., its Fourier coefficients are absolutely summable,*

$$\sum_{k=0}^{\infty} |a_k| < \infty.$$

Let A be generated by f. Then the spectra of $S^{-1}A$ are clustered around 1 for large n.

The main idea of their proof is to use an orthogonal transformation to transform $S - A$ into a Hankel matrix. Then Nehari's Theorem [35, p.120] is used to show that the limiting Hankel operator is compact. Using the theory of collectively compact sets of operators [3, pp.65–70], the spectra of the finite Hankel matrices are then shown to be clustered. However, this proof cannot be readily generalized to real-valued f, i.e., to Hermitian Toeplitz systems. Thus R. Chan in [11] developed a purely linear algebra technique to extend the results in Theorem 3.

Theorem 4 (R. Chan (1989) [11]) *Let f be a positive function in the Wiener class. Then the spectra of $S^{-1}A$ are clustered around 1 for large n.*

Proof: The approach is to decompose, for a given $\epsilon > 0$, the matrix $S - A$ into a sum of two matrices L and V, Here L is the n-by-n matrix obtained from $S-A$ by replacing the $(n-N)$ by $(n-N)$ leading principal submatrix of $S - A$ by the zero matrix and $V = S - A - L$. The leading $(n - N)$ by $(n - N)$ block of V is the leading $(n - N)$ by $(n - N)$ principal submatrix of $S - A$, hence this block is a Toeplitz matrix, and it is easy to see that the maximum absolute column sum of V is attained at the first column (or the $(n - N - 1)$-th column). Thus

$$||V||_1 = \sum_{k=m+1}^{n-N-1} |a_{k-n} - a_k| \leq \sum_{k=N+1}^{n-N-1} |a_k| < \epsilon.$$

Since V is Hermitian, we have $||V||_\infty = ||V||_1$. Thus

$$||V||_2 \leq (||V||_1 \cdot ||V||_\infty)^{\frac{1}{2}} < \epsilon.$$

Hence the spectrum of V lies in $(-\epsilon, \epsilon)$. By Cauchy Interlace Theorem, see for instance [45], we see that at most $2N$ eigenvalues of $S - A$ have absolute values exceeding ϵ. Using the fact that $S^{-1}A$ is similar to $S^{-1/2}AS^{-1/2}$, the result of Theorem 2 and

$$S^{-1/2}AS^{-1/2} = I + S^{-1/2}(A - S)S^{-1/2}, \tag{6.2.1}$$

the result follows. □

Using standard error analysis of the conjugate gradient method, we can then show that the convergence rate of the method is superlinear, see (6.1.8). If extra smoothness conditions are imposed on the generating function f, we can get more precise estimates on how $|||\mathbf{e_q}|||$ in (6.1.8) goes to zero.

Theorem 5 (Trefethen (1990) [76], Ku and Kuo (1993) [59, 60]) *Suppose f is a rational function of the form $f = p/q$ where p and q are polynomials of degrees μ and ν, respectively. Then the number of outlying eigenvalues of $S^{-1}A$ is exactly equal to $2\max\{\mu, \nu\}$. Hence, the method converges in at most $1 + 2\max\{\mu, \nu\}$ steps for large n. If however $f(z) = \sum_{j=0}^{\infty} a_j z^j$ is only analytic in a neighborhood of $|z| = 1$, then there exist constants c and $0 \leq r < 1$ such that*

$$\frac{|||\mathbf{e_{q+1}}|||}{|||\mathbf{e_0}|||} < c^q r^{q^2/4+q/2}. \tag{6.2.2}$$

The idea of Trefethen's proof is to use rational approximation to bound the singular values of the Hankel matrix considered in the proof of Theorem 3. It follows from (6.2.2) that

$$\frac{|||\mathbf{e_{q+1}}|||}{|||\mathbf{e_q}|||} \approx cr^q \to 0.$$

For generating functions f with Fourier coefficients a_j decaying at a slower rate, we have the following two theorems.

Theorem 6 (R. Chan (1989) [11]) *Let f be a ν-times differentiable function with $f^{(\nu)} \in L^1[-\pi,\pi]$ where $\nu > 1$ (i.e. $|a_j| \le \tilde{c}/j^{\nu+1}$ for some constant $\tilde{c}$.) Then there exists a constant c which depends only on f and ν, such that for large n,*

$$\frac{|||\mathbf{e_{2q}}|||}{|||\mathbf{e_0}|||} \le \frac{c^q}{((q-1)!)^{2\nu-2}}.$$

Theorem 6 was proved by using Cauchy's interlace theorem [45]. R. Chan and Yeung later used Jackson's theorem [33] in approximation theory to prove a stronger result than that in Theorem 6.

Theorem 7 (R. Chan and Yeung (1992) [28]) *Let f be a Lipschitz function of order ν, $0 < \nu \le 1$, or f has a continuous νth derivative, $\nu \ge 1$. Then there exists a constant c which depends only on f and ν, such that for large n,*

$$\frac{|||\mathbf{e_{2q}}|||}{|||\mathbf{e_0}|||} \le \prod_{k=2}^{q} \frac{c \log^2 k}{k^{2\nu}}.$$

Theorems 3–7 give the rate at which the error goes to zero in terms of the rate of decay of $|a_j|$. To see if solving Toeplitz systems by preconditioned conjugate gradient methods is more efficient than by fast direct Toeplitz solvers, Linzer [64] has performed tests for Toeplitz matrices with different condition numbers and coefficients with different decaying rate. His results show that the iterative methods have the edge if $|a_j|$ decays like $O(j^{-0.5})$ or faster for matrices with condition number about 10, and the rate of decay required increases to $O(j^{-2})$ when the condition number is about 10^5.

We also remark that Huckle [56] has compared the number of floating point operations for iterative methods with that of direct Toeplitz solvers and superfast Toeplitz solvers. He derived an upper bound of the number of PCG iterations such that the iterative method will be better than direct and superfast Toeplitz solvers. His finding shows that for positive definite Toeplitz systems, PCG methods are competitive for large matrices with small number of PCG iterations. In the indefinite or near-singular case, iterative methods may give a higher accuracy. In the nonsymmetric case, only classical $O(n^2)$ direct Toeplitz solvers are available and therefore the iterative methods will have the edge if a good preconditioner can be found, see §2.3.

<u>T. Chan's Preconditioner</u>

For an n-by-n Toeplitz matrix A, T. Chan's circulant preconditioner $c(A)$ is defined to be the minimizer of

$$||C - A||_F \tag{6.2.3}$$

over all n-by-n circulant matrices C, see T. Chan (1988) [30]. Here $||\cdot||_F$ denotes the Frobenius norm. In [30], the matrix $c(A)$ is called an *optimal* circulant preconditioner because it minimizes (6.2.3). The jth diagonals of $c(A)$ are shown to be equal to

$$c_j = \begin{cases} \dfrac{(n-j)a_j + ja_{j-n}}{n}, & 0 \le j < n, \\ c_{n+j}, & 0 < -j < n, \end{cases} \tag{6.2.4}$$

which are just the average of the diagonals of A, with the diagonals being extended to length n by a wrap-around. By using (6.1.5) and (6.2.4), we see that the eigenvalues $\lambda_k(c(A))$ of $c(A)$ are given by

$$\lambda_k(c(A)) = \sum_{j=-n+1}^{n-1} a_j \left(1 - \frac{|j|}{n}\right) e^{2\pi ijk/n}, \quad k = 0, \ldots, n-1. \tag{6.2.5}$$

When A is not a Toeplitz matrix, the circulant minimizer $c(A)$ of (6.2.3) can still be obtained easily by taking the arithmetic average of the entries of A, i.e., its diagonals are given by

$$c_\ell = \frac{1}{n} \sum_{j-k=\ell (\mathrm{mod}\ n)} a_{j,k}, \quad \ell = 0, \ldots, n-1, \tag{6.2.6}$$

see [78]. Therefore, T. Chan's preconditioner is particularly useful in solving *non-Toeplitz* systems arising from the numerical solutions of elliptic partial differential equations [14] and Toeplitz least squares problems arising from signal and image processing [19, 20, 21, 32, 67]. Convergence results for T. Chan's preconditioner have been established for these problems, see [23]. One good property of the T. Chan preconditioner is that it preserves the positive-definiteness of a given matrix.

Theorem 8 *If A is Hermitian positive definite, then $c(A)$ is Hermitian and positive definite. Moreover, we have*

$$\lambda_{\min}(A) \le \lambda_{\min}(c(A)) \le \lambda_{\max}(c(A)) \le \lambda_{\max}(A). \tag{6.2.7}$$

Proof: Tyrtyshnikov [78] first proved it for the real scalar field and R. Chan, Jin, and Yeung [17] generalized the result to the complex field. By (6.2.6), it is clear that $c(A)$ is Hermitian when A is Hermitian. As the Frobenius norm is a unitary-invariant norm, the minimizer of $||C - A||_F$

over all C of the form $C = F^*\Lambda F$, Λ a diagonal matrix, is attained at $F\Delta F^*$. Here Δ is a diagonal matrix with diagonal entries

$$\Delta_{j,j} = [FAF^*]_{j,j} \equiv \lambda_j, \quad j = 1, \ldots, n.$$

Suppose that $\lambda_j = \lambda_{\min}(c(A))$ and $\lambda_k = \lambda_{\max}(c(A))$. Let e_j and e_k denote the j-th and the k-th unit vectors respectively. Since A is Hermitian, we have

$$\lambda_{\max}(c(A)) = \lambda_k = \frac{e_k^* FAF^* e_k}{e_k^* e_k} \le \max_{x \ne 0} \frac{x^* FAF^* x}{x^* x} = \max_{x \ne 0} \frac{x^* Ax}{x^* x} = \lambda_{\max}(A).$$

Similarly,

$$\lambda_{\min}(A) = \min_{x \ne 0} \frac{x^* Ax}{x^* x} = \min_{x \ne 0} \frac{x^* FAF^* x}{x^* x} \le \frac{e_j^* FAF^* e_j}{e_j^* e_j} = \lambda_j = \lambda_{\min}(c(A)).$$

From the inequality above, we can easily see that $c(A)$ is positive definite when A is positive definite. □

We remark that Strang's preconditioner does not satisfy (6.2.7) even for Toeplitz matrices, see R. Chan and Yeung [27]. In addition, R. Chan and Wong [25] proved that for some Toeplitz matrices A, T. Chan's preconditioner $c(A)$ minimizes $\kappa(C^{-1}A)$ over all n-by-n non-singular circulant matrices C.

As for the performance of $c(A)$ as a preconditioner for the Toeplitz matrix A, R. Chan [12] proved that under the Wiener class assumptions in Theorem 4 (i.e., f is a positive function with absolutely summable Fourier coefficients), the spectra of $c(A) - A$ and $S - A$ are asymptotically the same as n tends to infinity, i.e., $\lim_{n\to\infty} ||c(A) - S||_2 = 0$. Hence, $c(A)$ works as well for Wiener class functions as S does.

Theorem 9 (R. Chan and Yeung (1992) [26]) *Let f be a positive function in $\mathbf{C}_{2\pi}$. Then the spectra of $c(A)^{-1}A$ are clustered around 1 for large n.*

Proof: Using Weierstrass' theorem to approximate 2π-periodic continuous generating function f by Wiener class function, for any given ϵ, there exists a trigonometric polynomial

$$p_M(\theta) = \sum_{k=-M}^{M} b_k e^{ik\theta}$$

such that

$$||f - p||_\infty \le \epsilon. \tag{6.2.8}$$

Here $||\cdot||_\infty$ is the supremum norm. Let A' be the Topelitz matrix generated by p. Then for all $n > 2M$, we write

$$c(A) - A = c(A - A') - A + A' + c(A') - A'.$$

Using (6.1.2), (6.2.8) and Theorem 8, we know that $\|A - A'\|_2$ and $\|c(A - A')\|_2$ are both less than or equal to ϵ. Since p is a trigonometric polynomial, the spectra of $c(A') - A'$ is clustered around zero. The result follows by considering (6.2.1) and the similar argument in Theorem 4. $\square$

However, the Weierstrass approach used in proving this theorem does not work for Strang's preconditioner. From Theorem 7, we see that the class of generating functions where Strang's preconditioner works is the class of 2π-periodic Lipschitz continuous functions. We will discuss this discrepancy in §2.2.

Huckle in 1992 [53] proposed a preconditioner that is an extension of the T. Chan's preconditioner. Let $1 \leq p \leq n$. Huckle's circulant preconditioner H is defined to be the circulant matrix with eigenvalues

$$\lambda_k(H) = \sum_{j=-p+1}^{p-1} a_j \left(1 - \frac{|j|}{p}\right) e^{2\pi ijk/n}, \quad k = 0, \ldots, n-1, \tag{6.2.9}$$

cf. (6.2.5). Thus, when $p = n$, it is the T. Chan preconditioner. Besides Wiener class functions, Huckle [52] has shown that H also works for generating functions with Fourier coefficients a_k that satisfy

$$\sum_{k=-\infty}^{\infty} |k||a_k|^2 < \infty.$$

Preconditioners by Embedding

Let the Toeplitz matrix A be embedded into a $2n$-by-$2n$ circulant matrix

$$\begin{bmatrix} A & B^* \\ B & A \end{bmatrix} \tag{6.2.10}$$

R. Chan's circulant preconditioner R is defined as $R = A + B$, see R. Chan (1989) [11]. We remark that T. Chan's and Huckle's preconditioners to A are just equal to R. Chan's preconditioner obtained from the Toeplitz matrix with diagonals $(1-|j|/n)a_j$ and $\max\{0, 1-|j|/p\}a_j$ respectively; i.e., the diagonals a_j of A are damped by the factor $(1 - |j|/n)$ and $\max\{0, 1 - |j|/p\}$ respectively, see (6.2.5) and (6.2.9).

Using the embedding (6.2.10), Ku and Kuo (1991) [58] constructed four different preconditioners K_i, $1 \leq i \leq 4$, based on different combinations of the matrices A and B. They are

$$K_1 = A + B; \quad K_2 = A - B; \quad K_3 = A + JB; \quad K_4 = A - JB$$

where J is the n-by-n anti-identity (reversal) matrix. We note that K_2, K_3, and K_4 are not circulant matrices.

Preconditioners by Minimization of Norms

Besides using the minimizer of $||C - A||_F$ as preconditioners for Toeplitz systems, minimizers of other approximations have also been proposed and used. For instance, Tyrtyshnikov's circulant preconditioner T (1992) [78] is defined to be the minimizer of

$$||I - C^{-1}A||_F \tag{6.2.11}$$

over all non-singular circulant matrices C. In [78], T is called the *super-optimal* circulant preconditioner because it minimizes (6.2.11) instead of (6.2.3) and is shown to be equal to

$$T = c(AA^*)c(A)^{-1},$$

see also [17]. Tyrtyshnikov [78] showed that for a general positive definite Toeplitz matrix A, T is also positive definite, cf. (6.2.7). Tismenetsky [74] independently proposed the same preconditioner.

Huckle [54] has considered the minimizer M by minimization of

$$||I - C^{-1/2}AC^{-1/2}||_F,$$

over all non-singular circulant matrices C. The constructions of T and M require $O(n \log n)$ operations, see R. Chan, Jin, and Yeung [17], Tismenetsky [74], and Huckle [54], respectively.

Finally, we compare the performance of these preconditioners (R, $\{K_i\}_{i=1}^4$, T, and M) with Strang's and T. Chan's preconditioners. It has been proved in R. Chan [11], R. Chan, Jin, and Yeung [18], Ku and Kuo [58], and Huckle [54] that under the same Wiener class assumptions, these circulant preconditioned systems have spectra that are asymptotically the same as Strang's and T. Chan's circulant preconditioned systems. In particular, all these preconditioned systems converge at the same rate for large n.

In the following, we illustrate the effectiveness of circulant preconditioners for Toeplitz systems by a numerical example. We use the continuous generating function

$$f(\theta) = \theta^4 + 1, \quad -\pi \le \theta \le \pi,$$

in the test. Table 1 shows the numbers of iterations required to solve non-preconditioned systems $A\mathbf{x} = \mathbf{b}$ and circulant preconditioned systems $C^{-1}A\mathbf{x} = C^{-1}\mathbf{b}$ for different preconditioners. The right hand side vector $\mathbf{b}$ is the vector of all ones. The zero vector is the initial guess. The stopping criterion is when the residual vector $\mathbf{r_q}$ at the qth iteration satisfies $||\mathbf{r_q}||_2/||\mathbf{r_0}||_2 < 10^{-7}$. All computations were done by Matlab.

In the table, I denotes that no preconditioner is used and H is Huckle's preconditioner with $p = n/2$ (see (6.2.9)). We see from Table 1 that the

n	I	S	$c(A)$	R	K_2	H	T
16	8	8	8	6	6	8	8
32	20	8	7	5	5	10	16
64	37	6	7	5	5	7	18
128	56	5	6	5	5	7	13
256	67	5	6	5	5	6	10
512	70	5	6	5	5	6	8

TABLE 1. Number of iterations for Different Preconditioned Systems.

number of iterations required for convergence for non-preconditioned systems is much greater than those for circulant preconditioned systems. Figure 2 depicts the spectra of the non-preconditioned matrix and the circulant preconditioned matrices for different circulant preconditioners. We note that the spectra of the circulant preconditioned matrices are indeed clustered around 1.

To illustrate that the circulant preconditioned conjugate gradient method is highly parallelizable, we test the performance of the algorithm on a DECmpp 12000/Sx parallel computer with 8192 processors for $n = 8$, 16, 32, ..., 4096. In Figure 3, we observe that the total time required for solving the Toeplitz system in Table 1 grows like $O(\log n)$.

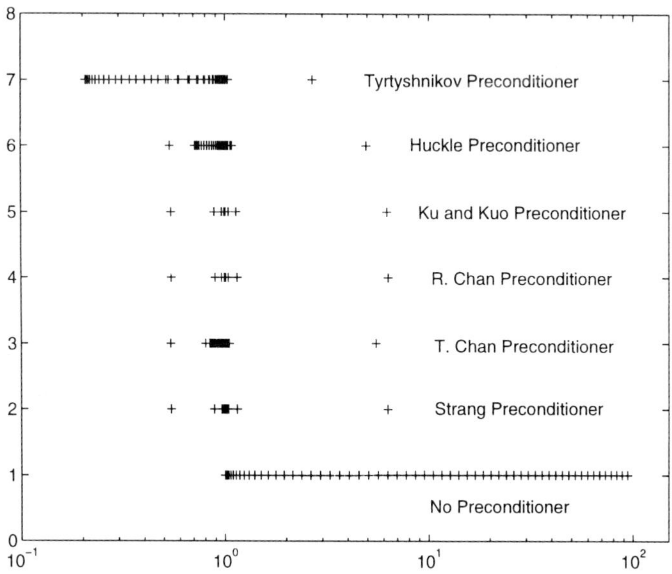

FIGURE 6.2. Spectra of Preconditioned Matrices for $n = 64$.

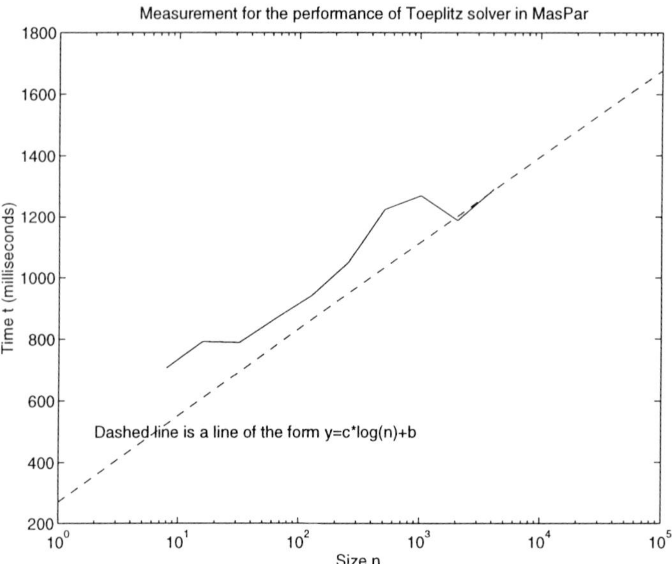

FIGURE 6.3. Time in milliseconds for the circulant PCG method.

6.2.2 Circulant Preconditioners from Kernel Functions

A unifying approach of constructing circulant preconditioners is given in R. Chan and Yeung [27], where it is shown that most of the above mentioned circulant preconditioners can be derived by using the convolution products of some well-known kernels with the generating function f.

In Theorem 2, we have noted that the eigenvalues of Strang's preconditioner S are equal to the partial sum of f evaluated at the points $\{2\pi j/n\}_{j=0}^{n-1}$. From Fourier analysis, see Zygmund [87, p.49] for instance, the partial sum of f is given by the convolution of f with the Dirichlet kernel $\hat{D}_{\lfloor \frac{n}{2} \rfloor}$ where

$$\hat{D}_k(\theta) = \frac{\sin(k+\frac{1}{2})\theta}{\sin\frac{1}{2}\theta}, \quad k = 1, 2, \ldots.$$

Thus the the eigenvalues of S can be expressed as in the form

$$\lambda_j(S) = (\hat{D}_{\lfloor \frac{n}{2} \rfloor} * f)(\frac{2\pi j}{n}), \quad 0 \le j < n,$$

where the convolution of the Dirichlet kernel with f is given by

$$(\hat{D}_{\lfloor \frac{n}{2} \rfloor} * f)(\theta) \equiv \frac{1}{2\pi} \int_{-\pi}^{\pi} \hat{D}_{\lfloor \frac{n}{2} \rfloor}(\theta - \phi) f(\phi) d\phi.$$

For the T. Chan's circulant preconditioner, by (6.2.5), the eigenvalues of $c(A)$ are the value of Cesàro summation process of order 1 for the Fourier series of f evaluated at the points $\{2\pi j/n\}_{j=0}^{n-1}$. This summation process is given by the convolution of f with the Fejér kernel $\hat{F}_n$, i.e. the eigenvalues of T. Chan's preconditioner $c(A)$ are given by

$$\lambda_j(c(A)) = (\hat{F}_n * f)(\frac{2\pi j}{n}), \quad 0 \le j < n,$$

where the Fejér kernels are given by

$$\hat{F}_k(\theta) = \frac{1}{k}\left(\frac{\sin\frac{k}{2}\theta}{\sin\frac{1}{2}\theta}\right)^2, \quad k = 1, 2, \cdots.$$

We see that Strang's and T. Chan's circulant preconditioners are constructed by using the Dirichlet and Fejér kernels, respectively. Similarly, the eigenvalues of R. Chan's preconditioner R and Huckle's preconditioner H are given by

$$\lambda_j(R) = (\hat{D}_{n-1} * f)(\frac{2\pi j}{n}), \quad 0 \le j < n$$

and

$$\lambda_j(H) = (\hat{F}_p * f)(\frac{2\pi j}{n}), \quad 0 \le j < n,$$

respectively. The idea can be applied to design other circulant preconditioners C from kernels $\hat{C}_n$ such as the von Hann kernel, Hamming kernel, and Bernstein kernel that are commonly used in function theory [83], signal processing [48], and time series analysis [69]. In [27], several circulant preconditioners were constructed using this approach.

We remark that the convolution products of these kernels with f are just smooth approximations of the generating function f itself. R. Chan and Yeung proved that if the convolution product is a good approximation of f, then the correspondingly constructed circulant matrix will be a good preconditioner.

Theorem 10 (R. Chan and Yeung (1992) [27]) *Let $f \in \mathbf{C}_{2\pi}$ be positive. Let $\hat{C}_n$ be a kernel such that $\hat{C}_n * f$ tends to f uniformly on $[-\pi, \pi]$. If C is the circulant matrix with eigenvalues given by*

$$\lambda_j(C) = (\hat{C}_n * f)(\frac{2\pi j}{n}), \quad 0 \le j < n, \tag{6.2.12}$$

then for large n, the circulants C and C^{-1} are uniformly bounded in ℓ_2 norm and the spectra of $C^{-1}A$ are clustered around 1 for large n.

Proof: Since $\hat{C}_n * f$ converges to f uniformly and $f_{\min} > 0$, there exists an $N > 0$, such that for all $n > N$,

$$|[f - f * \hat{C}_n](\frac{2\pi j}{n})| \le ||f - f * \hat{C}_n||_\infty < \frac{1}{2} f_{\min}, \quad 0 \le j < n.$$

Thus by (6.2.12), we have

$$\begin{aligned}\lambda_j(C) &= [f * \hat{C}_n - f](\frac{2\pi j}{n}) + f(\frac{2\pi j}{n}) \\ &\geq f_{\min} - [f - f * \hat{C}_n](\frac{2\pi j}{n}) \\ &\geq \frac{1}{2} f_{\min}, \quad 0 \leq j < n.\end{aligned}$$

Similarly, we can prove that $\lambda_j(C)$ is uniformly bounded above.

To prove the clustered spectra of $A - C$, we first rewrite $A - C$ as

$$A - C = \{A - c(A)\} + \{c(A) - C\},$$

where $c(A)$ is the T. Chan's circulant preconditioner. It suffices to show that

$$\lim_{n\to\infty} ||c(A) - C||_2 = 0. \tag{6.2.13}$$

Since $c(A)$ and C are both circulant matrices and hence can be diagonalized by the same Fourier matrix, we see that (6.2.13) is equivalent to

$$\lim_{n\to\infty} \max_{0\leq j<n} |\lambda_j(c(A)) - \lambda_j(C)| = 0.$$

Therefore, we have

$$\begin{aligned}\max_{0\leq j<n} |\lambda_j(c(A)) - \lambda_j(C)| &= \max_{0\leq j<n} |(\hat{F}_n * f)(\frac{2\pi j}{n}) - (\hat{C}_n * f)(\frac{2\pi j}{n})| \\ &\leq ||\hat{F}_n * f - \hat{C}_n * f||_\infty \\ &\leq ||\hat{F}_n * f - f||_\infty + ||f - \hat{C}_n * f||_\infty.\end{aligned}$$

Since $\hat{F}_n * f$ and $\hat{C}_n * f$ both converge to f uniformly, the spectra of $A - C$ is clustered around zero. The result follows by considering the same argument in Theorem 4. □

We have mentioned in §2.1 that Strang's and T. Chan's preconditioners are fundamentally different, where Strang's preconditioners work for Lipschitz continuous functions (Theorem 7) and T. Chan's preconditioners work for 2π-periodic continuous functions (Theorem 9). That can be explained by the associations of Strang's preconditioner with the Dirichlet kernel and T. Chan's preconditioner with the Fejér kernel. It is well-known in Fourier analysis that if f is 2π-periodic continuous (or respectively Lipschitz continuous), then the convolution product of the Fejér kernel (or respectively the Dirichlet kernel) with f will converge to f uniformly on $[-\pi, \pi]$, see Walker [83, p.79, p.52].

In addition, it is interesting to note that for a piecewise continuous function f, the convolution product with the Fejér kernel will no longer converge to f uniformly on $[-\pi, \pi]$. Therefore, for generating functions that are only piecewise continuous, we don't expect the spectra of $c(A)^{-1}A$ to be clustered around 1.

Theorem 11 (Yeung and R. Chan (1993) [85]) *Let f be a non-negative piecewise continuous function on $[-\pi, \pi]$. Then for any given $\epsilon > 0$, the number of eigenvalues of $c(A)^{-1}A$ that lie outside the interval $(1-\epsilon, 1+\epsilon)$ is at least of $O(\log n)$ for n sufficiently large. If moreover f is strictly positive, then the number of outlying eigenvalues is exactly of $O(\log n)$.*

The theorem is established by noting that $A - c(A)$ is orthogonally similar to a Hankel matrix. Then Widom's theorem [84], which gives an estimate of the eigenvalues of the Hilbert matrices, is used to estimate the number of outlying eigenvalues of $A - c(A)$. Numerical examples are given in [85] to verify that the convergence rate of the method will no longer be superlinear in general. In fact, the numbers of iterations required for convergence do increase like $O(\log n)$. These results have been extended by Tyrtyshnikov [79, 81]. In [79], he has established a generalized Szegö theorem that if f is in $\mathbf{L}_2$, then the singular values of A_n is equally distributed (in a generalized sense) as $|f(x)|$ (cf. (6.1.3)). He then used the result to prove that if f is in $\mathbf{L}_2$, then the number of outlying eigenvalues of the preconditioned system grows no more than $o(n)$. In [81], he further extended the results of circulant preconditioners for products of Toeplitz matrices.

6.2.3 Non-Hermitian Type Toeplitz Systems

In this subsection, we study Toeplitz matrices A generated by complex-valued functions. We note that such A are complex non-Hermitian matrices. In general the fast direct Toeplitz solvers are not applicable, and neither is the conjugate gradient method when applied to the system $A\mathbf{x} = \mathbf{b}$. For such matrices A, one can apply the conjugate gradient method to the normal equations $A^*A\mathbf{x} = A^*\mathbf{b}$. Another way of solving non-Hermitian Toeplitz systems is to employ some CG-like method [42] such as restarted GMRES [70] or TFQMR [40]. To speed up the CG-like methods, we can choose a matrix C such that the singular values of the preconditioned matrices $C^{-1}A$ are clustered.

Let us begin with skew-Hermitian type Toeplitz matrices, i.e., Toeplitz matrices of the form

$$A = \begin{bmatrix} a_0 & -\overline{a_1} & \cdots & -\overline{a_{n-2}} & -\overline{a_{n-1}} \\ a_1 & a_0 & -\overline{a_1} & & -\overline{a_{n-2}} \\ \vdots & a_1 & a_0 & \ddots & \vdots \\ a_{n-2} & & \ddots & \ddots & -\overline{a_1} \\ a_{n-1} & a_{n-2} & \cdots & a_1 & a_0 \end{bmatrix},$$

where a_0 is a real number. Obviously, $A = a_0 I + A_S$, where I is the identity matrix and A_S is a skew-Hermitian Toeplitz matrix. These systems or low rank perturbations of such systems often appear in solving discretized hyperbolic differential equations, see Buckley [9] and Holmgren and Otto [49, 50].

In [16], R. Chan and Jin used R. Chan's circulant preconditioners R defined in §2.1.3 to precondition these skew-Hermitian Toeplitz matrices. Under the Wiener class assumptions on the entries of the first column of A, they proved that the singular values of $R^{-1}A$ are clustered around 1.

For Toeplitz matrices A generated by a complex-valued function, R. Chan and Yeung [29] have proved that if the generating function is 2π-periodic continuous with no zeros on $[-\pi, \pi]$, then the spectra of the iteration matrices $(c(A)^{-1}A)^*(c(A)^{-1}A)$ are clustered around 1. From that they showed that if the condition number $\kappa(A)$ of A is of $O(n^\alpha)$, $\alpha > 0$, then the number of iterations required for convergence is at most $O(\alpha \log n)$. Hence the total complexity for solving non-Hermitian type Toeplitz systems is of $O(n \log^2 n)$. When $\alpha = 0$, i.e., A is well-conditioned, the method converges in $O(1)$ steps and the complexity is reduced to $O(n \log n)$.

Numerical results in [29] shows that the requirements on f, namely that f has no zeros and $\kappa(A) = O(n^\alpha)$ are indispensable in order to get the said convergence rate. Moreover, these two conditions are mutually exclusive. For instance, if $f(\theta) = e^{i\theta}$, then f has no zeros on $[-\pi, \pi]$ but A is singular for all n. On the other hand, if $f(\theta) = 4\sin^2\theta$, then A is just the 1-dimensional discrete Laplacian with $\kappa(A) = O(n^2)$.

We remark that Ku and Kuo [59, 60, 61] have also considered solving non-symmetric Toeplitz matrix systems by the preconditioned conjugate gradient method. In their papers, A is assumed to be generated by complex-valued rational functions in the Wiener class, which happens to be a subclass of the class of 2π-periodic continuous functions considered in [29].

6.2.4 $\{\omega\}$-Circulant Preconditioners

Circulant matrices belong to the class of $\{\omega\}$-circulant matrices which are defined as follows:

Definition 2 *Let $\omega = e^{i\theta_0}$ with $\theta_0 \in [-\pi, \pi]$. An n-by-n matrix W is said to be an $\{\omega\}$-circulant matrix if it has the spectral decomposition*

$$W = \Omega^* F^* \Lambda F \Omega.$$

Here $\Omega = \text{diag}\,[1, \omega^{-1/n}, \ldots, \omega^{-(n-1)/n}]$ and Λ is a diagonal matrix containing the eigenvalues of W.

Notice that $\{\omega\}$-circulant matrices are Toeplitz matrices with the first entry of each row obtained by multiplying the last entry of the preceding row by ω. In particular, $\{1\}$-circulant matrices are circulant matrices, while $\{-1\}$-circulant matrices are skew-circulant matrices. Huckle [53] and R. Chan and Jin [16] have used skew-circulant matrices as preconditioners for Toeplitz matrices and proved that under the Wiener class assumptions, the spectra of these preconditioned matrices are clustered around 1. Performances of general $\{\omega\}$-circulant matrices as preconditioners for Toeplitz matrices are discussed in R. Chan and Ng [22] and Huckle [55].

6.2.5 General Remarks on Circulant Preconditioners

In this section, we have discussed many different kinds of circulant preconditioners. We note from Theorem 10 that most of them can be derived from the convolution approach. Moreover, the theorem changes the problem of finding a preconditioner to a problem in approximation theory. In particular, using results in approximation theory, the theorem can give us a guideline as to which preconditioner is better for a given generating function. For example, for 2π-periodic continuous functions that are not in the Wiener class, T. Chan's preconditioner is better than the Strang preconditioner. Also if we use a positive kernel to construct the preconditioner, then the preconditioner retains positive definiteness of the given Toeplitz matrix.

We emphasize that the assumptions on the generating functions f in the theorems in this section are to simplify the arguments. The main thing required in the proof is not an explicit form of f but a bound on the rate of decay of the diagonals $\{a_j\}_{j=0}^{\infty}$, see the definition of Wiener class functions and also the statement of Theorem 6. Thus, for the circulant preconditioning methods to work, there is no need to know the exact form of f, but just an estimate of the decay rate of a_j. Along this line, we remark further that Zygmund [87, p.183] has shown that if the diagonals a_j are convex, i.e., the second-order differences $a_{j+1} - 2a_j + a_{j-1} \geq 0$ for all j, then f is non-negative. Moreover, if one of the second-order differences is positive, then f is positive.

6.3 Some Remarks

In 1986, Strang addressed the question of whether iterative methods can compete with direct methods for solving symmetric positive definite Toeplitz systems. The answer has turned out to be an unqualified yes. The conjugate gradient method coupled with a circulant preconditioner can solve a large class of n-by-n Toeplitz systems in $O(n \log n)$ operations, as compared to the $O(n \log^2 n)$ operations required by fast direct Toeplitz solvers. To conclude this chapter, we list the following remarks on other aspects of using conjugate gradient methods for solving Toeplitz systems. For details, see the recent survey paper [23].

1. From (6.1.4), we see that circulant matrices are precisely those matrices that can be diagonalized by the discrete Fourier transform, a

transform which has a fast algorithm for its computations. However, there are other transforms (for instance, the Hartley transform and the sine and cosine transforms) with fast algorithms too, see [82]. It is therefore natural to consider using these fast transforms to construct new classes of preconditioners for solving Toeplitz systems. It has been proved that under the Wiener class assumption on the Toeplitz matrices, the method converges superlinearly.

2. When the minimum of f is zero, the condition number of A is not uniformly bounded and the Toeplitz matrix A is ill-conditioned. Tyrtyshnikov has proved theoretically [80] that Strang's and T. Chan's preconditioners will fail in this case. Instead of finding other possible circulant preconditioners, R. Chan [13] resorted to using band-Toeplitz matrices as preconditioners. The motivation behind using band-Toeplitz matrices is to approximate the generating function f by trigonometric polynomials of fixed degree rather than by convolution products of f with some kernels. The advantage here is that trigonometric polynomials can be chosen to match the zeros of f, so that the preconditioned method still works when f has zeros.

3. Recently, the use of multigrid methods for solving Toeplitz systems has also been shown to be a successful methodology, see [38, 39, 15].

4. There are many applications of optimal transform based preconditioners to Toeplitz-related systems arising from partial differential equations, queueing problems, signal and image processing, integral equations, and time series analysis. Part of the motivation of using optimal transform based preconditioners is to exploit their fast inversion via their transform matrices. Applications of the method to partial differential equations, queueing networks, integral equations, image restoration, and time series analysis are also given. The results show that the method in some instances works better than traditional methods used specifically for these problems.

6.4 Acknowledgment

We thank Mark Mak for generating Figure 3.

6.5 References

[1] S. Akl, *The Design and Analysis of Parallel Algorithms*, Prentice-Hall, Englewood Cliffs, NJ, 1989.

[2] G. Ammar and W. Gragg, *Superfast Solution of Real Positive Definite Toeplitz Systems*, SIAM J. Matrix Anal. Appl., 9 (1988), pp. 61–76.

[3] P. Anselone, *Collectively Compact Operator Approximation Theory and Applications to Integral Equations*, Prentice-Hall, Englewood Cliffs, NJ, 1971.

[4] O. Axelsson and V. Barker, *Finite Element Solution of Boundary Value Problems, Theory and Computation*, Academic Press, Orlando, FL, 1984.

[5] G. Baxter, *Polynomials Defined By a Difference System*, J. Math. Anal. Appl., 2 (1961), pp. 223–263.

[6] D. Bertsekas and J. Tsitsiklis, *Parallel and Distributed Computation: Numerical Methods*, Prentice-Hall, Englewood Cliffs, NJ, 1989.

[7] R. Bitmead and B. Anderson, *Asymptotically Fast Solution of Toeplitz and Related Systems of Linear Equations*, Linear Algebra Appl., 34 (1980), pp. 103–116.

[8] R. Brent, F. Gustavson, and D. Yun, *Fast Solution of Toeplitz Systems of Equations and Computation of Padé Approximants*, J. Algo., 1 (1980), pp. 259–295.

[9] A. Buckley, *On the Solution of Certain Skew Symmetric Linear Systems*, SIAM J. Numer. Anal., 14 (1977), pp. 566–570.

[10] J. Bunch, *Stability of Methods for Solving Toeplitz Systems of Equations*, SIAM J. Sci. Stat. Comput., 6 (1985), pp. 349–364.

[11] R. Chan, *Circulant Preconditioners for Hermitian Toeplitz Systems*, SIAM J. Matrix Anal. Appl., 10 (1989), pp. 542–550.

[12] R. Chan, *The Spectrum of a Family of Circulant Preconditioned Toeplitz Systems*, SIAM J. Numer. Anal., 26 (1989), pp. 503–506.

[13] R. Chan, *Toeplitz Preconditioners for Toeplitz Systems with Nonnegative Generating Functions*, IMA J. Numer. Anal., 11 (1991), pp. 333–345.

[14] R. Chan and T. Chan, *Circulant Preconditioners for Elliptic Problems*, Numer. Linear Algebra Appl., 1 (1992), pp. 77–101.

[15] R. Chan, Q. Chang and H. Sun, *Multigrid Method for Ill-Conditioned Symmetric Toeplitz Systems*, SIAM J. Sci. Comput., to appear.

[16] R. Chan and X. Jin, *Circulant and Skew-circulant Preconditioners for Skew-Hermitian Type Toeplitz Systems*, BIT, 31 (1991), pp. 632–646.

[17] R. Chan, X. Jin, and M. Yeung, *The Circulant Operator in the Banach Algebra of Matrices*, Linear Algebra Appl., 149 (1991), pp. 41–53

[18] R. Chan, X. Jin, and M. Yeung, *The Spectra of Super-Optimal Circulant Preconditioned Toeplitz Systems*, SIAM J. Numer. Anal., 28 (1991), pp. 871–879.

[19] R. Chan, J. Nagy, and R. Plemmons, *Circulant Preconditioned Toeplitz Least Squares Iterations*, SIAM J. Matrix Anal. Appl., 15 (1994), pp. 80–97.

[20] R. Chan, J. Nagy, and R. Plemmons, *FFT-Based Preconditioners for Toeplitz-Block Least Squares Problems*, SIAM J. Numer. Anal., 30 (1993), pp. 1740–1768.

[21] R. Chan, J. Nagy, and R. Plemmons, *Displacement Preconditioner for Toeplitz Least Squares Iterations*, Elec. Trans. Numer. Anal., 2 (1994), pp. 44–56.

[22] R. Chan and M. Ng, *Toeplitz Preconditioners for Hermitian Toeplitz Systems*, Linear Algebra Appl., 190 (1993), pp. 181–208.

[23] R. Chan and M. Ng, *Conjugate Gradient Methods for Toeplitz Systems*, SIAM Review, to appear.

[24] R. Chan and G. Strang, *Toeplitz Equations by Conjugate Gradients with Circulant Preconditioner*, SIAM J. Sci. Stat. Comput., 10 (1989), pp. 104–119.

[25] R. Chan and C. Wong, *Best-Conditioned Circulant Preconditioners*, Linear Algebra Appl., 218 (1995), pp. 205–212.

[26] R. Chan and M. Yeung, *Circulant Preconditioners for Toeplitz Matrices with Positive Continuous Generating Functions*, Math. Comp., 58 (1992), pp. 233–240.

[27] R. Chan and M. Yeung, *Circulant Preconditioners Constructed from Kernels*, SIAM J. Numer. Anal., 29 (1992), pp. 1093–1103.

[28] R. Chan and M. Yeung, *Jackson's Theorem and Circulant Preconditioned Toeplitz Systems*, J. Approx. Theory, 70 (1992), pp. 191–205.

[29] R. Chan and M. Yeung, *Circulant Preconditioners for Complex Toeplitz Matrices*, SIAM J. Numer. Anal., 30 (1993), pp. 1193–1207.

[30] T. Chan, *An Optimal Circulant Preconditioner for Toeplitz Systems*, SIAM J. Sci. Stat. Comput., 9 (1988), pp. 766–771.

[31] T. Chan and P. Hansen, *A Look-Ahead Levinson Algorithm for General Toeplitz Systems*, IEEE Trans. Signal Process., 40 (1992), pp. 1079–1090.

[32] T. Chan and J. Olkin, *Circulant Preconditioners for Toeplitz-block Matrices*, Numer. Algo., 6 (1994), pp. 89–101.

[33] E. Cheney, *Introduction to Approximation Theory*, McGraw-Hill, New York, 1966.

[34] C. Chui and A. Chan, *Application of Approximation Theory Methods to Recursive Digital Filter Design*, IEEE Trans. Acoust., Speech, Signal Process., 30 (1982), pp. 18–24.

[35] C. Chui and G. Chen, *Signal Processing and Systems Theory*, Selected Topics, Springer-Verlag, Berlin, 1992.

[36] J. Cooley and J. Tukey, *An Algorithm for the Machine Calculation of Complex Fourier Series*, Math. Comp., 19 (1965), pp. 297–301.

[37] P. Davis, *Circulant Matrices*, John Wiley & Sons, New York, 1979.

[38] G. Fiorentino and S. Serra, *Multigrid Methods for Toeplitz Matrices*, Calcolo, 28 (1991), pp. 283–305.

[39] G. Fiorentino and S. Serra, *Multigrid Methods for Symmetric Positive Definite Block Toeplitz Matrices with Nonnegative Generating Functions*, SIAM J. Sci. Comput., to appear.

[40] R. Freund, *A Transpose-free Quasi-minimal Residual Algorithm for Non-Hermitian Linear Systems*, SIAM J. Sci. Comput., 14 (1993), pp. 470–482.

[41] R. Freund, *A Look-ahead Bareiss Algorithm for General Toeplitz Matrices*, Numer. Math., 68 (1994), pp. 35–69.

[42] R. Freund, G. Golub, and N. Nachtigal, *Iterative Solution of Linear Systems*, Acta Numerica, 1 (1992), pp. 57–100.

[43] R. Freund and H. Zha, *Formally Biorthogonal Polynomials and a Look-ahead Levinson Algorithm for General Toeplitz Systems*, Linear Algebra Appl., 188,189 (1993), pp. 255–303.

[44] I. Gohberg and I. Fel'dman, *Convolution Equations and Projection Methods for Their Solution*, Transl. Math. Monographs, Vol. 41, Amer. Math. Soc., Providence, RI, 1974.

[45] G. Golub and C. Van Loan, *Matrix Computations*, 2nd ed., The Johns Hopkins University Press, Baltimore, MD, 1989.

[46] U. Grenander and G. Szegö, *Toeplitz Forms and Their Applications*, 2nd ed., Chelsea Publishing, New York, 1984.

[47] M. Gutknecht, *Stable Row Recurrences for the Padé Table and Generically Superfast Look-ahead Solvers for Non-Hermitian Toeplitz Systems*, Linear Algebra Appl., 188,189 (1993), pp. 351–421.

[48] R. Hamming, *Digital Filters*, 3rd ed., Prentice Hall, Englewood Cliffs, NJ, 1989.

[49] S. Holmgren and K. Otto, *A Comparison of Preconditioned Iterative Methods for Nonsymmetric Block-tridiagonal Systems of Equations*, Report No. 123 (revised), Dept. of Scientific Computing, Uppsala Univ., Uppsala, Sweden, 1990.

[50] S. Holmgren and K. Otto, *Iterative Solution Methods and Preconditioners for Block-tridiagonal Systems of Equations*, SIAM J. Matrix Anal. Appl., 13 (1992), pp. 863–886.

[51] F. de Hoog, *A New Algorithm for Solving Toeplitz Systems of Equations*, Linear Algebra Appl., 88/89 (1987), pp. 123–138.

[52] T. Huckle, *Circulant/Skew Circulant Matrices as Preconditioners for Hermitian Toeplitz Systems*, in Proc. IMACS Conf. on Iterative Methods in Linear Algebra, Brussels, 1991.

[53] T. Huckle, *Circulant and Skew Circulant Matrices for Solving Toeplitz Matrix Problems*, SIAM J. Matrix Anal. Appl., 13 (1992), pp. 767–777.

[54] T. Huckle, *Some Aspects of Circulant Preconditioners*, SIAM J. Sci. Comput., 14 (1993), pp. 531–541.

[55] T. Huckle, *Iterative Methods for Toeplitz-like Matrices*, Report SCCM-94-05, Computer Science Dept., Stanford Univ., Stanford, CA, 1994.

[56] T. Huckle, *Fast Transforms for Tridiagonal Linear Equations*, BIT, 34 (1994), pp. 99–112.

[57] A. Jain, *Fundamentals of Digital Image Processing*, Prentice-Hall, Englewood Cliffs, NJ, 1989.

[58] T. Ku and C. Kuo, *Design and Analysis of Toeplitz Preconditioners*, IEEE Trans. Signal Process., 40 (1992), pp. 129–141.

[59] T. Ku and C. Kuo, *Spectral Properties of Preconditioned Rational Toeplitz Matrices*, SIAM J. Matrix Anal. Appl., 14 (1993), pp. 146–165.

[60] T. Ku and C. Kuo, *Spectral Properties of Preconditioned Rational Toeplitz Matrices: the Nonsymmetric Case*, SIAM J. Matrix Anal. Appl., 14 (1993), pp. 521–544.

[61] T. Ku and C. Kuo, *A Minimum-phase LU Factorization Preconditioner for Toeplitz Matrices*, SIAM J. Sci. Stat. Comput., 13 (1992), pp. 1470-1487.

[62] N. Levinson. *The Wiener RMS (Root Mean Square) Error Criterion in Filter Design and Prediction*, J. Math. and Phys., 25 (1946), pp. 261–278.

[63] E. Linzer, *Arithmetic Complexity of Iterative Toeplitz Solvers*, manuscript, 1993.

[64] E. Linzer, *Extended Circulant Conditioning of Toeplitz Systems*, Tech. Report R.C, IBM Research, Yorktown Heights, NY, October 1992.

[65] D. Luenberger, *Linear and Nonlinear Programming*, 2nd ed., Addison & Wesley, Reading, MA, 1984.

[66] M. Morf, *Doubling Algorithms for Toeplitz and Related Equations*, in Proc. IEEE Intl. Conf. on Acoust., Speech, and Signal Process., 3 (1980), pp. 954–959.

[67] M. Ng and R. Plemmons, *Fast RLS Adaptive Filtering by FFT-based Conjugate Gradient Iterations*, SIAM J. Sci. Comput., (1996), to appear.

[68] J. Olkin, *Linear and Nonlinear Deconvolution Problems*, Ph.D. thesis, Rice Univ., Houston, TX, 1986.

[69] M. Priestley, *Spectral Analysis and Time Series*, Academic Press, London, 1981.

[70] Y. Saad and M. Schultz, *GMRES: A Generalized Minimal Residual Algorithm for Solving Nonsymmetric Linear Systems*, SIAM J. Sci. Stat. Comput., 7 (1986), pp. 856–869.

[71] G. Strang, *A Proposal for Toeplitz Matrix Calculations*, Stud. Appl. Math., 74 (1986), pp. 171–176.

[72] G. Strang and A. Edelman, *The Toeplitz-Circulant Eigenvalue Problem $Ax = \lambda Cx$*, in Oakland Conf. on PDE and Appl. Math., L. Bragg and J. Dettman, eds., Longman Sci. Tech., New York, 1987.

[73] P. Swarztrauber, *Multiprocessor FFTs*, Parallel Comput., 5 (1987), pp. 197–210.

[74] M. Tismenetsky, *A Decomposition of Toeplitz Matrices and Optimal Circulant Preconditioning*, Linear Algebra Appl., 154–156 (1991), pp. 105 121.

[75] O. Toeplitz, *Zur Theorie der quadratischen und bilinearen Formen von unendlichvielen Veränderlichen. I. Teil: Theorie der L-Formen.*, Math. Annal., 70 (1911), pp. 351–376.

[76] N. Trefethen, *Approximation Theory and Numerical Linear Algebra*, in Algorithms for Approximation II, J. Mason and M. Cox, eds., Chapman and Hall, London, 1990, pp. 336–360.

[77] W. Trench, *An Algorithm for the Inversion of Finite Toeplitz Matrices*, SIAM J. Appl. Math., 12 (1964), pp. 515–522.

[78] E. Tyrtyshnikov, *Optimal and Superoptimal Circulant Preconditioners*, SIAM J. Matrix Anal. Appl., 13 (1992), pp. 459–473.

[79] E. Tyrtyshnikov, *A Unifying Approach to Some Old and New Theorems on Distribution and Clustering*, Linear Algebra. Appl., 232 (1996), pp. 1–43.

[80] E. Tyrtyshnikov, *Circulant Preconditioners with Unbounded Inverses*, Linear Algebra Appl., 216 (1995), pp. 1–24.

[81] E. Tyrtyshnikov, *Influence of Matrix Operations on the Distribution of Eigenvalues and Singular Values of Toeplitz Matrices*, Linear Algebra Appl., 207 (1994), pp. 225–249.

[82] C. Van Loan, *Computational Frameworks for the Fast Fourier Transform*, SIAM, Philadelphia, PA, 1992.

[83] J. Walker, *Fourier Analysis*, Oxford University Press, New York, 1988.

[84] H. Widom, *Hankel Matrices*, Trans. Amer. Math. Soc., 121 (1966), pp. 1–35.

[85] M. Yeung and R. Chan, *Circulant Preconditioners for Toeplitz Matrices with Piecewise Continuous Generating Functions*, Math. Comp., 61 (1993), pp. 701–718.

[86] S. Zohar, *The Solution of a Toeplitz Set of Linear Equations*, J. Assoc. Comput. Mach., 21 (1974), pp. 272–276.

[87] A. Zygmund, *Trigonometric Series*, Cambridge University Press, London, 1968.

7

Iterative Methods for Problems in Computational Fluid Dynamics

Howard C. Elman*
David J. Silvester†
Andrew J. Wathen‡

We discuss iterative methods for solving the algebraic systems of equations arising from linearization and discretization of primitive variable formulations of the incompressible Navier-Stokes equations. Implicit discretization in time leads to a coupled but linear system of partial differential equations at each time step, and discretization in space then produces a series of linear algebraic systems. We give an overview of commonly used time and space discretization techniques, and we discuss a variety of algorithmic strategies for solving the resulting systems of equations. The emphasis is on preconditioning techniques, which can be combined with Krylov subspace iterative methods. In many cases the solution of subsidiary problems such as the discrete convection-diffusion equation and the discrete Stokes equations plays a crucial role. We examine iterative techniques for these problems and show how they can be integrated into effective solution algorithms for the Navier-Stokes equations.

7.1 Introduction

Our objective is to compute solutions of incompressible flow problems modelled by the Navier-Stokes equations in a flow domain $\Omega \subset \mathbb{R}^d$ ($d = 2$ or

---

*Department of Computer Science and Institute for Advanced Computer Studies, University of Maryland, College Park, MD 20742, USA, e-mail: elman@cs.umd.edu. This work was supported by the U. S. National Science Foundation under grants ASC-8958544 and DMS-9423133.

†Department of Mathematics, University of Manchester Institute of Science and Technology, Manchester M601QD, UK, email: djs@lanczos.ma.umist.ac.uk

‡Oxford University Computing Laboratory, Wolfson Building, Parks Road, Oxford OX13QD, UK, email: Andy.Wathen@comlab.ox.ac.uk

3) with a piecewise smooth boundary $\partial\Omega$:

$$\frac{\partial \mathbf{u}}{\partial t} + \mathbf{u}\cdot\nabla\mathbf{u} - \nu\nabla^2\mathbf{u} + \nabla p = 0 \qquad \text{in } \mathcal{W} \equiv \Omega \times (0,T) \tag{7.1.1}$$

$$\nabla\cdot\mathbf{u} = 0 \qquad \text{in } \mathcal{W}. \tag{7.1.2}$$

together with boundary and initial conditions of the form

$$\mathbf{u}(\mathbf{x},t) = \mathbf{g}(\mathbf{x},t) \qquad \text{on } \overline{\mathcal{W}} \equiv \partial\Omega \times [0,T]; \tag{7.1.3}$$

$$\mathbf{u}(\mathbf{x},0) = \mathbf{u}_0(\mathbf{x}) \qquad \text{in } \Omega. \tag{7.1.4}$$

Our notation is standard: $\mathbf{u}$ is the fluid velocity, p is the pressure, $\nu > 0$ is a specified viscosity parameter (in a non-dimensional setting it is the inverse of the Reynolds number), and $T > 0$ is some final time. The initial velocity field $\mathbf{u}_0$ will be assumed to satisfy the incompressibility constraint, that is, $\nabla\cdot\mathbf{u}_0 = 0$. The boundary velocity field satisfies $\int_{\partial\Omega}\mathbf{g}\cdot\mathbf{n}\,ds = 0$ for all time t, where $\mathbf{n}$ is the unit vector normal to $\partial\Omega$. We also assume that the pressure solution is uniquely specified e.g. by insisting that its mean value is zero.

If $\mathbf{g}$ is independent of t then the usual objective is simply to compute steady-state solutions of (7.1.1)–(7.1.2). In other cases however, time-accuracy is important and the requirements of the time discretisation will be more demanding; specifically, an accurate and unconditionally stable time-discretisation is necessary to adaptively change the timestep to reflect the dynamics of the underlying flow. Two classes of time discretisation scheme are described below, operator splitting methods and linearised implicit methods.

7.1.1 Operator splitting Methods

One attractive approach ensuring stability and high accuracy is to decouple the convection and incompressibility operators using an "alternating-direction" splitting; see Glowinski & Dean [5]. Assuming uniform timesteps $\Delta\mathrm{t} = T/n$ for ease of exposition, the simplest two-stage (Peaceman-Rachford) scheme [39] is given below.

Algorithm 7.1.1 *Given* $\mathbf{u}^0$, $\theta \in [0,1]$, $\alpha \in (0,1)$, $\beta \in (0,1)$, *find* $\mathbf{u}^1$,$\mathbf{u}^2$, …, $\mathbf{u}^n$ *via*

$$\begin{aligned}\frac{\mathbf{u}^{n+\theta}-\mathbf{u}^n}{\theta\Delta\mathrm{t}} - \alpha\nu\nabla^2\mathbf{u}^{n+\theta} + \mathbf{u}^*\cdot\nabla\mathbf{u}^{n+\theta} &= \beta\nu\nabla^2\mathbf{u}^n - \nabla p^n \quad \text{in } \Omega,\\ \mathbf{u}^{n+\theta} &= \mathbf{g}^{n+\theta} \quad \text{on } \partial\Omega.\end{aligned} \tag{7.1.5}$$

$$\begin{aligned}\frac{\mathbf{u}^{n+1}-\mathbf{u}^{n+\theta}}{(1-\theta)\Delta\mathrm{t}} - \beta\nu\nabla^2\mathbf{u}^{n+1} + \nabla p^{n+1} &= \alpha\nu\nabla^2\mathbf{u}^{n+\theta} - \mathbf{u}^*\cdot\nabla\mathbf{u}^{n+\theta}\\ \nabla\cdot\mathbf{u}^{n+1} &= 0 \quad \text{in } \Omega,\\ \mathbf{u}^{n+1} &= \mathbf{g}^{n+1} \quad \text{on } \partial\Omega.\end{aligned} \tag{7.1.6}$$

In practice the choice of parameters is restricted; the splitting of the diffusive terms must be done consistently i.e. $\alpha + \beta = 1$, and the "frozen" velocity in the convective term must be divergence free i.e. $\nabla \cdot \mathbf{u}^* = 0$, otherwise the skew symmetry of the convective term will not be preserved. Another important consideration with regard to the choice of $\mathbf{u}^*$ is the linearity, or otherwise, of the equation systems that must be solved at each time level. In particular, the natural choice of $\mathbf{u}^* = \mathbf{u}^n$ gives a linear method but in this case the accuracy is only first order. Second order accuracy can be achieved by setting $\mathbf{u}^* = \mathbf{u}^{n+\theta}$, but in this case a nonlinear convection-diffusion problem (7.1.5) must be solved at every time level, in addition to the generalised Stokes problem (7.1.6).

The Peaceman-Rachford splitting method has one drawback (see [5] and [46]), namely, that it is not asymptotically stable when applied to the standard model problem with an exponentially decaying solution. Thus we can expect any implementation of Algorithm 7.1.1 to perform poorly if the time-step is not small enough when the underlying flow exhibits fast transient behaviour. In addition, the methodology is not well suited to computing steady-state flow solutions by "pseudo-timestepping" with large timesteps. Motivated by these observations, Glowinski [5] proposed a three-stage variant of the Peaceman-Rachford scheme which has all the good features of the original method whilst retaining stability in the asymptotic limit $\Delta \mathrm{t} \to \infty$. The resulting method is commonly referred to as "Le θ-scheme".

Algorithm 7.1.2 *Given* $\mathbf{u}^0$, $\theta \in (0, 1/2)$, $\alpha \in (0,1)$, $\beta \in (0,1)$, *find* $\mathbf{u}^1$, $\mathbf{u}^2$, ..., $\mathbf{u}^n$ *via*

$$\begin{aligned} \frac{\mathbf{u}^{n+\theta} - \mathbf{u}^n}{\theta \Delta \mathrm{t}} - \alpha\nu\nabla^2 \mathbf{u}^{n+\theta} + \nabla p^{n+\theta} &= \beta\nu\nabla^2\mathbf{u}^n - \mathbf{u}^n \cdot \nabla \mathbf{u}^n \\ \nabla \cdot \mathbf{u}^{n+\theta} &= 0 \qquad \text{in } \Omega, \\ \mathbf{u}^{n+\theta} &= \mathbf{g}^{n+\theta} \quad \text{on } \partial\Omega. \end{aligned} \tag{7.1.7}$$

$$\begin{aligned} \frac{\mathbf{u}^{n+1-\theta} - \mathbf{u}^{n+\theta}}{(1-2\theta)\Delta \mathrm{t}} - \beta\nu\nabla^2 \mathbf{u}^{n+1-\theta} + \ldots & \\ \mathbf{u}^* \cdot \nabla \mathbf{u}^{n+1-\theta} = \alpha\nu\nabla^2\mathbf{u}^{n+\theta} - \nabla p^{n+\theta} & \quad \text{in } \Omega, \\ \mathbf{u}^{n+1-\theta} = \mathbf{g}^{n+1-\theta} \quad \text{on } \partial\Omega. & \end{aligned} \tag{7.1.8}$$

$$\begin{aligned} \frac{\mathbf{u}^{n+1} - \mathbf{u}^{n+1-\theta}}{\theta \Delta \mathrm{t}} - \alpha\nu\nabla^2 \mathbf{u}^{n+1} + \ldots & \\ \nabla p^{n+1} = \beta\nu\nabla^2\mathbf{u}^{n+1-\theta} - \mathbf{u}^* \cdot \nabla \mathbf{u}^{n+1-\theta} & \\ \nabla \cdot \mathbf{u}^{n+1} = 0 \qquad \text{in } \Omega, & \\ \mathbf{u}^{n+1} = \mathbf{g}^{n+1} \quad \text{on } \partial\Omega. & \end{aligned} \tag{7.1.9}$$

The nonlinear scheme $\mathbf{u}^* = \mathbf{u}^{n+1-\theta}$ is considered in [5], and requires the solution of two generalised Stokes problems (7.1.7),(7.1.9), and one nonlinear convection-diffusion equation (7.1.8) at each time step. In this case, either choosing $\alpha{=}\beta = 1/2$ or else setting $\theta = 1-1/\sqrt{2}$ with $\alpha+\beta = 1$ gives second order accuracy as $\Delta t \to 0$. In particular, setting $\theta = 1 - 1/\sqrt{2}$, $\alpha = (1-2\theta)/(1-\theta)$ and $\beta = \theta/(1-\theta)$ gives a method which is second order accurate in time, unconditionally stable, has good asymptotic properties and has commonality between the coefficient matrices at the various substages, see [5]. The unconditional stability of the scheme in an incompressible Navier-Stokes setting was established by Klouček & Rys [31].

In [46] a linear θ-scheme is developed which retains second order accuracy (setting $\mathbf{u}^* = \mathbf{u}^{n+\theta}$ in Algorithm 7.1.2 reduces the accuracy to first order). The key here is to use an appropriate combination of the two convection matrices when freezing the velocity in Algorithm 7.1.2, viz:

$$\mathbf{u}^* = \frac{2\theta - 1}{\theta}\mathbf{u}^n + \frac{1-\theta}{\theta}\mathbf{u}^{n+\theta}. \tag{7.1.10}$$

In this case a linear convection-diffusion problem (7.1.8) must be solved at each time level. Furthermore, the results in [46] show that the accuracy of the resulting method is not compromised. In contrast, making the choice $\mathbf{u}^* = \mathbf{u}^{n+\theta}$ badly impinges on accuracy as $\Delta t \to 0$.

Summarising the discussion of splitting methods; both the two-stage Algorithm 7.1.1 in the case of small Δt, and the three-stage Algorithm 7.1.2 in general, give accurate time discretisation of the Navier-Stokes problem (7.1.1)–(7.1.4). In sections 3 and 4 that follow we describe how the component Stokes and (linear) convection-diffusion sub-problems may be solved efficiently using contemporary preconditioned Krylov subspace iteration methods. The spatial discretisations of the convection-diffusion and Stokes subproblems which arise above are discussed in section 2.

If ν is small then an efficient alternative to operator-splitting is to use the "characteristics" of the associated hyperbolic problem (looking backwards in time to ensure stability), see Douglas & Russell [6]. Using this approach a single Stokes problem of the form (7.1.6) must be solved at every time-level so the discussion in section 4 is also relevant to this class of methods.

7.1.2 Linearised Implicit Methods

The simplest time-stepping approach for the Navier-Stokes equations is a simple one-stage finite difference discretisation. A generic (and unconditionally stable) algorithm (cf. Algorithms 7.1.1 and 7.1.2) is given below.

Algorithm 7.1.3 *Given* $\mathbf{u}^0$, $\theta \in [1/2, 1]$, *find* $\mathbf{u}^1$, $\mathbf{u}^2$, $\ldots$, $\mathbf{u}^n$ *via*

$$\begin{aligned} \frac{(\mathbf{u}^{n+1} - \mathbf{u}^n)}{\Delta t} + \mathbf{u}^* \cdot \nabla \mathbf{u}^{n+\theta} &- \nu\nabla^2\mathbf{u}^{n+\theta} + \nabla p^{n+\theta} = 0 \\ \nabla \cdot \mathbf{u}^{n+\theta} &= 0 \qquad \text{in } \Omega, \\ \mathbf{u}^{n+\theta} &= \mathbf{g}^{n+\theta} \quad \text{on } \partial\Omega. \end{aligned} \tag{7.1.11}$$

Here $\mathbf{u}^{n+\theta} = \theta\mathbf{u}^{n+1} + (1-\theta)\mathbf{u}^n$ and $p^{n+\theta} = \theta p^{n+1} + (1-\theta)p^n$. Note that p^0 is required if $\theta \neq 1$ so the Algorithm 7.1.3 is not self-starting in general. In this case an approximation to p^0 must be computed explicitly by manipulation of the continuum problem, or alternatively it must be approximated by taking one (very small) step of a self-starting algorithm (e.g. with $\theta = 1$ above).

Algorithm 7.1.3 contains the well known nonlinear schemes of backward Euler and Crank-Nicolson. These methods are given by ($\mathbf{u}^{n+\theta} = \mathbf{u}^{n+1}$, $\mathbf{u}^* = \mathbf{u}^{n+1}$), ($\mathbf{u}^{n+\theta} = \mathbf{u}^{n+\frac{1}{2}}$, $\mathbf{u}^* = \mathbf{u}^{n+\frac{1}{2}}$), and are first and second order accurate respectively. In either case, a nonlinear problem must be solved at every time-level. As a result neither of these methods is to be recommended if time-accuracy is needed. A well known linearisation strategy is to set $\mathbf{u}^* = \mathbf{u}^n$ above. This does not affect the stability properties of the time-discretisation, but it does reduce the Crank-Nicolson accuracy to first order as $\Delta t \to 0$ (the first order accuracy of backward Euler is unchanged). To retain second order accuracy in a linear scheme the Simo-Armero scheme [45] given by setting $\mathbf{u}^{n+\alpha} = \mathbf{u}^{n+\frac{1}{2}}$ with $\mathbf{u}^* = (3\mathbf{u}^n - \mathbf{u}^{n-1})/2$ in Algorithm 7.1.3 is recommended.

Using linearised backward Euler (or the Simo-Armero scheme) a frozen-coefficient Navier-Stokes problem (or *Oseen* problem) arises at each discrete time step. In contrast to the operator splitting case, the Oseen methodology is primarily of interest when solving steady-state problems—the linearised backward Euler method is uniquely well suited to pseudo-timestepping since it inherits the long term asymptotic dissipative behaviour of (7.1.1)–(7.1.2), see [45] for details. Alternatively, attacking the steady state version of (7.1.1)–(7.1.2) directly introduces a (steady-state) Oseen system at every iterative level. In section 5 we consider techniques for solving such Oseen problems using preconditioned Krylov subspace methods.

7.2 Spatial discretisation

In this section, the spatial discretisation of the sub-problems arising from the operator splitting methods in section 1.1 are discussed. For simplicity, we only consider the *steady-state* limit of the linearised convection-diffusion and Stokes sub-problems here; for example, as would arise from setting $\Delta t \to \infty$ in (7.1.5) and (7.1.6) respectively.

7.2.1 *The linearised convection-diffusion problem*

The problem addressed here is the following: Given some convective velocity field (or "wind") $\mathbf{w} \in \mathbb{R}^d$ such that $\nabla \cdot \mathbf{w} = 0$, find a scalar variable u (the transported quantity) satisfying

$$-\nu\nabla^2 u + \mathbf{w} \cdot \nabla u = f \qquad \text{in } \Omega, \tag{7.2.1}$$

with a boundary condition $u(\mathbf{x}) = g(\mathbf{x})$ on $\partial\Omega$. In practice, for example when solving (7.1.5), the "wind" is not actually pointwise divergence-free. Our discussion is still relevant in such cases—our starting point is then an equivalent formulation of the momentum conservation equations (7.1.1), with the convection term expressed in *skew-symmetric form*, see [45] for details.

Simple finite difference methods are often appropriate when spatially discretising the model problem (7.2.1)), especially if the geometry is straightforward and "fast solution" is the goal. Alternatively, if the flow domain is irregular or if adaptive refinement via a posteriori error control is to be included, then finite element spatial approximation is best. The theory underlying finite element approximation of (7.2.1) is summarised for completeness below. For further details, see for example, Quarteroni & Valli [40].

The weak formulation of (7.2.1) is defined in terms of the Sobolev space $H_0^1(\Omega)$ (the set of functions with derivatives in $L^2(\Omega)$ and which are zero on $\partial\Omega$). Defining the space $X \equiv H_0^1(\Omega)$, it is easy to see that the solution u satisfies

$$a(u,v) = (f,v) \qquad \forall v \in X, \tag{7.2.2}$$

where $a(\cdot,\cdot)$ is the bilinear form $a(u,v) = \nu(\nabla u, \nabla v) + (\mathbf{w}\cdot\nabla u, v)$, and $(\cdot,\cdot)$ denotes the usual scalar $L^2(\Omega)$ inner product.

Since Ω is bounded and $\mathbf{w}$ is divergence-free, the bilinear form $a(\cdot,\cdot)$ is coercive and bounded over X

$$a(u,u) = \quad \nu\,\|\nabla u\|^2 \qquad \forall u \in X, \tag{7.2.3}$$

$$|a(u,v)| \leq \quad C_{\mathbf{w}}\,\|\nabla u\|\,\|\nabla v\| \qquad \forall u \in X, \forall v \in X, \tag{7.2.4}$$

and the continuity constant $C_{\mathbf{w}}$ is given by

$$C_{\mathbf{w}} = \nu + C_\Omega \|\mathbf{w}\|_{L^\infty(\Omega)}$$

where C_Ω is the Poincaré constant associated with Ω. Existence and uniqueness of the solution to (7.2.1) then follows from the Lax-Milgram lemma.

To generate a discrete system we take a finite dimensional subspace $X_h \subset X$, where h is a representative mesh parameter, and enforce (7.2.2) over X_h. Specifically, we look for a function u_h such that $u_h = g_h$ on $\partial\Omega$, which solves

$$a(u_h,v) = (f_h,v) \qquad \forall v \in X_h, \tag{7.2.5}$$

where f_h is the $L^2(\Omega)$ orthogonal projection of f into X_h, and g_h is typically the interpolant of of the boundary data g.

Since we are using a conforming approximation, u_h is also uniquely defined, and if $g = 0$, (7.2.3) and (7.2.4) imply the following a priori error estimate

$$\|\nabla(u-u_h)\| \leq \frac{C_{\mathbf{w}}}{\nu} \inf_{v\in X_h} \|\nabla(u-v)\|. \tag{7.2.6}$$

Although the finite element approximation in (7.2.6) is of optimal order as $h \to 0$, the stability clearly depends on the ratio $C_{\mathbf{w}}/\nu$. In general, oscillatory solutions are observed if the characteristic "mesh Peclet number" is large, i.e.

$$P_e \equiv \frac{h\|\mathbf{w}\|}{2\nu} > 1,$$

for example, if there are any boundary layers which are not resolved by the mesh. In general, when convection dominates, the discrete solution "inherits" instability from the associated solution of (7.2.2).

An alternative to adaptive mesh refinement is to "ignore" physical boundary layers, and to stabilise the discrete problem; e.g. using some form of *upwinded* discretisation. In a finite element setting this is conveniently achieved using a Petrov-Galerkin framework [27, 28] with a "shifted" (non-conforming) test space, say,

$$a(u_h, v + \delta\mathbf{w}\cdot\nabla v) = (f_h, v + \delta\mathbf{w}\cdot\nabla v) \qquad \forall v \in X_h, \tag{7.2.7}$$

where δ is an appropriately chosen stabilisation/upwinding parameter, see below. Taking a standard element-wise evaluation of the non-conforming term, and using a linear P_1 (or Q_1) approximation space, the formulation (7.2.7) simplifies to the so-called *streamline-diffusion* method

$$b(u_h, v) \equiv a(u_h, v) + (\mathbf{w}\cdot\nabla u_h, \delta\mathbf{w}\cdot\nabla v) = (f_h, v + \delta\mathbf{w}\cdot\nabla v) \qquad \forall v \in X_h. \tag{7.2.8}$$

This formulation clearly has better stability properties than the original since there is additional coercivity in the local flow direction,

$$b(u, u) = \nu\,\|\nabla u\|^2 + \delta\|\mathbf{w}\cdot\nabla u\|^2 \quad \forall u \in X_h. \tag{7.2.9}$$

Another appealing feature of the stabilised formulation (7.2.7) is that it is *consistent*—the exact solution of the differential equation (7.2.1) satisfies (7.2.7). This means that high order approximations (P_k or Q_k for $k \geq 2$) can be used without compromising accuracy.

Returning now to the choice of δ, it is possible to show that the solution of (7.2.7) satisfies the "best possible" error estimate (for any degree of polynomial approximation), under the assumption that δ in (7.2.7) is of the form

$$\delta = \frac{\alpha h}{\|\mathbf{w}\|} \quad \text{for all } P_e > 1. \tag{7.2.10}$$

Here $\alpha > 0$ is a "tuning parameter", and h is the usual representative mesh parameter. For a more complete discussion, and a review of the error analysis of the streamline diffusion method, see [28]. Note that if the discretised problem is diffusion-dominated (i.e. $P_e \leq 1$) then the corresponding "best" choice above is $\delta = 0$, in which case (7.2.7) reduces to the standard Galerkin formulation (7.2.5).

In practice, determining an appropriate choice of α in (7.2.10) is crucial. (This issue will also arise when we consider stabilised Stokes formulations below). There are two aspects to consider here: firstly, it very easy to over-stabilise giving smooth but inaccurate solutions, secondly, the performance of iterative solvers applied to (7.2.8) will clearly be influenced by the choice of parameter. This second aspect will be an issue in section 3, where some experiments are presented for a model problem with constant "wind" $\mathbf{w}$, solving (7.2.5) and (7.2.7), using uniform grids of bilinear finite elements. (The associated software is freely available, see section 6 for details.) If the wind is constant, then an optimal value is known (from Fourier analysis) $\alpha^* = 1/2(1 - 1/P_e)$, which minimises the contraction rate of iterative solvers applied to (7.2.8), see [17] for further details. Note that $\alpha^* \to 0$ as $P_e \to 1$ so that "stabilising" the standard method is likely to adversely affect the convergence of iterative solvers if the discrete problem is diffusion dominated.

7.2.2 The Stokes problem

Here we consider the following problem: find the velocity vector $\mathbf{u} \in \mathbb{R}^d$ and the scalar p (the "pressure") satisfying

$$-\nabla^2\mathbf{u} + \nabla p = f \quad \text{in } \Omega \qquad (7.2.11)$$

$$\nabla \cdot \mathbf{u} = 0 \quad \text{in } \Omega, \qquad (7.2.12)$$

with specified velocity boundary conditions

$$\mathbf{u}(\mathbf{x}) = \mathbf{g}(\mathbf{x}) \qquad \text{on } \partial\Omega. \qquad (7.2.13)$$

Note that in (7.2.11)–(7.2.12) the viscosity coefficient ν has been incorporated into the definition of the forcing function and the pressure.

The theory underlying the solution of (7.2.11)–(7.2.13) using finite element methods is outlined below. For full details see Girault & Raviart [21]. The weak formulation of (7.2.11)–(7.2.12) is defined in terms of the Sobolev spaces $H_0^1(\Omega)$ and $L_0^2(\Omega)$ (the set of functions in $L^2(\Omega)$ with zero mean value on Ω). Defining a velocity space $\mathbf{X} \equiv (H_0^1(\Omega))^d$ and a pressure space $M \equiv L_0^2(\Omega)$, it is easy to see that the solution $(\mathbf{u}, p)$ of (7.2.11)–(7.2.12) satisfies

$$(\nabla\mathbf{u}, \nabla\mathbf{v}) - (p, \nabla\cdot\mathbf{v}) = (\mathbf{f}, \mathbf{v}) \qquad \forall \mathbf{v} \in \mathbf{X} \qquad (7.2.14)$$

$$(\nabla\cdot\mathbf{u}, q) = 0 \qquad \forall q \in M, \qquad (7.2.15)$$

where $(\cdot,\cdot)$ denotes the usual vector or scalar $L^2(\Omega)$ inner product. Since Ω is bounded and connected there exists a constant κ satisfying the continuous *inf-sup* condition:

$$\sup_{\mathbf{w}\in\mathbf{X}} \frac{(p, \nabla\cdot\mathbf{w})}{\|\mathbf{w}\|_{\mathbf{X}}} \geq \kappa\|p\|_M \quad \forall p \in M. \qquad (7.2.16)$$

Existence and uniqueness of solution follows, see [21].

To generate a discrete system we take finite dimensional subspaces $\mathbf{X}_h \subset \mathbf{X}$ and $M_h \subset L^2(\Omega)$, where h is a representative mesh parameter, and enforce (7.2.14)–(7.2.15) over the discrete subspaces (again specifying that functions in M_h have zero mean to ensure uniqueness). Specifically, we look for functions $\mathbf{u}_h$ and p_h such that

$$(\nabla \mathbf{u}_h, \nabla \mathbf{v}) - (p_h, \nabla \cdot \mathbf{v}) = (\mathbf{f}_h, \mathbf{v}) \quad \forall \mathbf{v} \in \mathbf{X}_h \tag{7.2.17}$$

$$(\nabla \cdot \mathbf{u}_h, q) = 0 \quad \forall q \in M_h. \tag{7.2.18}$$

Here, $\mathbf{f}_h$ is the $(L^2(\Omega))^d$ orthogonal projection of $\mathbf{f}$ into $\mathbf{X}_h$.

The well-posedness of (7.2.17)–(7.2.18) is not automatic since we do not have an internal approximation (i.e. functions satisfying (7.2.18) do not necessarily satisfy (7.2.15)). A sufficient condition for the existence and uniqueness of the solution to (7.2.17)–(7.2.18) is that the following *discrete inf-sup* condition is satisfied: there exists a constant γ independent of h such that

$$\sup_{\mathbf{w} \in \mathbf{X}_h} \frac{(p, \nabla \cdot \mathbf{w})}{\|\nabla \mathbf{w}\|} \geq \gamma \|p\| \quad \forall p \in M_h. \tag{7.2.19}$$

Note that the semi-norm $\|\nabla \mathbf{w}\|$ in (7.2.19) is equivalent to the norm $\|\mathbf{w}\|_{\mathbf{X}}$ used in (7.2.16) for functions $\mathbf{w} \in \mathbf{X}$. In the case $\mathbf{g} = \mathbf{0}$ the condition (7.2.19) also guarantees optimal approximation in the sense of the error estimate

$$\|\nabla(\mathbf{u} - \mathbf{u}_h)\| + \|p - p_h\| \leq C\left(\inf_{\mathbf{v} \in \mathbf{X}_h} \|\nabla(\mathbf{u} - \mathbf{v})\| + \inf_{q \in M_h} \|p - q\|\right). \tag{7.2.20}$$

Note that the constant C in (7.2.20) is inversely proportional to the inf-sup constant γ in (7.2.19).

The simplest example of an unstable method is the computationally convenient equal-order velocity/pressure approximation based on a single grid. The problem is that the pressure space is too rich compared to the velocity space in this case. The simplest way of constructing an equal order approximation such that (7.2.19) is uniformly satisfied is to introduce two grids: for example in $\mathbb{R}^2$ starting from a coarse grid of rectangles, a refined grid can be constructed by joining the mid-points of the edges. The condition (7.2.19) is then satisfied by taking a C^0 piecewise bilinear function on the coarse mesh for the pressure approximation, and a C^0 piecewise bilinear function on the fine mesh for each of the velocity components. Numerical results presented in sections 4 and 5 were generated using this approach—henceforth referred as the Q_1–iso–Q_2 method.

To construct the matrix analogue of (7.2.17)–(7.2.18) it is convenient to introduce discrete operators $\mathcal{A} : \mathbf{X}_h \mapsto \mathbf{X}_h$ and $\mathcal{B} : \mathbf{X}_h \mapsto M_h$ defined via

$$(\mathcal{A}\mathbf{v}_h, \mathbf{w}_h) = (\nabla \mathbf{v}_h, \nabla \mathbf{w}_h) \quad \forall \mathbf{v}_h, \mathbf{w}_h \in \mathbf{X}_h, \tag{7.2.21}$$

$$(\mathcal{B}\mathbf{v}_h, q_h) = (\nabla \cdot \mathbf{v}_h, q_h) \quad \forall \mathbf{v}_h \in \mathbf{X}_h, \forall q_h \in M_h, \tag{7.2.22}$$

so that $\mathcal{B}^*$ is the adjoint of $\mathcal{B}$, i.e. $(\mathbf{v}_h, \mathcal{B}^* q_h) = (\mathcal{B}\mathbf{v}_h, q_h)$. With these definitions the discrete problem (7.2.17)–(7.2.18) can be rewritten as a matrix system:

$$\begin{pmatrix} \mathcal{A} & \mathcal{B}^* \\ \mathcal{B} & 0 \end{pmatrix} \begin{pmatrix} \mathbf{u}_h \\ p_h \end{pmatrix} = \begin{pmatrix} \mathbf{f}_h \\ 0 \end{pmatrix}. \tag{7.2.23}$$

Furthermore, the inf-sup inequality (7.2.19) simplifies to

$$\gamma \|p_h\| \le \sup_{\mathbf{w}_h \in \mathbf{X}_h} \frac{(\mathcal{B}\mathbf{w}_h, p_h)}{(\mathcal{A}\mathbf{w}_h, \mathbf{w}_h)^{1/2}} \qquad \forall p_h \in M_h. \tag{7.2.24}$$

It is instructive to express the inf-sup condition in terms of the actual finite element matrices that arise in practice. To this end, let us explicitly introduce the finite element basis sets, say,

$$\mathbf{X}_h = \text{span}\{\phi_i\}_{i=1}^n, \qquad M_h = \text{span}\{\psi_j\}_{j=1}^m; \tag{7.2.25}$$

and associate the functions $\mathbf{u}_h$, p_h, $\mathbf{f}_h$ with the vectors $u \in \mathbb{R}^n$, $p \in \mathbb{R}^m$ and $f \in \mathbb{R}^n$ of generalised coefficients, $\mathbf{u}_h = \sum_{i=1}^n u_i \phi_i$ etc. Defining the $n \times n$ "vector-stiffness matrix" $A_{ij} = (\nabla\phi_i, \nabla\phi_j)$ and also the $m \times n$ "divergence matrix" $B_{ij} = -(\nabla . \phi_j, \psi_i)$, gives the finite element version of (7.2.23):

$$\begin{pmatrix} A & B^t \\ B & 0 \end{pmatrix} \begin{pmatrix} u \\ p \end{pmatrix} = \begin{pmatrix} f \\ 0 \end{pmatrix}. \tag{7.2.26}$$

Moreover, introducing the $m \times m$ pressure "mass matrix" $Q_{ij} = (\psi_i, \psi_j)$; leads to the finite element version of (7.2.19) or (7.2.24): for all $p \in \mathbb{R}^m$,

$$\gamma (p^t Q p)^{1/2} \le \max_u \frac{p^t B u}{(u^t A u)^{1/2}} \tag{7.2.27}$$

$$= \max_{w = A^{1/2} u} \frac{p^t B A^{-1/2} w}{(w^t w)^{1/2}} \tag{7.2.28}$$

$$= (p^t B A^{-1} B^t p)^{1/2}, \tag{7.2.29}$$

since the maximum is attained when $w = A^{-1/2} B^t p$. Thus, we have a characterisation of the inf-sup constant:

$$\gamma^2 = \min_{p \ne 0} \frac{p^t B A^{-1} B^t p}{p^t Q p}. \tag{7.2.30}$$

In simple terms it is precisely the square root of the smallest eigenvalue of the Schur complement preconditioned by the pressure mass matrix: $Q^{-1} B A^{-1} B^t$.

The discrete inf-sup condition is extremely restrictive. The problem is that the simplest conforming finite element methods such as Q_1–P_0 (trilinear/bilinear velocity with constant pressure) are not stable in the sense that pressure vectors $p \in M_h$ can be constructed for which the inf-sup constant tends to zero under uniform refinement. This type of instability can be difficult to detect in practice since the associated discrete systems are non-singular, (so that each of the discrete problems are uniquely solvable), however they become rapidly ill-conditioned as $h \to 0$.

The simplest way of getting such low-order methods to work in practice is to relax the discrete incompressibility condition (7.2.18). An efficient approach is the following *local* stabilisation method, which is based on controlling the jumps in pressure across element boundaries within an appropriate macroelement subdidvision, $\mathcal{M}$ say, as follows

$$(\nabla \mathbf{u}_h, \nabla \mathbf{v}) - (p_h, \nabla \cdot \mathbf{v}) = (\mathbf{f}_h, \mathbf{v}) \qquad \forall \mathbf{v} \in \mathbf{X}_h \qquad (7.2.31)$$

$$(\nabla \cdot \mathbf{u}_h, q) - \beta \sum_{\substack{\mathrm{m}\in\mathcal{M}\\ e\in\Gamma_\mathrm{m}}} h_\mathrm{m} \int_e [\![p_h]\!]_e [\![q]\!]_e \mathrm{d}s = 0 \qquad \forall q \in M_h. \qquad (7.2.32)$$

In (7.2.32), Γ_m is the set of all edges/faces in the *interior* of the m'th macroelement, β is a positive stabilisation parameter (see below) and h_m is a local measure of the macroelement's size, see [30]. Of course, if stability is to be achieved then the number of elements in each macroelement must be sufficiently large—if every macroelement contained just one element there are no internal jump terms (i.e. $\Gamma_\mathrm{m} = \emptyset$), and (7.2.31)–(7.2.32) degenerates to the unstabilised formulation. In the motivating paper [30], it is rigorously established that as long as $\mathcal{M}$ is constructed so that each macroelement is topologically equivalent to a reference macroelement having a velocity node on every edge (or every face in three-dimensions), then there exists a minimal parameter value β_0 such that the formulation (7.2.31)–(7.2.32) is stable; i.e. there exists a constant γ_s bounded away from zero independently of h such that the following "inf-sup like" condition is satisfied

$$\sup_{\mathbf{w}\in\mathbf{X}_h} \frac{(p, \nabla \cdot \mathbf{w})}{\|\nabla \mathbf{w}\|} \geq \sqrt{2}\,\gamma_s \|p\| - (\beta \sum_{\substack{\mathrm{m}\in\mathcal{M}\\ e\in\Gamma_\mathrm{m}}} h_\mathrm{m} \int_e [\![p]\!]_e^2 \mathrm{d}s)^{\frac{1}{2}} \quad \forall p \in M_h. \qquad (7.2.33)$$

As a result, if $\beta \geq \beta_0$ then an optimal error estimate can be established in the case $\mathbf{g} = \mathbf{0}$ (see [30])

$$\|\nabla(\mathbf{u} - \mathbf{u}_h)\| + \|p - p_h\| \leq Ch \qquad (7.2.34)$$

where C is a constant independent of h and β (it depends only on β_0). Note that the same estimate (7.2.34) characterises the approximation accuracy of the Q_1–iso–Q_2 method above (with a different constant C).

Using a stabilised formulation of the form (7.2.31)–(7.2.32) leads to the following matrix system

$$\begin{pmatrix} A & B^t \\ B & -\beta S \end{pmatrix} \begin{pmatrix} u \\ p \end{pmatrix} = \begin{pmatrix} f \\ 0 \end{pmatrix}. \qquad (7.2.35)$$

where A and B are as defined in (7.2.26), and S corresponds to the pressure stabilisation term in (7.2.32). Furthermore we have an explicit representation of the stability constant γ_s in (7.2.33)

$$\gamma_s^2 = \min_{p\neq 0} \frac{p^t B A^{-1} B^t p + \beta p^t S p}{p^t Q p}, \tag{7.2.36}$$

which is the analogue of (7.2.30) in the unstabilised case.

One of the features of (7.2.31)–(7.2.32) is that if the discrete incompressibility constraints are added together, then the jump terms sum to zero in each macroelement (a specific example is given below). This is crucially important to the success of the method since it implies that the local incompressibility of the original method is retained after stabilisation (albeit over macroelements). The major potential limitation of this approach is that stability is only guaranteed if the stabilisation parameter β is bigger than the critical value β_0. Fortunately this does not cause any difficulty in practice, since an over-estimate of the critical parameter is easily computed if the extremal eigenvalues of the Schur complement and the stabilisation matrix are known; specifically, it is shown in [43] that $\beta_* \geq \beta_0$ if $\beta_* = \Gamma^2/\Theta^2$ with

$$\Gamma^2 = \max_{p\neq 0} \frac{p^t B A^{-1} B^t p}{p^t Q p}, \tag{7.2.37}$$

$$\Delta^2 = \max_{p\neq 0} \frac{p^t S p}{p^t Q p}. \tag{7.2.38}$$

A simple estimate of Γ is well known (see [18]): a Cauchy-Schwarz argument yields

$$\frac{|(\operatorname{div}\mathbf{v}, p)|^2}{\|\mathbf{v}\|_X^2 \|p\|_M^2} \leq \frac{\|\operatorname{div}\mathbf{v}\|^2}{\|\nabla \mathbf{v}\|^2} \leq d, \tag{7.2.39}$$

so for example in $\mathbb{R}^2$ we have $\sqrt{2} \geq \Gamma$. In practice, this estimate (which holds for all mixed approximations) seems to be pessimistic. In particular, in the case of the Q_1–P_0 approximation, numerical computations on quasi-uniform Cartesian grids of rectangular elements suggest that that $\Gamma \to 1$ from below, as $h \to 0$.

Using a macroelement stabilisation, Δ in (7.2.38) can be computed locally. To illustrate this, consider the case of a uniform grid of $j \times j$ square Q_1–P_0 elements of side h. If j is even then local stabilisation can be based on 2×2 macroelements, and with an appropriate local numbering, the stabilisation matrix S is block diagonal with identical 4×4 blocks of the following form

$$S_{\mathcal{M}} = h^2 \begin{pmatrix} 2 & -1 & 0 & -1 \\ -1 & 2 & -1 & 0 \\ 0 & -1 & 2 & -1 \\ -1 & 0 & -1 & 2 \end{pmatrix}. \tag{7.2.40}$$

As a result the eigenvalues of S are $0, 2h^2, 2h^2, 4h^2$ (each with multiplicity equal to $j^2/4$). Furthermore, since the pressure is piecewise constant the mass matrix Q is diagonal with entries equal to h^2. Hence, $\Delta^2 = 4$ in (7.2.38), and $\Gamma^2 = 1$ in (7.2.37) so that a "good" parameter value is easily deduced, namely $\beta = 1/4$. This is important since it allows the possibility of constructing usable software built around Q_1–P_0 for discretising Stokes problems (see section 6). Some numerical results using this software/methodology are described in section 4.

Finally, we note that the discretisation of the Oseen problem (7.1.11), which arises using the linearised implicit time-stepping methods (see section 1.2) can be done using the Stokes methodology described above, and will give good results if the flow is diffusion-dominated in the sense that boundary layers are properly resolved by the mesh. Some numerical results using stabilised Q_1–P_0 are described in section 5. Generalising the streamline-diffusion approximations of the transport terms (cf. section 2.1) is also possible, although the characterisation of appropriate stabilisation parameters is much more difficult to do automatically in the Oseen case.

7.3 Solution methods for the discrete convection-diffusion equation

Discretization of the convection-diffusion equation (7.2.1) using finite differences or finite elements (via (7.2.5) or (7.2.8)) leads to a linear system of equations

$$Fu = f \tag{7.3.1}$$

where u and f are vectors in $\mathbb{R}^n$. F is a nonsymmetric matrix of the form

$$\nu A + N.$$

Here $A = -\Delta_h$, the discrete Laplacian, for the usual finite difference or Galerkin discretizations, or in the case of streamline upwinding, $A = -\Delta_h + A_w$ where A_w corresponds to the stabilizing term of (7.2.8). N is a skew-symmetric matrix, the discrete convection operator.

We will emphasize splitting methods for (7.3.1), that is, representations of the coefficient matrix in the form

$$F = Q - R$$

where Q is the nonsingular *splitting matrix*. Such a splitting can be used to produce a stationary iteration

$$u^{(k+1)} = Q^{-1}(Ru^{(k)} + f) \tag{7.3.2}$$

where $u^{(0)}$ is an arbitrary initial guess, or Q can be used as a preconditioner for (7.3.1) in combination with Krylov subspace methods. The classical analysis of the stationary method (7.3.2) proceeds as follows; see [1, 48, 52] for comprehensive presentations of these results and [34, 47] for concise overviews. Let $e^{(k)} = u - u^{(k)}$ denote the error at the kth step of the stationary iteration (7.3.2). Then

$$e^{(k)} = (Q^{-1}R)^k e^{(0)},$$

and for any consistent norm $\|\cdot\|$,

$$\|e^{(k)}\| \leq \|(Q^{-1}R)^k\| \; \|e^{(0)}\|. \tag{7.3.3}$$

Analysis is based on the fact that

$$\lim_{k\to\infty} \|(Q^{-1}R)^k\|^{1/k} = \rho(Q^{-1}R) \tag{7.3.4}$$

where ρ denotes the spectral radius. The iteration is convergent if and only if $\rho(Q^{-1}R) < 1$, and roughly speaking, the error decreases in magnitude by a factor of $\rho(Q^{-1}R)$ at each step. Consequently, ρ is referred to as the convergence factor. The effectiveness of Krylov subspace methods depends in large part on the existence of a polynomial that takes on the values 1 at the origin and is small on the eigenvalues of $Q^{-1}A = I - Q^{-1}R$ [1, 41]; thus, it is also desirable to make $\rho(Q^{-1}R)$ as small as possible for Krylov subspace methods.

7.3.1 *Analysis of convergence factors*

We first consider versions of the classical Jacobi, Gauss-Seidel and successive over-relaxation (SOR) iterative methods and discuss analytic bounds on convergence factors for these methods. Throughout our discussion, we will use the two-dimensional version of (7.2.1); see [1, 34, 47, 48, 52] for general presentations. Let Ω denote the unit square $(0,1) \times (0,1)$ and assume the discretisation is performed on a uniform grid using finite differences or linear or trilinear finite elements. If the grid is ordered with a natural left-to-right bottom-to-top ordering, then the resulting matrix F has block tridiagonal form in which the block diagonal is a tridiagonal matrix. Figure 7.1 shows an example for a 6×6 grid ordered by horizontal lines. The nonzero structure of the matrix for a nine-point operator on this grid is shown on the right. This structure would arise from a bilinear finite element discretisation; for finite differences or linear finite elements (with unidirectional triangles), the off-diagonal blocks would be diagonal or bidiagonal, respectively. Figure 7.2 shows an alternative *line red-black ordering* and the structure of the corresponding matrix. Variants based on vertical orderings are defined analogously.

For any of the line orderings, let $F = D - L - U$ where D denotes the block diagonal of F, $-L$ denotes the lower triangular matrix consisting of

FIGURE 7.1. Natural horizontal line ordering and nonzero structure of matrix.

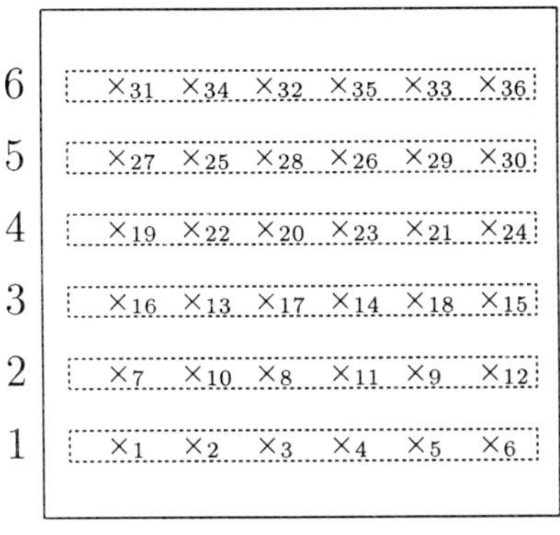

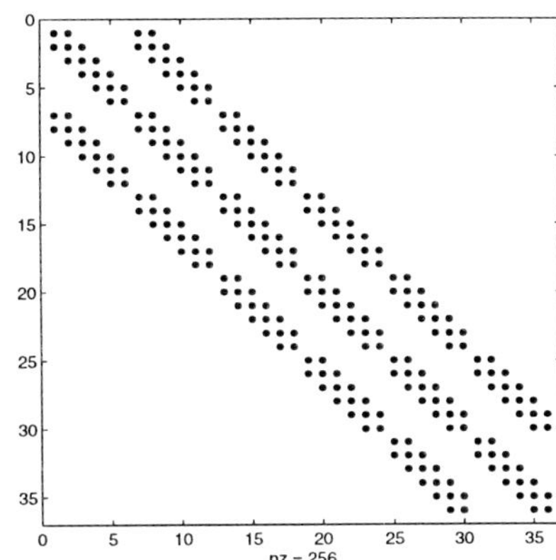

FIGURE 7.2. Horizontal line red-black ordering and nonzero structure of matrix.

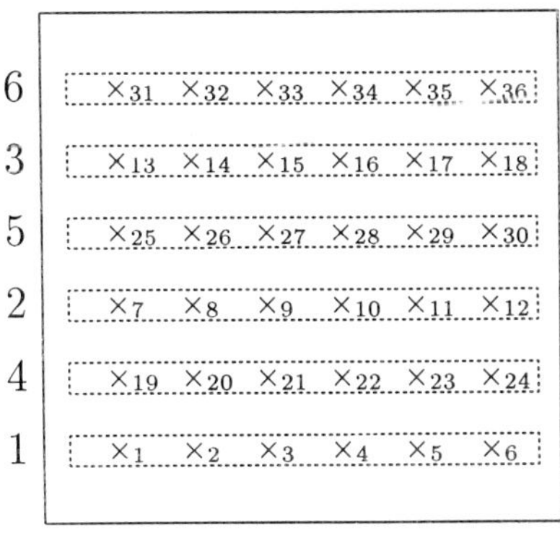

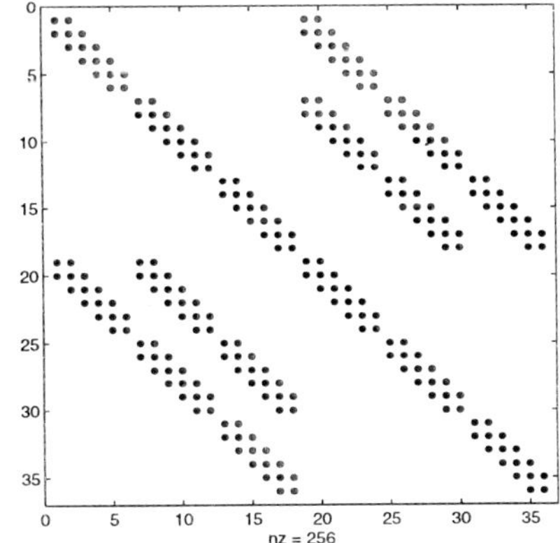

entries below the block diagonal D, and $-U$ is the analogous upper triangular matrix. The classical stationary methods are defined by the following splittings:

line Jacobi:	$Q = D,$	$R = D - F;$
line Gauss-Seidel:	$Q = D - L,$	$R = U;$
line SOR:	$Q = \frac{1}{\omega}(D - \omega L),$	$R = \frac{1}{\omega}\left[(1-\omega)D + \omega U\right].$

The matrix F arising from these orderings is block consistently ordered [52]. Consequently, the spectral radii of the line Jacobi and line Gauss-Seidel iteration matrices are related by

$$\rho((D-L)^{-1}U) = \rho(D^{-1}C)^2 \qquad (7.3.5)$$

where $C = D - F$. Moreover, if the Jacobi matrix $D^{-1}C$ has real eigenvalues and its spectral radius is less than one, then the spectral radius of the SOR iteration matrix $\mathcal{L}_\omega = (D - \omega L)^{-1}[(1-\omega)D + \omega U]$ is minimized by $\omega^* = \frac{2}{1+\sqrt{1-\rho(D^{-1}C)^2}}$ and

$$\rho(\mathcal{L}_{\omega^*}) = \omega^* - 1. \qquad (7.3.6)$$

Thus the key to the analysis is to bound $\rho(D^{-1}C)$. We treat some specific cases individually.

The constant coefficient problem $\nu = 1$ and $\mathbf{w} = (\sigma, \tau)$ in (7.2.1) is the starting point for much of the analysis. Any finite difference discretisation produces a five-point operator which can be represented by a "computational molecule"

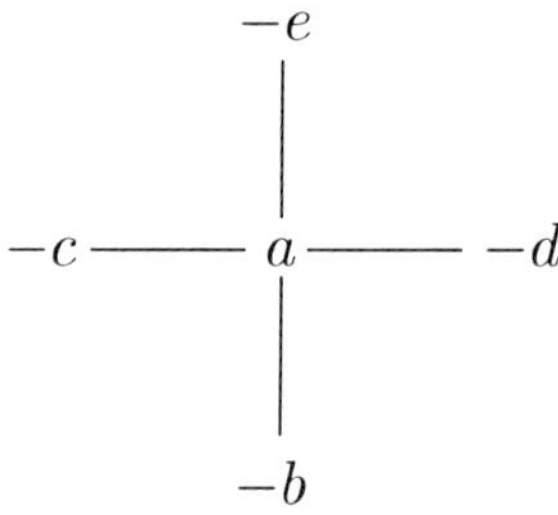

(7.3.7)

In this, case, we can give an exact expression for $\rho(D^{-1}C)$. The proof depends on the fact that the block diagonal matrix D can be symmetrized using a diagonal similarity transformation, see [11].

Theorem 7.3.1 *If $cd \geq 0$, then the spectral radius of the block Jacobi iteration matrix for the horizontal line ordering is*

$$\frac{2\sqrt{be}\cos(\pi h)}{a - 2\sqrt{cd}\cos(\pi h)}.$$

If $be \geq 0$, then the spectral radius of the block Jacobi iteration matrix for the vertical line ordering is

$$\frac{2\sqrt{cd}\cos(\pi h)}{a - 2\sqrt{be}\cos(\pi h)}.$$

The conditions in this theorem are satisfied if all of the off-diagonal entries b, c, d and e of F are greater than equal to zero. This is the case, for example, if centred finite differences are used to discretise the first derivatives in (7.2.1) on a fine enough mesh, or if upwind differencing is used [11, 12]. For example, if centred differences are used on a uniform $n \times n$ grid with $h = 1/(n+1)$, then with $\gamma = \sigma h/2$, $\delta = \tau h/2$, Theorem 7.3.1 is equivalent to the following result.

Corollary 7.3.1 *For centred differences, if $|\gamma| < 1$ then the spectral radius of the block Jacobi iteration matrix for the horizontal line ordering is*

$$\frac{\sqrt{|1-\delta^2|}\cos(\pi h)}{2 - \sqrt{1-\gamma^2}\cos(\pi h)}.$$

If $|\delta| < 1$, then the spectral radius of the block Jacobi iteration matrix for the vertical line ordering is

$$\frac{\sqrt{|1-\gamma^2|}\cos(\pi h)}{2 - \sqrt{1-\delta^2}\cos(\pi h)}.$$

Figure 7.3 shows some examples of spectral radii of Gauss-Seidel iteration matrices for centred difference discretisations and various parameters. Larger values of γ (respectively δ) correspond to increased convection in the horizontal (vertical) direction. These results indicate that it is advantageous to orient the grid lines in directions orthogonal to the dominant direction of flow, i.e., to perform the Gauss-Seidel sweep in the direction of flow.

For finite difference operators with computational molecules of type (7.3.7), the grid points and equations can be ordered with a red-black ordering so that every equation centred at a "red" point depends only on "black" unknowns, and every equation centred at a "black" point depends only on "red" unknowns. In matrix notation, this process corresponds to ordering the rows and columns of F so that the problem (7.3.1) has the form

$$\begin{pmatrix} F_{11} & F_{12} \\ F_{21} & F_{22} \end{pmatrix} \begin{pmatrix} u^{(r)} \\ u^{(b)} \end{pmatrix} = \begin{pmatrix} f^{(r)} \\ f^{(b)} \end{pmatrix}$$

where F_{11} and F_{22} are nonsingular diagonal matrices. Decoupling of the red points $u^{(r)}$ is equivalent to producing a smaller system

$$\hat{F}u^{(b)} = \hat{f}$$

FIGURE 7.3. Spectral radii of line Gauss-Seidel iteration matrices for various parameters.

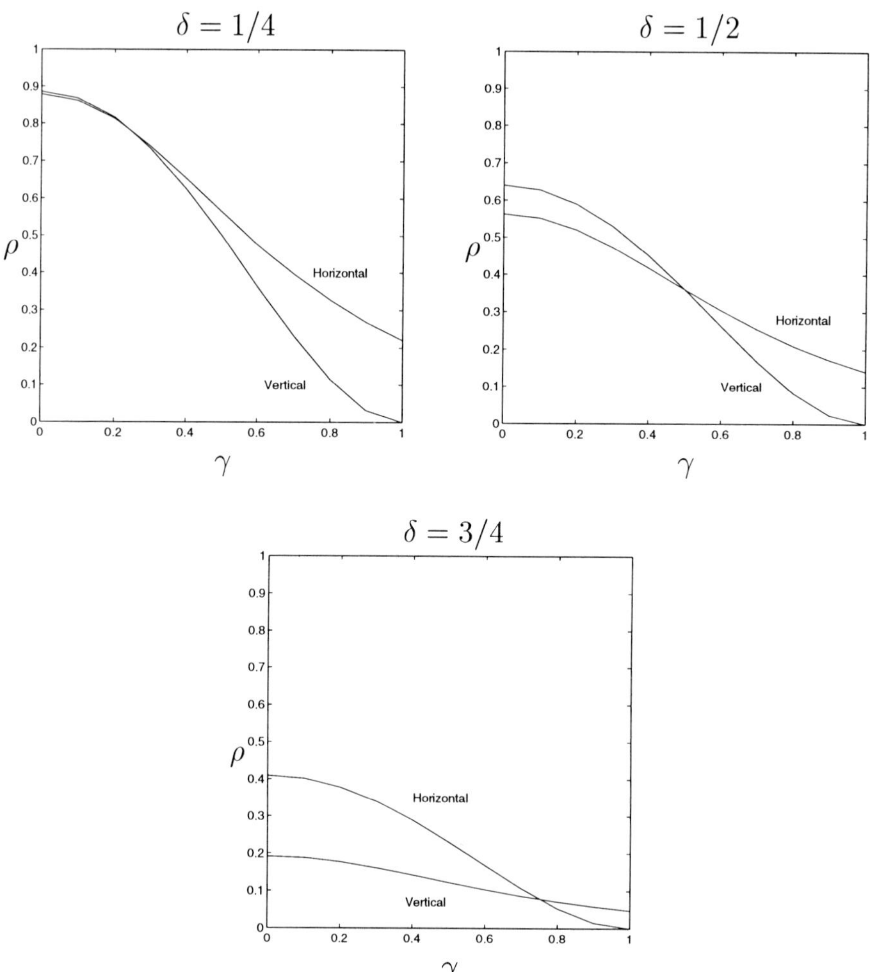

FIGURE 7.4. Horizontal two-line orderings of the reduced grid.

Line						
3	×13	×14	×15	×16	×17	×18
2	×7	×8	×9	×10	×11	×12
1	×1	×2	×3	×4	×5	×6

Line						
2	×7	×8	×9	×10	×11	×12
3	×13	×14	×15	×16	×17	×18
1	×1	×2	×3	×4	×5	×6

where $\hat{F} = F_{22} - F_{21}F_{11}^{-1}F_{12}$ and $\hat{f} = f^{(b)} - F_{21}F_{11}^{-1}f^{(r)}$.

$\hat{F}$ is also a sparse matrix, and for appropriate orderings of the reduced grid $\hat{F}$ is block consistently ordered. Two examples, a natural *two-line* ordering and a red-black *two-line* ordering, are shown in figure 7.4. As above, analysis of the Gauss-Seidel and SOR methods uses (7.3.5) and (7.3.6). For the orderings of figure 7.4, we have the following bounds for the block Jacobi iteration matrices. The proof again depends on finding an appropriate symmetrization operator for $\hat{F}$; see [11, 12, 13, 14] for proofs and extensive discussions of other problems and orderings.

Theorem 7.3.2 *If $be \geq 0$ and $cd \geq 0$, then the spectral radius of the two-line Jacobi iteration matrix is bounded by*

$$\frac{2\,be\,\cos 2\pi h + 4\sqrt{bcde}\cos\pi h}{\begin{array}{l}[a^2 - 2(\sqrt{cd}+\sqrt{be})^2 - 2cd + \\ \qquad 4\sqrt{bcde}\,(1-\cos\pi h) + 4cd\,(1-\cos^2\pi h)]\end{array}} + o(h^2).$$

Corollary 7.3.2 *For centred differences, if $|\gamma| < 1$ and $|\delta| < 1$, then the spectral radius of the two-line block Jacobi iteration matrix for the reduced system is bounded by*

$$\frac{(1-\delta^2)\cos 2\pi h + 2\sqrt{(1-\gamma^2)(1-\delta^2)}\,\cos\pi h}{\begin{array}{l}[8 - (\sqrt{1-\gamma^2}+\sqrt{1-\delta^2})^2 - (1-\gamma^2) + \\ 2\sqrt{(1-\gamma^2)(1-\delta^2)}\,(1-\cos\pi h) + 2(1-\gamma^2)\,(1-\cos^2\pi h)]\end{array}} + o(h^2).$$

These bounds are typically stronger than those above for the unreduced system. Results for vertical two-line orderings can be established in the same way.

The results above are derived from properties of the matrices D and C of the block Jacobi splitting. An alternative approach due to Parter [36] and Parter and Steuerwalt [38] based more closely on the differential operators reveals asymptotic convergence rates as $h \to 0$. (See also [37].) Let $\mathcal{F}$ denote the differential operator on the left side of (7.2.1), and assume the discretization matrix F is scaled so that F/h^2 approximates $\mathcal{F}$ with truncation error $o(1)$ at all mesh points of Ω not next to the boundary, and $O(1)$ at points next to $\partial\Omega$. Let $F = Q - R$ be a splitting.

Theorem 7.3.3 *Suppose the following conditions hold for all small h:*
1. $\rho(Q^{-1}R) < 1$.
2. $\rho(Q^{-1}R)$ is an eigenvalue of $Q^{-1}R$.
3. $\|R\|_2$ is bounded independent of h.
4. There is a smooth function q satisfying $q(x,y) \geq q_0 > 0$ on $\bar{\Omega}$, such that

$$(Ru, v) = (qu, v) + E \tag{7.3.8}$$

where in (7.3.8), q refers to the vector of mesh values, and $E = he_1(u,v) + h^2e_2(u,v)$ depends on σ and τ.[1] *Then as $h \to 0$, $\rho(Q^{-1}R) = 1 - \Lambda_0 h^2 + o(h^2)$, where Λ_0 is the smallest eigenvalue of the problem*

$$\mathcal{F}u = \Lambda qu \quad \text{in } \Omega, \qquad u = 0 \text{ on } \partial\Omega. \tag{7.3.9}$$

This result is very easy to apply to the constant coefficient problem. The mesh function q of (7.3.8) is a constant obtained by inspection as the sum of the entries of the computational molecule that define R. For example, for the Jacobi splittings, $q = 2$ for the one-line ordering of the full system and $q = 3/4$ for the two-line ordering of the reduced system. If the minimal eigenvalue of (7.3.9) is known, then the asymptotic convergence factor is identified. On the unit square,

$$\Lambda_0 = \frac{1}{q}\left(\frac{\sigma^2}{4} + \frac{\tau^2}{4} + 2\pi^2\right).$$

For the line Jacobi splittings we have discussed here, the convergence factors are

$$\begin{array}{ll} 1 - \left(\dfrac{\sigma^2}{8} + \dfrac{\tau^2}{8} + \pi^2\right) & \text{one-line ordering, full system} \\ 1 - \left(\dfrac{\sigma^2}{3} + \dfrac{\tau^2}{3} + \dfrac{8}{3}\pi^2\right) & \text{two-line ordering, reduced system.} \end{array}$$

The first of these expressions agrees with the asymptotic convergence factor obtained from Corollary 7.3.1 and the second one is slightly stronger than that obtained from Corollary 7.3.2. Note that Corollaries 7.3.1 and 7.3.2 provide insight into the nonsymptotic regime. Other examples of the use of this methodology are given in [12, 36, 38].

---

[1] Here e_1 is a function of first order differences in u and v and e_2 is a function of second order differences; see [38] for a more precise statement.

Finally, we note that another popular splitting method for discrete convection-diffusion equations is based on incomplete LU (ILU) factorization of the coefficient matrix. Recall that a nonsingular M-matrix B is one for which $B_{ij} \le 0$ for $i \ne j$ and $B^{-1} \ge 0$ [48]. It is well-known [32] that for any such B there is a unique ILU factorization $Q = LU$ such that L is unit lower triangular, U is upper triangular, $l_{ij} = 0$ and $u_{ij} = 0$ for $(i,j) \notin \mathcal{N}$, and $[Q - B]_{ij} = 0$ for $(i,j) \in \mathcal{N}$, where $\mathcal{N}$ is an index set containing all diagonal indices (i,i). It can be shown [3, 13, 51] that if

$$B = Q_1 - R_1 = Q_2 - R_2,$$

where $Q_1 = L_1U_1$ and $Q_2 = L_2U_2$ are incomplete factorizations such that the set of matrix indices for which $L_1 + U_1$ is permitted to be nonzero is contained in the set of indices for which $L_2 + U_2$ is permitted to be nonzero, then $\rho(Q_2^{-1}R_2) \le \rho(Q_1^{-1}R_1)$. For the examples arising from finite differences that we have considered, both F and $\hat{F}$ are nonsingular M-matrices for a fine enough mesh. Let $Q_1 = Q$ obtained by the ILU(0) factorization (i.e., the index set $\mathcal{N}$ equals the nonzero set of of the coefficient matrix) with error matrix R, and let $Q_2 = D$ from the block Jacobi splitting. It follows that

$$\rho(Q^{-1}R) \le \rho(D^{-1}C).$$

Thus, all the bounds obtained above for the block Jacobi method carry over to the ILU(0) factorization.

7.3.2 Ordering effects

We now turn to some issues associated with the underlying flow and the effects of ordering of the discrete grid. As noted in the discussion following Corollary 7.3.1, some of the analysis depends on the orientation of lines in the grid. Once that orientation is fixed, however, there is no dependence on ordering of unknowns. For example, none of results above depend on whether a "natural" or "red-black" ordering is used, and all of them are independent of the sign of the coefficients of the convection terms, which determine the direction of flow. Indeed, for a natural ordering in which relaxation is performed in a direction *opposite* the direction of flow, the bounds on convergence factor are identical.[2]

In practice, the performance of relaxation methods is sensitive to ordering. As might be expected from intuition, it is better to relax in the direction of flow than in the opposite direction, and performance for orderings such as red-black that don't bear a clear relation to flow direction is somewhere in between these extremes. The difference between the analytic results and these performance characteristics stems from the difference between (7.3.3) and (7.3.4). The expression (7.3.4) provides insight

[2] Theorem 7.3.3 has no dependence even on line orientation.

FIGURE 7.5. Four orderings for a one-dimensional grid, for $\sigma > 0$ and $n = 8$.

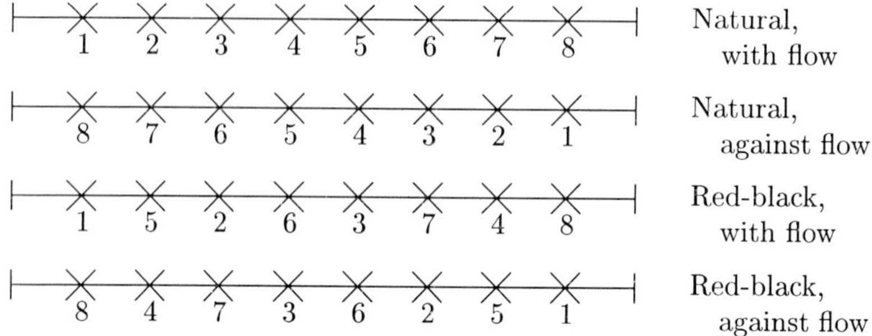

into asymptotic behaviour as the number of iterations becomes large, but it provides no information about transient behaviour displayed before the limiting value is approached.

Many aspects of this issue can be understood from the one-dimensional version of (7.2.1)

$$-u'' + \sigma u' = f$$

on the unit interval $(0, 1)$ with Dirichlet boundary conditions and $\sigma > 0$. Let n denote the number of interior mesh points of a uniform grid. Finite difference and linear finite element discretization lead to a linear system (7.3.1) in which, for a natural ordering, the coefficient matrix F is tridiagonal of order n, with constant values on its three interior bands. Assume that F is normalized to have unit diagonal, so that it can be represented as

$$F = tri\,[-b,\, 1,\, -c]\,.$$

In addition, assume $b + c = 1$ (needed for a consistent discretisation) and $b > 0$, $c > 0$ (for a nonoscillatory solution [25]). We say that the discrete problem is *convection-dominated* if b is large, i.e., close to 1. The Gauss-Seidel iteration matrix is $\mathcal{L}_1 = (I - L)^{-1}U$, where L and U are the strict lower triangular and upper triangular parts of F.

Figure 7.5 shows examples of four different orderings for $n = 8$. There are two natural orderings, together with two red-black orderings induced by the natural orderings. Figure 7.6 shows a representative example of the behaviour of relaxation for convection-dominated problems that reveals the limitations of the standard analysis. The figure plots $\|e^{(k)}\|_1$, on a logarithmic scale, against the iteration count k, for the Gauss-Seidel method corresponding to the four ordering schemes. Here, $n = 32$ and $b = 7/8$. The initial guess is a normally distributed random vector with mean 0 and variance 1, and the right hand side and solution are identically zero. The spectral radius for each of the orderings is $\rho(\mathcal{L}_1) = .434$. Figure 7.7 shows the norms $\|\mathcal{L}_1^k\|_1$. In both figures, the highlighted values correspond to $k = n - 1$ and $k = n/2 - 1$ for the natural ordering against the flow and

FIGURE 7.6. l_1-norms of the errors in Gauss-Seidel iteration, for $n = 32$ and $b = 7/8$.

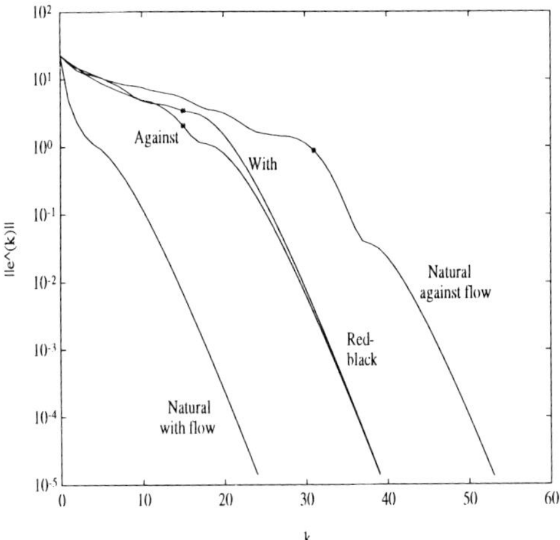

red-black orderings, respectively. It is evident that the norms are closely correlated with the performance of the solution algorithm, and that the spectral radius reveals nothing about the transient behaviour.

The iteration matrices arising from different orderings will be distinguished as follows. For the left-to-right natural ordering, inducing a relaxation sweep oriented with the flow, the iteration matrix is $F = (I - L)^{-1}U$, where

$$L = tri\ [b, 0, 0], \quad U = tri\ [0, 0, c].$$

The red-black ordering induced by this natural ordering gives rise to the coefficient matrix $F = I - L_{RB} - U_{RB}$ where

$$L_{RB} = \begin{pmatrix} 0 & 0 \\ B & 0 \end{pmatrix}, \quad U_{RB} = \begin{pmatrix} 0 & C \\ 0 & 0 \end{pmatrix},$$

and

$$B = tri\ [0, b, c], \quad C = tri\ [b, c, 0],$$

of dimensions $\lceil n/2 \rceil \times \lfloor n/2 \rfloor$ and $\lfloor n/2 \rfloor \times \lceil n/2 \rceil$ respectively. The iteration matrix is given by

$$F_{RB} = (I - L_{RB})^{-1}U_{RB} = (I + L_{RB})U_{RB} = \begin{pmatrix} 0 & C \\ 0 & BC \end{pmatrix}.$$

For sweeps oriented against the flow, rather than reversing the ordering, it is equivalent to use the left-to-right natural ordering and perform an "upper-triangular" sweep, i.e., with the iteration matrix $G = (D - U)^{-1}L$.

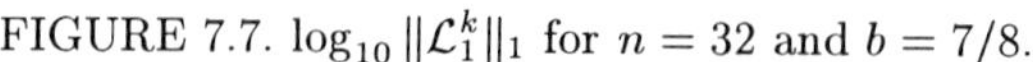

FIGURE 7.7. $\log_{10} \|\mathcal{L}_1^k\|_1$ for $n = 32$ and $b = 7/8$.

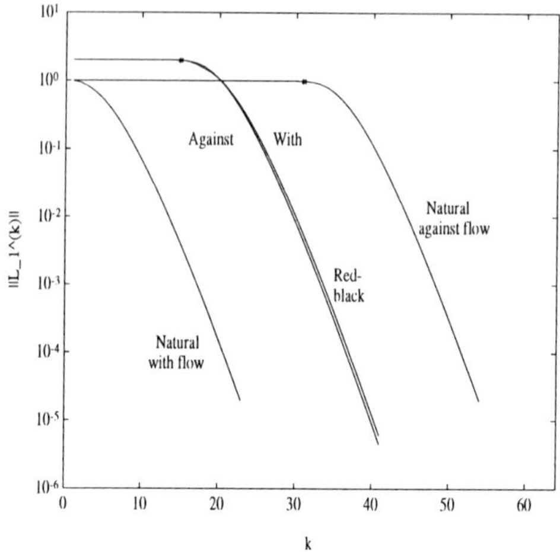

We summarize an analysis for the one-dimensional problem below. Proofs and descriptions of additional numerical experiments are given in [9]. There are three results: *lower bounds* on the values of both $\|G^k\|_1$ and $\|F_{RB}^k\|_1$, and *upper bounds* on the values of $\|F^k\|_1$. Essentially the same lower bounds apply in the l_∞-norm, and the upper bounds can be generalized to any l_p-norm.

Theorem 7.3.4 *The norm $\|G^k\|_1$ for Gauss-Seidel iteration with sweeps against the flow is bounded below for $k < n$ by*

$$\|G^k\|_1 \geq (1-c^2)^{k-1}(1-c^{n-(k-1)}).$$

Theorem 7.3.5 *For problems whose order n is divisible by four, the norm $\|F_{RB}^k\|_1$ for Gauss-Seidel iteration associated with the red-black ordering induced by a left-to-right natural ordering is bounded below as follows:*

$$\|F_{RB}^k\|_1 \geq 2 - \psi(k,c) \quad \textit{for } k \leq n/2 - 1.$$

where $\psi(k,c)$ is zero for $k < n/4$ and close to zero for $n/4 \leq k < n/2$. (See [9] for a precise definition.)

Theorem 7.3.6 *The norm $\|F^k\|_1$ for Gauss-Seidel iteration with sweeps that follow the flow is bounded above by*

$$\|F^k\|_1 \leq 1 - b^{n+k-2}\left(\sum_{j=0}^{k-1}\binom{n+k-3+j}{j}c^j\right). \tag{7.3.10}$$

FIGURE 7.8. Comparison of $||G^k||_1$ with lower bounds, for $n = 32$ and $c = 1/8$.

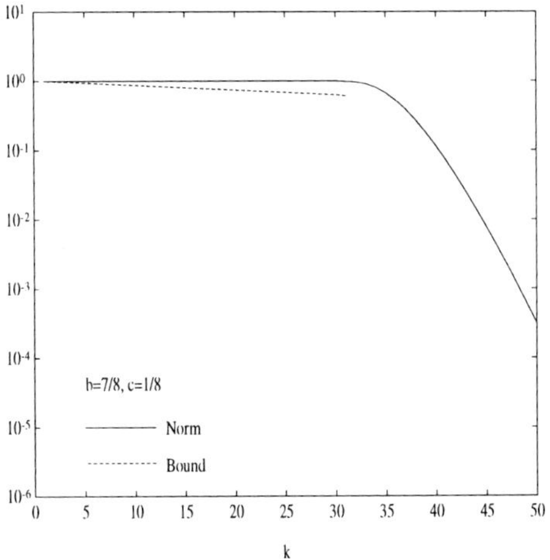

FIGURE 7.9. Comparison of $||F_{RB}^k||_1$ with lower bounds, for $n = 32$ and $c = 1/8$.

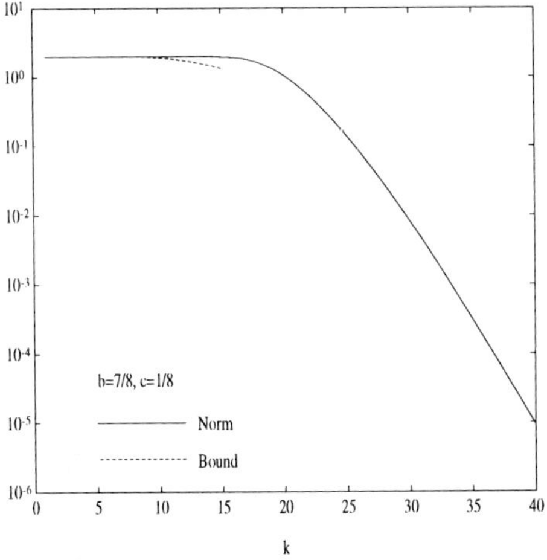

FIGURE 7.10. Comparison of $\|F^k\|_1$ with upper bounds, for $n = 32$ and $b = 7/8$.

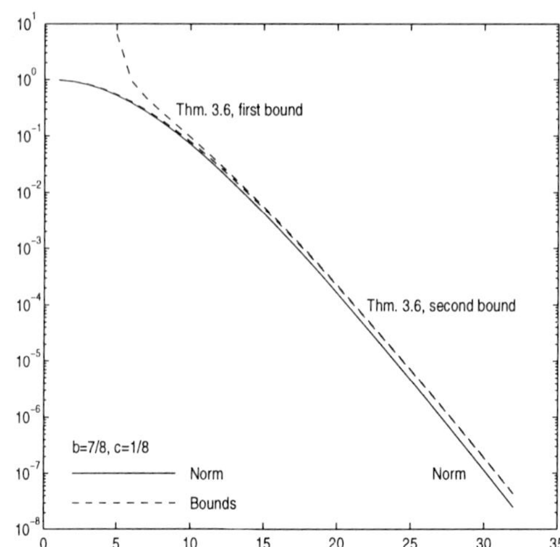

For $k > (n-3)c/(1-2c)$, the following simpler upper bound holds:

$$\|F^k\|_1 \leq \frac{k\binom{n+2k-3}{k} b^{n-2}(bc)^k}{k(1-2c)-(n-3)c}. \tag{7.3.11}$$

Figures 7.8, 7.9 and 7.10 plot norms and the bounds from each of these results, for the problem used for Figs. 7.6 and 7.7. The results of Theorems 7.3.4 and 7.3.5 indicate that if b is near 1 then the norms of the iteration matrices for sweeping against the flow and the red-black ordering are close to one for $n-1$ and $n/2-1$ steps, respectively. Consequently, these orderings incur a latency in which little reduction in the error is obtained. In contrast, Theorem 7.3.6 shows that the norm of the iteration matrix is small for sweeping with the flow.

It is possible to generalize the upper bounds of Theorem 7.3.6 to line relaxation methods applied to two-dimensional problems. In this case, the angle between the flow direction and sweep direction plays a role in both performance and bounds; details can be found in [10].

7.3.3 Discussion

We now discuss some practical issues associated with solving the convection-diffusion equation. We first note some limitations of the analysis cited in sections 3.1 and 3.2, namely, the results apply only to constant coefficient problems and they are limited to finite difference discretisations. The latter restriction is not transparent for Theorem 7.3.3 but requirement (2) of this result is typically established using the fact that the coefficient matrix is

	Galerkin			Streamline Upwinding		
ν	Hor.	Vert.	Hor. R/B	Hor.	Vert.	Hor. R/B
1/10	240	262	257	–	–	–
1/25	84	110	98	–	–	–
1/50	29	57	43	31	59	45
1/100	15	46	31	15	44	29
1/200	div.	div.	div.	6	37	22
1/250	div.	div.	div.	9	38	23

TABLE 7.1. Iterations for bilinear finite elements applied to the convection-diffusion equation.

an M-matrix and applying the Perron-Frobenius theory; see [38], p. 1185. Standard finite element discretisations of the convection-diffusion equation do not produce M-matrices, and we know of little analysis for anything other than finite differences. (Cf. [10].)

Despite these limitations, we have found that the analysis above gives a good indication of the behaviour of splitting methods for simple flows or other discretisations. Examples demonstrating this for a semicircular flow are given in [13], which also contains some analysis for variable coefficient problems. As an example of behaviour for other discretisations, we consider two versions of bilinear finite elements applied to the problem (7.2.1) with constant convection coefficients $v = (-\frac{\sqrt{2}}{2}, \frac{\sqrt{2}}{2})$ on the unit square, $f = 0$, and Dirichlet boundary conditions $u(x,1) = u(y,0) = 0$, u(1,y)=1, $u(x,0) = 0$ for $x \leq 1/2$ and $u(x,0) = 1$ for $x > 1/2$. We discretise on a uniform $n \times n$ element grid using either a pure Galerkin method (7.2.5) or a streamline upwind Petrov-Galerkin method (7.2.8). These problems were generated using the MATLAB code described in section 6.

Table 7.1 shows the iteration counts needed by three variants of line relaxation (horizontal, vertical and horizontal red-black) to solve the discrete problems with $n = 32$, with stopping criterion

$$\|f - Fu^{(k)}\|_2/\|f\|_2 \leq 10^{-6}.$$

Note that the flow direction forms a $-45°$ angle with the horizontal axis. Consequently, the direction of horizontal line relaxation contains a large component in the direction of flow whereas vertical line relaxation is essentially sweeping against the flow. The results show that as the flow becomes stronger (i.e. as the viscosity ν decreases), the differences among the methods are essentially as predicted in section 3.2: sweeping against the flow incurs a latency of approximately n steps and the red-black ordering incurs a latency of approximataly $n/2$ steps. Moreover, the iteration counts decrease dramatically as convection becomes more dominant, as the results of section 3.1 predict.

We also note, however, that the use of these ideas for more complex flows and on large-scale parallel computers lead to some open questions. For example, for the circular flow arising in the driven cavity problem (see section 5), there are portions of the domain where neither a horizontal or vertical line orientation produces a sweep in the direction of flow, and it may be necessary to use more sophisticated strategies to handle such flows. We expect complex three-dimensional flows to add additional difficulties. On parallel architectures, it is known that red-black and multi-color orderings lead to higher parallel efficiencies than natural orderings. However, the latency associated with red-black orderings shows that these reorderings may have limitations that need to be overcome to produce effective solution methods.

7.4 Solution methods for the discrete Stokes equations

The stability issue associated with mixed approximation of the Stokes problem is of central importance when it comes to finding fast and reliable iterative solution methods. In this section we will develop the theory and present computational results for only one class of method, namely those based on preconditioned Minimum Residual iteration, but see [8] for a comparison of various competitive techniques. The relevant theory for any of the alternative approaches is based on the key result (7.2.30) which is a direct consequence of the stability required to ensure accuracy properties of the underlying approximation.

In our examples, we concentrate on two particular mixed finite elements: the stable Q_1–*iso*–Q_2 and locally stabilised $Q_1 - P_0$ approximations. Our aim is to illustrate the general structure for stable and stabilised mixed spaces with these convenient and popular choices. Since we wish to concentrate on the stability issue, we consider here the steady state Stokes problem as mentioned in section 2. For consideration of the additional issue arising with time-dependent problems see [4].

7.4.1 Statement of the problem

As in section 2, we can express the discrete Stokes problem as

$$\mathcal{A}x := \begin{pmatrix} A & B^t \\ B & -\beta S \end{pmatrix} \begin{pmatrix} u \\ p \end{pmatrix} = \begin{pmatrix} f \\ 0 \end{pmatrix} \tag{7.4.1}$$

where A is the discrete vector-Laplacian, B is the discrete gradient so that its adjoint B^t is the discrete negative divergence and S is the stabilisation matrix with β being the non-negative stabilisation parameter. The vector u contains the velocity coefficients in terms of the selected basis and p correspondingly for the pressure. Certainly A is symmetric and it will also be positive definite with the usual Dirichlet boundary conditions, B will be full rank except that the vector $p = (1, 1, \ldots, 1)^t$ representing hydrostatic (constant) pressure will be in the null space (unless it is explicitly removed) and S will be symmetric and positive semi-definite ($S = 0$ in the case of an unstabilised approximation). Any body forces are represented in the vector f.

Employing the Sylvester Law of Inertia ([22] pp. 274), the congruence transform

$$\begin{pmatrix} A & B^t \\ B & -\beta S \end{pmatrix} = \begin{pmatrix} I & 0 \\ BA^{-1} & I \end{pmatrix} \begin{pmatrix} A & 0 \\ 0 & -\beta S - BA^{-1}B^t \end{pmatrix} \begin{pmatrix} I & A^{-1}B^t \\ 0 & I \end{pmatrix}$$

reveals that $\mathcal{A}$ has n_u positive eigenvalues and n_p negative eigenvalues for a stable or stabilised method. This observation follows directly from the discrete stability condition (7.2.36).

The indefiniteness of the Stokes system is thus clear. Note that if the mesh size h is reduced and the discrete problem size correspondingly increased, both the number of positive and negative eigenvalues increases: some authors refer to such systems as being highly (or strongly) indefinite. It is the solution of such linear systems which we address in this section.

7.4.2 The MINRES method

There are two applicable Krylov subspace iterative methods for such symmetric and indefinite systems: SYMMLQ and MINRES, both based on the symmetric Lanczos procedure and both due to Paige and Saunders [35]. Here we concentrate on the MINRES methods since it possesses a minimisation property. We comment that to our knowledge, SYMMLQ has not been tried on discrete Stokes problems.

In the generic context of solving the symmetric and indefinite matrix system

$$\mathcal{A}x = b,$$

MINRES is characterised by the following. The kth iterate x_k lies in the (affine) Krylov subspace

$$x_0 + \text{span}\{r_0, \mathcal{A}r_0, \mathcal{A}^2 r_0, \ldots, \mathcal{A}^{k-1} r_0\}$$

where $r_0 = b - \mathcal{A}x_0$ is the initial residual. Correspondingly for the kth residual vector we have

$$r_k \in r_0 + \text{span}\{\mathcal{A}r_0, \mathcal{A}^2 r_0, \ldots, \mathcal{A}^k r_0\},$$

the defining condition being that $\|r_k\|_2$ is minimum from this space. Thus if Π_k is the set of all real polynomials of degree less than or equal to k

then $r_k = p(\mathcal{A})r_0$, with $p(0) = 1$ and p being optimal in the above sense. Employing a spectral (eigenvector) expansion

$$r_0 = \sum \alpha_i v_i \quad , \quad \mathcal{A} v_i = \lambda_i v_i$$

we have

$$r_k = p(\mathcal{A}) \sum \alpha_i v_i = \sum \alpha_i p(\lambda_i) v_i$$

so that

$$\begin{aligned} \|r_k\|_2 &= \min_{p\in\Pi_k, p(0)=1} \|\sum \alpha_i p(\lambda_i) v_i\|_2 \\ &= \min_{p\in\Pi_k, p(0)=1} \left(\sum \alpha_i^2 p(\lambda_i)^2 v_i^t v_i\right)^{\frac{1}{2}} \\ &\leq \min_{p\in\Pi_k, p(0)=1} \max_i |p(\lambda_i)| \left(\sum \alpha_i^2 v_i^t v_i\right)^{\frac{1}{2}} \end{aligned}$$

or

$$\frac{\|r_k\|_2}{\|r_0\|_2} \leq \min_{p\in\Pi_k, p(0)=1} \max_{\lambda\in\Lambda(\mathcal{A})} |p(\lambda)|$$

where $\Lambda(\mathcal{A})$ denotes the eigenvalue spectrum. Note that the orthogonality of the eigenvectors which is a consequence of the symmetry of $\mathcal{A}$ is important here. Also if one were interested in positive definite symmetric matrices $\mathcal{A}$, this convergence estimate would be the same as that for the 'classical' Conjugate Gradient method (which requires fewer operations per iteration) except that $\|r_k\|_2$ would be replaced by $\sqrt{r_k^t \mathcal{A}^{-1} r_k} = \sqrt{(x - x_k)^t \mathcal{A}\,(x - x_k)} \stackrel{\text{def}}{=} \|x - x_k\|_{\mathcal{A}}$.

In order to achieve rapid convergence, preconditioning will be as important here as in the symmetric and positive definite case. Also it is desirable to ensure that any preconditioner does not destroy the underlying symmetry of the original problem else more general non-symmetric iterative methods such as GMRES ([42] or see the paper by Van der Vorst in this volume) would have to be employed. Such methods are generally less efficient than their symmetric counterparts (see for example [16]). In order to preserve symmetry in the preconditioned system we employ a symmetric and positive definite preconditioner $\mathcal{M}$ which for theoretical purposes only we factor as $\mathcal{M} = \mathcal{M}^{\frac{1}{2}} \mathcal{M}^{\frac{1}{2}}$. (A Cholesky factorisation could equally be used). We are then interested in applying MINRES to the preconditioned system

$$\mathcal{M}^{-\frac{1}{2}} \mathcal{A} \mathcal{M}^{-\frac{1}{2}} (\mathcal{M}^{\frac{1}{2}} x) = \mathcal{M}^{-\frac{1}{2}} b$$

or

$$\widetilde{\mathcal{A}}\, \widetilde{x} = \widetilde{b}$$

say. Now the corresponding preconditioned residual is

$$\widetilde{r} = \widetilde{b} - \widetilde{\mathcal{A}}\, \widetilde{x} = \mathcal{M}^{-\frac{1}{2}} (b - \mathcal{A} x) = \mathcal{M}^{-\frac{1}{2}} r$$

so that

$$\|\widetilde{r_k}\|_2^2 = \widetilde{r_k}^t \widetilde{r_k} = r_k^t \mathcal{M}^{-1} r_k = \|r_k\|_{\mathcal{M}^{-1}}^2.$$

The preconditioned MINRES convergence estimate therefore becomes

$$\frac{\|r_k\|_{\mathcal{M}^{-1}}^2}{\|r_0\|_{\mathcal{M}^{-1}}^2} \leq \min_{p\in\Pi_k, p(0)=1} \max_{\lambda\in\Lambda(\mathcal{M}^{-1}\mathcal{A})} |p(\lambda)| := \widehat{\rho}_k. \tag{7.4.2}$$

Note that the use of a positive definite preconditioner was necessary as $\|\cdot\|_{\mathcal{M}^{-1}}$ does not define a norm for indefinite $\mathcal{M}$. A consequence is that preconditioning can not alter the inertia of the original system since $\mathcal{M}^{-\frac{1}{2}}\mathcal{A}\mathcal{M}^{-\frac{1}{2}}$ is a congruence transform and the Sylvester Law of Inertia applies. That is, any symmetric and indefinite matrix preconditioned by a positive definite matrix is necessarily left with the same number of positive and negative eigenvalues. The role of preconditioning in this case is therefore to cluster both the positive and the negative eigenvalues so that the polynomial approximation error $\widehat{\rho}_k$ in (7.4.2) is small for low number of iterations, k.

A second point is that (unlike in the case of the conjugate gradient method) reduction of the residual in the preconditioned MINRES algorithm is in a norm which is dependent on the preconditioner. Thus one must be careful not to select a preconditioner which simply distorts this norm. We will return to this point later.

At each MINRES iteration we will require the solution of a system of equations with the preconditioner as coefficient matrix. Thus from the point of view of practicality, this must be readily achieved.

7.4.3 Preconditioning

The convergence estimate (7.4.2) shows that convergence depends on the eigenvalues of the preconditioned system: our goal now is to estimate these eigenvalues. In particular for a partial differential equation problem such as the Stokes problem, we are interested in the rate of MINRES convergence for large discrete problems, i.e. for discretisations on fine meshes which lead to very large dimensional matrix systems. It is therefore appropriate to consider how the rate of convergence depends on the representative mesh-size, h, as $h \to 0$. The best case will be if the number of iteration required to achieve convergence to a given tolerance does not depend on h.

Since this preserves the underlying block structure of the coefficient matrix, we are interested in block diagonal preconditioning matrices of the form

$$\begin{pmatrix} P & 0 \\ 0 & M \end{pmatrix} \tag{7.4.3}$$

where both P and M are symmetric and positive definite. The eigenvalues we wish to estimate are therefore the eigenvalues λ of

$$\begin{pmatrix} A & B^t \\ B & -\beta S \end{pmatrix}\begin{pmatrix} u \\ p \end{pmatrix} = \lambda \begin{pmatrix} P & 0 \\ 0 & M \end{pmatrix}\begin{pmatrix} u \\ p \end{pmatrix}.$$

We readily see that if $P = A$ then $\lambda = 1$ is an eigenvalue of multiplicity $n_u - n_p$ corresponding to any eigenvector $[u, 0]^t$ with $Bu = 0$. (The multiplicity comes simply from the size of the right null space of the rectangular matrix B). In the stable case ($S = 0$), if also $M = BA^{-1}B^t$, the remaining eigenvalues satisfy

$$(1-\lambda)Au = -B^t p \qquad \text{and} \qquad Bu = \lambda BA^{-1}B^t p$$

or by eliminating u,

$$(\lambda^2 - \lambda + 1)BA^{-1}B^t p = 0.$$

Thus since the assumed inf-sup stability in this case ensures that $BA^{-1}B^t$ is positive definite, we deduce that $\lambda = 1/2 \pm \sqrt{5}/2$ are the remaining eigenvalues each with multiplicity n_p. This is an ideal situation from the point of view of convergence of MINRES: since the preconditioned matrix $\mathcal{M}^{-\frac{1}{2}}\mathcal{A}\mathcal{M}^{-\frac{1}{2}}$ has only three distinct eigenvalues the convergence bound (7.4.2) will be zero for $k = 3$ as there is a cubic polynomial with these three roots. That is, MINRES will terminate with the exact solution after three iterations regardless of the size of the discrete problem.

Unfortunately use of the Schur complement $BA^{-1}B^t$ in the preconditioner is not desirable since it is in general a dense matrix which is not easy to construct let alone to invert (or rather solve a system) at each MINRES iteration. But this is where the discrete inf-sup stability condition (7.2.30) and (7.2.37) provides the key: the pressure mass matrix Q is spectrally equivalent to $BA^{-1}B^t$ and so we lose little by selecting $M = Q$. The analysis with this choice is similar to the above: we have

$$\begin{aligned} Au + B^t p &= \lambda Au \\ Bu &= \lambda Qp. \end{aligned}$$

The case $\lambda = 1$ arises with the same eigenvectors (and thus multiplicity) as above, and for $\lambda \neq 1$ eliminating u using the first of these equations gives

$$BA^{-1}B^t p = \lambda(\lambda - 1)Qp.$$

Thus for each eigenvalue μ of $Q^{-\frac{1}{2}}BA^{-1}B^tQ^{-\frac{1}{2}}$ there are a pair of eigenvalues

$$\lambda = \frac{1}{2} - \frac{1}{2}\sqrt{1+4\mu} < 0 \quad \text{and} \quad \lambda = \frac{1}{2} + \frac{1}{2}\sqrt{1+4\mu} > 0$$

of the original problem. Now since discrete inf-sup stability and boundedness imply $\gamma^2 \le \mu \le \Gamma^2$, we see that

$$\lambda \in \left[\frac{1-\sqrt{1+4\Gamma^2}}{2}, \frac{1-\sqrt{1+4\gamma^2}}{2}\right] \cup \{1\} \cup \left[\frac{1+\sqrt{1+4\gamma^2}}{2}, \frac{1+\sqrt{1+4\Gamma^2}}{2}\right]$$

for every eigenvalue. That is, the multiple eigenvalue $\lambda = 1$ is retained and the remaining eigenvalues are pairwise symmetric about $\frac{1}{2}$ and lie in small intervals which are uniformly bounded and uniformly bounded away from the origin. In this situation the convergence of MINRES will not take only three iterations, but nevertheless it will be fast and (crucially) will be independent of the size of the discrete problem. For an unstable approximation we have $\gamma = 0$ (or $\gamma \to 0$ under mesh refinement), so the negative eigenvalues would not be bounded away from the origin and poor convergence results.

Before proceding to more general theory, we motivate other approximations which will preserve the effective form of this 'ideal' preconditioner but which lead to a more practical overall preconditioner.

It is apparent that preconditioning with $\mathcal{M}$ as above requires at each MINRES iteration the solution of two systems of equations of size n_u and n_p and with coefficient matrices P and M respectively. The 'ideal' choice $P = A$ thus requires an exact solution of a Poisson equation for each of the velocity components since A is the vector Laplacian coming from approximation of the viscous terms. Conveniently there has been much analysis of preconditioners for the Laplacian (see for example the papers by Xu and Chan in this volume).

We do not need to use an inner preconditioned conjugate gradient iteration to effect an exact solution, but are in a position to simply take P to be a domain decomposition or multilevel preconditioner for example. That is we simply let P be a preconditioner for the Laplacian. By applying a suitable scaling to P if necessary we will assume that

$$\alpha \le \frac{u^t A u}{u^t P u} \le 1 \qquad \text{for all } u. \tag{7.4.4}$$

If we use a powerful preconditioner such as a multigrid cycle, then α will be near 1 independently of the discrete problem size (usually expressed in terms of inverse powers of the mesh size parameter, h) and we might expect that only a few more MINRES iterations will be required than if we made the more expensive choice $P = A$. If we use a weaker preconditioner such as diagonal scaling for which $\alpha = O(h^2)$ then more MINRES iterations will be needed for convergence.

It is a much simpler matter to approximate the ideal choice of M further by approximating the pressure mass matrix Q without significantly affecting the convergence of MINRES. The simplest choice $M = \text{diag}(Q)$ is proved to be a good approximation to Q in [49]. Specifically we will assume

$$\theta^2 \le \frac{p^t Q p}{p^t M p} \le \Theta^2 \qquad \text{for all } p. \tag{7.4.5}$$

Using a continuous P_1 pressure approximation for example, replacing the mass matrix Q by its diagonal is very convenient computationally, and furthermore (7.4.5) is satisfied in this case with $\theta = 1/\sqrt{2}$ and $\Theta = \sqrt{2}$.

7.4.4 Eigenvalue bounds

Our analysis proceeds with the assumptions (7.4.4), (7.4.5) and the further assumption of boundedness of the stabilisation matrix S:

$$\frac{p^t S p}{p^t Q p} \leq \Delta^2 \qquad \text{for all } p. \tag{7.4.6}$$

Using the locally stabilised $Q_1 - P_0$ mixed approximation described in section 2, we choose $M = Q$ in (7.4.5) since Q is a diagonal matrix and in this case we know that $\Delta = 2$ on a uniform mesh.

For the eigenvalue analysis it is convenient to consider the symmetrically preconditioned system:

$$\begin{aligned} \mathcal{M}^{-\frac{1}{2}} \mathcal{A} \mathcal{M}^{-\frac{1}{2}} &= \begin{pmatrix} P^{-\frac{1}{2}} A P^{-\frac{1}{2}} & P^{-\frac{1}{2}} B^t M^{-\frac{1}{2}} \\ M^{-\frac{1}{2}} B P^{-\frac{1}{2}} & -\beta M^{-\frac{1}{2}} S M^{-\frac{1}{2}} \end{pmatrix} \\ &= \begin{pmatrix} \widetilde{A} & \widetilde{B}^t \\ \widetilde{B} & -\beta \widetilde{S} \end{pmatrix} = \widetilde{\mathcal{A}}. \end{aligned} \tag{7.4.7}$$

In the following, we denote by $\sigma_{\max}$ the largest singular value of $\widetilde{B}$ (i.e. the largest eigenvalue of $\widetilde{B}\widetilde{B}^t$).

Lemma 4.1. *All negative eigenvalues λ of $\widetilde{\mathcal{A}}$ satisfy*

$$\frac{1}{2}\left(\alpha - \beta\Delta^2\Theta^2 - \sqrt{\left(\alpha + \beta\Delta^2\Theta^2\right)^2 + 4\sigma_{\max}^2}\right) \leq \lambda \tag{7.4.8}$$

and

$$\lambda \leq \frac{1}{2}\left(\alpha - \sqrt{\alpha^2 + 4\gamma^2\theta^2\alpha}\right) \tag{7.4.9}$$

and all positive eigenvalues λ of $\widetilde{\mathcal{A}}$ satisfy

$$\alpha \leq \lambda, \tag{7.4.10}$$

and

$$\lambda \leq \frac{1}{2}\left(1 + \sqrt{1 + 4\sigma_{\max}^2}\right). \tag{7.4.11}$$

Proof. If λ is an eigenvalue of $\widetilde{\mathcal{A}}$ then there are vectors u, p not both zero satisfying

$$\widetilde{A}u + \widetilde{B}^t p = \lambda u \tag{7.4.12}$$

$$\widetilde{B}u - \beta\widetilde{S}p = \lambda p. \tag{7.4.13}$$

If $\lambda > 0$ then $u \neq 0$ since otherwise (7.4.13) implies $p = 0$ as $\widetilde{S}$ is positive semi-definite. If $\lambda < 0$ then $p \neq 0$ since otherwise (7.4.12) implies $u = 0$ as $\widetilde{A}$ is positive definite.

Taking the scalar product of (7.4.12) with u and the scalar product of (7.4.13) with p and subtracting gives

$$u^t \widetilde{A} u + \beta p^t \widetilde{S} p = \lambda\, u^t u - \lambda\, p^t p$$

which using (7.4.4) and the positive semi-definiteness of $\beta \widetilde{S}$ gives

$$(\alpha - \lambda) u^t u \leq -\lambda p^t p$$

leading to (7.4.10) for positive λ since $u \neq 0$ in this case.

Further for $\lambda > 0$, substituting for p from (7.4.13) into the scalar product of u with (7.4.12) gives

$$u^t \widetilde{A} u + \frac{1}{\lambda} u^t \widetilde{B}^t \left(I + \frac{\beta}{\lambda} \widetilde{S} \right)^{-1} \widetilde{B} u = \lambda u^t u$$

where the stated matrix inverse certainly exists because $\beta, \lambda > 0$ and $\widetilde{S}$ is positive semi-definite. Moreover the maximum eigenvalue of $(I + \frac{\beta}{\lambda} \widetilde{S})^{-1}$ is 1 thus

$$\lambda u^t \widetilde{A} u + u^t \widetilde{B}^t \widetilde{B} u \geq \lambda^2 u^t u$$

from which follows

$$0 \geq \lambda^2 - \lambda - \sigma_{\max}^2.$$

This gives (7.4.11).

For $\lambda < 0$, $\widetilde{A} - \lambda I$ is invertible, so we can take the scalar product of (7.4.13) with p and substitute for u from (7.4.12) to obtain

$$p^t \widetilde{B} (\widetilde{A} - \lambda I)^{-1} \widetilde{B}^t p + \beta p^t \widetilde{S} p = -\lambda p^t p. \tag{7.4.14}$$

Considering (7.4.14), if $\lambda < 0$ is an eigenvalue of $\widetilde{A}$ then

$$p^t \widetilde{B} \widetilde{A}^{-\frac{1}{2}} (I - \lambda \widetilde{A}^{-1})^{-1} \widetilde{A}^{-\frac{1}{2}} \widetilde{B}^t p + \beta p^t \widetilde{S} p = -\lambda p^t p$$

where $p \neq 0$. Because the eigenvalues of $(I - \lambda \widetilde{A}^{-1})^{-1}$ are

$$(1 - \lambda/\alpha)^{-1} \leq \ldots \leq (1 - \lambda)^{-1},$$

we have

$$(1 - \lambda/\alpha)^{-1} p^t \widetilde{B} \widetilde{A}^{-1} \widetilde{B}^t p + \beta p^t \widetilde{S} p \leq -\lambda p^t p,$$

and since $0 \leq (1 - \lambda/\alpha)^{-1} \leq 1$ there follows

$$(1 - \lambda/\alpha)^{-1} \left(p^t \widetilde{B} \widetilde{A}^{-1} \widetilde{B}^t p + \beta p^t \widetilde{S} p \right) \leq -\lambda p^t p.$$

Using (7.4.7) to express this in terms of the blocks of the original unpreconditioned Stokes matrix (7.4.1) this is

$$(1 - \lambda/\alpha)^{-1} p^t M^{-\frac{1}{2}} (B A^{-1} B^t + \beta S) M^{-\frac{1}{2}} p \leq -\lambda p^t p.$$

Now using the stability property (7.2.36) this implies

$$\gamma^2(1-\lambda/\alpha)^{-1}p^tM^{-\frac{1}{2}}QM^{-\frac{1}{2}}p \le -\lambda p^tp$$

which by employing (7.4.5) further implies

$$\gamma^2\theta^2(1-\lambda/\alpha)^{-1}p^tp \le -\lambda p^tp.$$

Since $p \neq 0$ this gives

$$0 \le \lambda^2 - \alpha\lambda - \alpha\gamma^2\theta^2$$

from which (7.4.9) easily follows.

To derive (7.4.8) we use (7.4.6) and (7.4.5) in (7.4.14) to obtain

$$(\alpha-\lambda)^{-1}\sigma_{\max}^2 + \beta\Delta^2\Theta^2 \ge -\lambda$$

or

$$0 \ge \lambda^2 + (\beta\Delta^2\Theta^2 - \alpha)\lambda - \sigma_{\max}^2 - \beta\Delta^2\Theta^2\alpha$$

which yields the result. □

It is convenient to remove $\sigma_{\max}$ from these bounds since estimates for this quantity are not readily available.

Lemma 4.2.

$$\sigma_{\max} \le \Gamma\Theta \tag{7.4.15}$$

Proof. For all p we have

$$\begin{aligned} p^t\widetilde{B}\widetilde{B}^tp &= p^tM^{-\frac{1}{2}}BP^{-1}B^tM^{-\frac{1}{2}}p \\ &\le p^tM^{-\frac{1}{2}}BA^{-1}B^tM^{-\frac{1}{2}}p \end{aligned}$$

using (7.4.4). So given that (7.2.37) holds in the stable or stabilised case we have

$$\begin{aligned} p^t\widetilde{B}\widetilde{B}^tp &\le \Gamma^2\, p^tM^{-\frac{1}{2}}QM^{-\frac{1}{2}}p \\ &\le \Gamma^2\, \Theta^2\, p^tp \end{aligned}$$

where we have further used (7.4.5). We have thus proved

$$\sigma_{\max}^2 \le \Gamma^2\, \Theta^2$$

and hence (7.4.15). □

Employing Lemma 4.2, the bounds (7.4.8) and (7.4.11) become

$$\frac{1}{2}\left(\alpha - \beta\Delta^2\Theta^2 - \sqrt{(\alpha+\beta\Delta^2\Theta^2)^2 + 4\Gamma^2\,\Theta^2}\right) \le \lambda \tag{7.4.16}$$

and

$$\lambda \le \frac{1}{2}\left(1 + \sqrt{1+4\Gamma^2\,\Theta^2}\right). \tag{7.4.17}$$

Regarding (7.4.16),(7.4.9),(7.4.10) and (7.4.17) as the best bounds which we can estimate, we are now in a position to find an upper bound on the convergence rate of the preconditioned MINRES algorithm by considering the approximation problem in (7.4.2). Before doing so let us just point out the dependencies of the relevant quantities α, γ, Γ, θ, Θ and Δ as well as the stabilisation parameter β (which arises only in a stabilised formulation):

- α: depends on how well the preconditioning block P approximates the discrete Laplacian A
- γ: stability constant—bounded above zero independently of the mesh
- Γ: boundedness constant: $\Gamma \leq \sqrt{d}$ for any domain $\Omega \subset \mathbb{R}^d$ (see (7.2.39))
- θ, Θ: positive constants independent of problem size even for the simple choice $M = \mathrm{diag}(Q)$. For such a choice of the preconditioning block M, these constants are tabulated in ([49]) for many different finite elements types
- Δ: upper bound on the stabilisation matrix S—an $O(1)$ constant.
- β: positive stabilisation parameter optimally chosen to be just large enough to achieve stability (see [43]).

Any of these parameters may depend on the geometry of the domain and/or the computational grid, *BUT it is only α which can depend explicity on the size of the discrete problem.* That is, the only way that mesh-size dependence arises in the eigenvalues bounds for the preconditioned Stokes coefficient matrix $\widetilde{\mathcal{A}}$ is through a dependence of α on the representative mesh-size, h. Therefore, provided that some simple approximation of the pressure mass matrix is used so that (7.4.5) is satisfied and provided a suitable stable or stabilised formulation is employed so that (7.2.30) or (7.2.36) and (7.4.6) hold then the convergence of the preconditioned MINRES algorithm will be essentially determined by the quality of the Laplacian preconditioner, P. Let us illustrate with a few examples.

Example 1: $P = \mathrm{diag}(A)$ and M is any suitable choice (such as $\mathrm{diag}(Q)$) which satisfies (7.4.5).

For this case we have $\alpha = ch^2 + \mathrm{O}(h^4)$ for some constant c independent of h (see for example [2], pp. 240). By considering the leading asymptotic term for small h it is apparent that the eigenvalue bounds (7.4.16),(7.4.9),(7.4.10) and (7.4.17) define a pair of eigenvalue inclusion intervals of the form

$$\Lambda(\mathcal{M}^{-\frac{1}{2}}\mathcal{A}\mathcal{M}^{-\frac{1}{2}}) \subset [-a, -bh] \cup [ch^2, d]. \tag{7.4.18}$$

The constants a, b, c and d are defined in terms of γ, Γ, θ, Θ, Δ and β by the above formulae *but* they do not depend on h. (c is exactly as above because of the simple form of (7.4.10)). We demonstrate the asymptotic manipulation for the least obvious bound (7.4.9):

$$\begin{aligned}\lambda &\leq \frac{1}{2}\left(ch^2 + O(h^4) - \sqrt{(ch^2 + O(h^4))^2 + 4\gamma^2\theta^2(ch^2 + O(h^4))}\right)\\ &= \frac{1}{2}\left(ch^2 + O(h^4) - 2c^{\frac{1}{2}}\theta\gamma h\left(1 + O(h^4)\right)^{\frac{1}{2}}\right)\\ &= -c^{\frac{1}{2}}\gamma\theta h + \frac{1}{2}ch^2 + O(h^4).\end{aligned}$$

The important point to note here is that as $h \to 0$ the negative eigenvalues approach the origin at only half the rate at which the positive eigenvalues can approach from above.

h	λ^-_{max}	λ^-_{min}	λ^+_{min}	λ^+_{max}
1/8	−0.7547	−0.1556e0	0.2747e0	2.0640
1/16	−0.7701	−0.9500e-1	0.7444e-1	2.1347
1/32	−0.7740	−0.5253e-1	0.1902e-1	2.1531
1/64	−0.7749	−0.2770e-1	0.4783e-2	2.1577
1/128	−0.7752	−0.1427e-1	0.1198e-2	2.1589

TABLE 7.2. Extreme eigenvalues: $Q_1 - P_0$ element with diagonal preconditioning

In table 7.2 we show the results of eigenvalue computations on the diagonally preconditioned Stokes coefficient matrix as above for a driven cavity flow problem (see section 6 for associated software). We show the extreme eigenvalues of $\mathcal{M}^{-\frac{1}{2}}\mathcal{A}\mathcal{M}^{-\frac{1}{2}}$ for a sequence of regular grids refined by bisection and using the locally stabilised $Q_1 - P_0$ element with the 'optimal' stabilisation parameter value $\beta = 0.058$ (see [43]). The driven cavity problem was solved on only half of the flow domain by using the natural symmetry about the centreline. The most positive and most negative eigenvalues (λ^+_{max} and λ^-_{max} respectively) clearly approach constant values as h is reduced, the negative eigenvalue nearest to the origin (λ^-_{min}) is approximately halved and the smallest positive eigenvalue (λ^+_{min}) reduces by approximately a quarter as h is halved: these results therefore show that (7.4.18) is descriptive and is not just providing crude bounds.

We note that example 1 is illustrative of the generic situation: if P is chosen such that $\alpha = O(h^r)$ and M is an appropriate approximation of the pressure mass matrix then

$$\Lambda(\mathcal{M}^{-\frac{1}{2}}\mathcal{A}\mathcal{M}^{-\frac{1}{2}}) \subset [-a, -bh^{r/2}] \cup [ch^r, d]. \qquad (7.4.19)$$

That is the negative eigenvalues always approach the origin at half of the rate of the positive eigenvalues. The analysis given here therefore applies

h	$\mu_{\min}$	$\mu_{\max}$	$\lambda_{\min}(P^{-1}A)$	$\lambda_{\max}(P^{-1}A)$
1/8	0.1686	0.9340	0.8519	1.0000
1/16	0.1655	0.9862	0.8220	1.0000
1/32	0.1642	0.9967	0.8090	1.0000

TABLE 7.3. Extreme eigenvalues of $Q^{-1}BA^{-1}B^t$ and $P^{-1}A$: $Q_1-iso-Q_2$ element with multigrid preconditioning

to a wide range of Laplacian preconditioners including, for example, the modified incomplete cholesky factorisation ([32], [26]) for which $r = 1$.

Example 2: P is a multigrid cycle for A (see for example the paper by Xu in this volume) and $M = Q$ (or some approximation).

This is actually the easiest situation from the view point of the analysis as α is bounded away from zero independently of h. In table 7.3 we give the computed extremal eigenvalues of $P^{-1}A$ for our test problem employing the stable $Q_1-iso-Q_2$ element on a sequence of refined meshes. The preconditioner P represents a single multigrid V-cycle with an 'optimally' damped Jacobi smoother. It is apparent that $\alpha \approx 0.8$ in this situation. Also tabulated in 7.3 are the extreme eigenvalues, $\mu_{\min}$ and $\mu_{\max}$ of $Q^{-1}BA^{-1}B^t$: these show that $\gamma \approx 0.16$ and $\Gamma \approx 1$ for this element. It follows that the bounds (7.4.16),(7.4.9),(7.4.10) and (7.4.17) are all independent of h.

An interesting point arises with the use of spectrally equivalent preconditioners such as in this example, namely convergence of MINRES occurs in norm which is naturally associated with the problem, see [44].

7.4.5 The rate of convergence of MINRES

In the case of a spectrally equivalent Laplacian preconditioner P such as a multigrid cycle, we may simply note that since all of the eigenvalues are bounded away from infinity and away from the origin independently of the mesh-size h, then $\hat{\rho}_k$ in (7.4.2) is also independent of h. The convergence of MINRES in this case should therefore be independent of problem size. This is clearly displayed in table 7.4 where we present some preconditioned MINRES iteration counts.

The problem is again the leaky lid driven cavity but solved on only half of the domain by using the natural symmetry. For these results the stable $Q_1-iso-Q_2$ element was used and the convergence criterion was a reduction by 10^{-6} in the $\mathcal{M}^{-1}$-norm of the residual. The preconditioner A_{MG1} represents a single multigrid V-cycle: as above an 'optimally' damped Jacobi smoother was employed. Iteration and total flop counts using the 'ideal' block preconditioner $P = A$, the diagonally scaled MINRES method of example 1 above, and the block preconditioner based on a Modified Incomplete Cholesky factorisation (MIC) are also included for comparison. Note that the cost of the incomplete factorisation is not included in the flop

h	$P = A$ $M = Q$	$P = A_{MG1}$ $M = Q$	$P = A_{MIC}$ $M = Q$	$P = \text{diag}(A)$ $M = \text{diag}(Q)$
1/8	23	27 (0.44)	28 (0.19)	41 (0.23)
1/16	25	28 (1.97)	38 (1.22)	94 (2.25)
1/32	27	30 (8.10)	53 (7.69)	206(20.63)
1/64	27	31 (36.33)	78 (54.89)	427(175.04)

TABLE 7.4. MINRES iterations (Megaflops): Q_1–*iso*–Q_2 element

counts given in the table; preconditioning is via sparse upper and lower triangular matrix solves in this case. The use of $P = A$ is expensive in operation counts since a full factorisation is needed in this case, so only the MINRES iteration counts are included for comparison. The computations were done on a Sun Sparcstation-10 using MATLAB 4.1.

Note that use of $P = A$ or of the more practical multigrid cycle as a preconditioner for the Laplacian does indeed imply that the number of MINRES iterations does not depend on the discrete problem size. Use of the multigrid preconditioner rather than the 'ideal' choice $P = A$ is seen to increase the number of iterations only slightly: it is nearly ideal, but much more efficient overall. Indeed the multigrid preconditioner gives an 'optimal' Stokes solver: the total number of floating point operation increases by a factor of approximately four each time the grid is refined to create four times as many discrete variables. This is a very desirable property.

It remains to analyse the convergence of MINRES for preconditioners such as those in the two right hand columns of table 7.4 above which do not involve multigrid or some other spectrally equivalent Laplacian preconditioner.

Having estimated the eigenvalue spectrum in the form $\Lambda(\widetilde{\mathcal{A}}) \subset E$ where E comprises two intervals of the form $[-a, -b] \cup [c, d]$ with a, b, c and d being positive, our attention therefore turns to the approximation problems

$$\begin{aligned} \widehat{\rho}_k &= \min_{p\in\Pi_k, p(0)=1} \max_{\lambda\in\Lambda(\widetilde{\mathcal{A}})} |p(\lambda)| && (7.4.20) \\ &\leq \min_{p\in\Pi_k, p(0)=1} \max_{x\in E} |p(x)| := \rho_k. && (7.4.21) \end{aligned}$$

We know from (7.4.2) that $\widehat{\rho}_k$ bounds the relative reduction in the MINRES residual after k iterations; if little is lost in the inequality above then it is more tractable to deal with the approximation problem (7.4.21) on intervals rather than (7.4.20) on the discrete eigenvalue set. A rapidly decreasing sequence ρ_k will still indicate fast convergence.

In fact, when a single number is desired to represent convergence, it is convenient to consider the *asymptotic convergence factor*

$$\rho := \lim_{k\to\infty} \rho_k^{1/k}$$

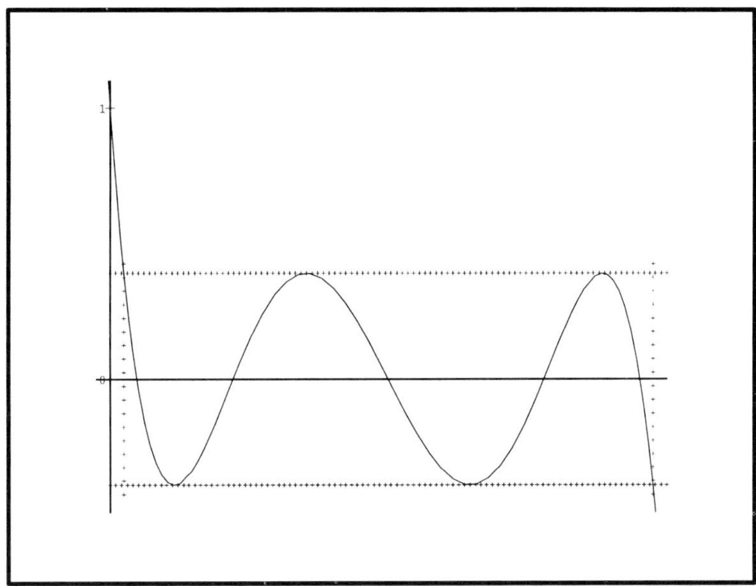

FIGURE 7.11. Optimal polynomial on a single interval

which represents a bound on the average contraction in the residual per iteration.

Firstly we require some results from Approximation Theory to characterise polynomials $p \in \Pi_k, p(0) = 1$ which solve the minimax problem (7.4.21) for different sets E (see for example [33] for these results). Note that $\{\rho_k\}$ must be a decreasing (non-negative) sequence as each successive iteration simply increases the allowable degree of p by one. Existence and uniqueness of the solution is known and a characterisation is expressed in terms of the number of points in the set E at which ρ_k is attained.

Let us consider first the simpler problem when $E = [c, d]$ with $c > 0$ such as would arise if $\widetilde{A}$ were symmetric and *positive definite.* In this case the optimal polynomial $p \in \Pi_k$ satisfies $|p(x_j)| = \rho_k$ for $k + 1$ distinct points $a = x_0 < x_1 < \ldots < x_{k-1} < x_k = b$. Moreover $p(x_j) = -p(x_{j-1})$ for $j = 1, 2, \ldots, k$. It is then a straightforward matter to see that p must be as sketched in figure 7.11.

We now take the unusual step of writing down an ordinary differential equation initial value problem which must be satisfied by the polynomial p. Noting that $p = \pm\rho_k$ at the points $x_1, \ldots, x_{k-1}$ where the derivative p' vanishes as well as at the endpoints of the interval $[c, d]$ we have

$$k^2(p^2(x) - \rho_k^2) = (p'(x))^2(x - c)(x - d) \tag{7.4.22}$$

where the constant scaling term k^2 comes from equating the leading coefficient (of x^{2k}) on both sides of this equation. The 'initial' value is $p(0) = 1$.

The nonlinear ordinary differential equation (7.4.22) can now be differentiated to give

$$2k^2pp' = 2p'p''(x-c)(x-d) + (p')^2(2x-c-d)$$

so the common factor p' can be cancelled to reveal the linear ordinary differential equation

$$(x-c)(x-d)p'' + (x-(c+d)/2)p' - k^2p = 0. \tag{7.4.23}$$

This is the classical Chebyshev equation (see for example [24], pp. 1033) the solutions of which are the well known Chebyshev polynomials $T_k(x) = \cos k\phi,\ x = \cos\phi$ suitably shifted to the interval $[c,d]$ and scaled to satisfy the side condition $p(0) = 1$. (This may be discovered by seeking a series solution).

This is a rather unusual way to show the well-known result that the solution of the polynomial approximation problem (7.4.21) is

$$p(x) = T_k\left(\frac{2x-c-d}{d-c}\right) \Big/ T_k\left(\frac{c+d}{c-d}\right)$$

(see for example [2]).

Using the definition in terms of the cosine it follows that $-1 \leq T_k \leq 1$ for the relevant argument and so the preconditioned MINRES convergence estimate (7.4.2) becomes

$$\frac{\|r_k\|}{\|r_0\|} \leq \rho_k = 1\Big/T_k\left(\frac{c+d}{c-d}\right). \tag{7.4.24}$$

If c and/or d are defined asymptotically in terms of h then use can be made of the asymptotics of Chebyshev polynomials to give asymptotic formulae for ρ_k in terms of h. For example if $c = O(h^r), d = O(1)$ then

$$\lim_{k\to\infty} \rho_k^{1/k} = 1 - O(h^{r/2})$$

(see [2]).

We use this non-standard approach here because it is actually more general since it extends to various situations where $\mathcal{M}^{-\frac{1}{2}}\mathcal{A}\mathcal{M}^{-\frac{1}{2}}$ is indefinite.

The first indefinite case we consider is $E = [-d,-c] \cup [c,d]$. We say that a symmetric matrix $\widetilde{\mathcal{A}}$ with $\lambda \in \Lambda(\widetilde{\mathcal{A}}) \Rightarrow -\lambda \in \Lambda(\widetilde{\mathcal{A}})$ is 'symmetrically indefinite': such a matrix necessarily leads to consideration of an inclusion set of this form. In this case the optimal polynomials p in (7.4.21) must inherit the symmetry of the inclusion set and so must be of the form sketched in figure 7.12. We see that p' vanishes at the origin as well as at the points

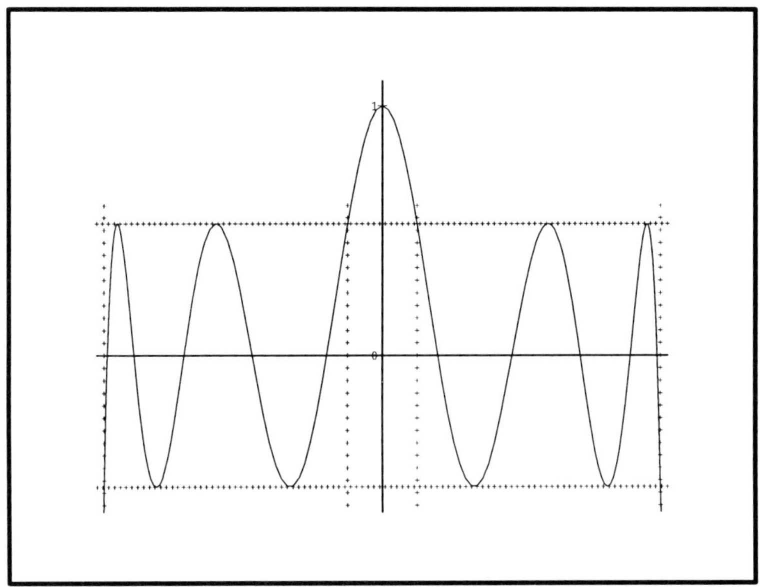

FIGURE 7.12. Optimal polynomial on two intervals symmetric about the origin

where p attains $\pm\rho_k$, so in a similar manner to the above we obtain the ordinary differential equation

$$\begin{aligned} k^2x^2(p^2(x) - \rho_k^2) &= (p'(x))^2(x-c)(x-d)(x+c)(x+d) \\ &= (p'(x))^2(x^2-c^2)(x^2-d^2) \end{aligned}$$

which is necessarily satisfied by the optimal polynomial.

Making the change of variable $y = x^2$ and setting $p(x) = q(y)$ so that $p'(x) = 2xq'(y)$ we obtain

$$k^2x^2(q^2(y) - \rho_k^2) = 4x^2(q'(y))^2(y-c^2)(y-d^2)$$

or

$$(k/2)^2(q^2(y) - \rho_k^2) = (q'(y))^2(y-c^2)(y-d^2).$$

This is precisely in the form of (7.4.22) and so proceding as above we obtain q as the Chebyshev polynomial of degree $k/2$ shifted to the interval $[c^2, d^2]$ and scaled to satisfy $q(0) = 1$:

$$q(y) = T_{k/2}\left(\frac{2y - c^2 - d^2}{d^2 - c^2}\right) \Big/ T_{k/2}\left(\frac{c^2 + d^2}{c^2 - d^2}\right).$$

Thus in terms of the optimal polynomial p of degree k we have

$$p(x) = T_{k/2}\left(\frac{2x^2 - c^2 - d^2}{d^2 - c^2}\right) \Big/ T_{k/2}\left(\frac{c^2 + d^2}{c^2 - d^2}\right)$$

and so for symmetrically indefinite systems the MINRES convergence estimate is

$$\frac{\|r_k\|}{\|r_0\|} \le \rho_k = 1 \Big/ T_{k/2}\left(\frac{c^2 + d^2}{c^2 - d^2}\right). \qquad (7.4.25)$$

In a partial differential equation situation where $c = O(h^r), d = O(1)$ as above, we would thus have

$$\lim_{k\to\infty} \rho_k^{1/k} = 1 - O(h^r)$$

It is instructive to compare this with the convergence that would be achieved by an iterative method such as MINRES (or Conjugate Gradients) applied to the symmetric and positive definite 'normal equations' (NE), $\widetilde{\mathcal{A}}^2\,\widetilde{x} = \widetilde{\mathcal{A}}\,\widetilde{b}$. For this system $\Lambda(\widetilde{\mathcal{A}}^2) \subset [c^2, d^2]$ so that we can estimate convergence using (7.4.24) to obtain

$$\frac{\|r_k^{NE}\|}{\|r_0^{NE}\|} \le \rho_k^{NE} = 1 \Big/ T_k\left(\frac{c^2 + d^2}{c^2 - d^2}\right). \qquad (7.4.26)$$

Comparing (7.4.25) with (7.4.26) we see that we can expect that MINRES for the original indefinite symmetric problem will take twice the number of iterations as MINRES (or the more efficient Conjugate Gradient method) for the normal equations to achieve the same reduction of residual. Since for the normal equations *two* matrix-vector multiplies will be required at each iteration compared to only one for the indefinite system so that the normal equation method will be twice as expensive per iteration, we deduce that there is essentially nothing to choose between these two approaches in this case. That is, the iterative solution of 'symmetrically indefinite' systems using a method such as MINRES is no better than the much more generally applicable normal equations approach. A more precise statement of this result is given by Freund [19].

When proposing the use of preconditioned MINRES for a class of indefinite systems it is therefore important to show that the eigenvalues are *not* symmetric about the origin. For the Stokes problem the results of the previous section establish a precise non-symmetry in the eigenvalues: for non-optimal preconditioners, the negative eigenvalues approach the origin at half of the rate of the positive eigenvalues under mesh refinement.

In this situation, the approach employing ordinary differential equations as above can still be employed to derive a convergence estimate, though the details are rather more involved (see [50]). We quote only the result: If the eigenvalues of $\widetilde{\mathcal{A}}$ are contained in a set of the form

$$[-a, -bh^{r/2}] \cup [ch^r, d]$$

then

$$\lim_{k\to\infty} \rho_k^{1/k} = 1 - O(h^{3r/4}).$$

That is the convergence of MINRES on the Stokes problem is at a rate precisely half way between that achived for a symmetric positive definite problem such as the Laplacian and that achieved for the corresponding normal equations.

7.5 Solution methods for the discrete Oseen equations

In this section we examine methods for solving the steady-state Navier-Stokes equations that combine and build on the techniques of sections 3 and 4. The methods are designed for the steady-state Oseen equations ($\Delta t \to \infty$ in (7.1.11)). These equations also arise from a nonlinear iteration for solving the Navier-Stokes equations in which $\mathbf{u}^*$ represents the iterate from a given step and the solution $\mathbf{u}$ is the iterate for the next step. See [29] for a convergence analysis.

Discretisation leads to a matrix problem

$$\begin{pmatrix} F & B^t \\ B & -\beta S \end{pmatrix} \begin{pmatrix} u \\ p \end{pmatrix} = \begin{pmatrix} f \\ 0 \end{pmatrix}, \tag{7.5.1}$$

where u and p now represent discrete versions of velocity and pressure, respectively. F is a discrete vector convection-diffusion operator and B represents the coupling between the discrete velocity u and the pressure p. For simplicity of presentation we only present results for the unstabilised case $S = 0$.

7.5.1 *Preconditioning I: Convection-diffusion solves*

We first describe two preconditioning techniques developed in [7] that generalise the methods of section 4 essentially by replacing the approximation P to the vector Laplacian operator in (7.4.3) with an approximation to the vector convection-diffusion operator F. It is easiest to describe the ideas using the exact operator F. Thus, consider the block diagonal preconditioner

$$\begin{pmatrix} F & 0 \\ 0 & \frac{1}{\nu}Q \end{pmatrix} \tag{7.5.2}$$

where Q is the pressure mass matrix. As in section 4, the eigenvalues of the preconditioned system are the solutions of the generalised eigenvalue problem

$$\begin{pmatrix} F & B^t \\ B & 0 \end{pmatrix} \begin{pmatrix} u \\ p \end{pmatrix} = \lambda \begin{pmatrix} F & 0 \\ 0 & \frac{1}{\nu}Q \end{pmatrix} \begin{pmatrix} u \\ p \end{pmatrix}.$$

These are given by $\lambda = 1$ or

$$\lambda = \frac{1 \pm \sqrt{1+4\mu}}{2}$$

where μ comes from the generalised eigenvalue problem for the Schur complement system,

$$BF^{-1}B^t p = \mu \left(\frac{1}{\nu}Q\right) p. \tag{7.5.3}$$

The following result provides a bound on μ.

Theorem 7.5.1 *The eigenvalues of the generalised Schur complement problem* (7.5.3) *for the Oseen operator are contained in a rectangular box in the right half plane of the form*

$$\left[\frac{\gamma^2\nu^2}{\delta^2+\nu^2}, \Gamma^2\right] \times i \left[\frac{\Gamma^2}{2}, \frac{\Gamma^2}{2}\right].$$

where γ *and* Γ *are as in* (7.2.30) *and* (7.2.37), *and* $\delta = \rho(A^{-1}N)$.

This is proved [7] by bounding the eigenvalues of the symmetric part of $BF^{-1}B^t$ (with respect to $\frac{1}{\nu}Q$) and the skew-symmetric part of $BF^{-1}B^t$, and then applying Bendixson's theorem ([47], p. 418). But γ and Γ are independent of the mesh size h of the discretisation. Moreover, since N and A are first-order and second-order operators, respectively, δ is also independent of h [15]. Consequently, the box containing the generalised eigenvalues of (7.5.3) are independent of the discretisation mesh size. A bound on the eigenvalues of the preconditioned Oseen operator is an immediate consequence.

Corollary 7.5.1 *The eigenvalues of the discrete Oseen operator* (7.5.1) *preconditioned by* (7.5.2) *consist of* $\lambda = 1$ *of multiplicity* $n_u - n_p$, *together with four sets consisting of points of the form* $1 + (a \pm bi)$ *and* $-a \pm bi$. *These sets can be enclosed in two rectangular regions that are symmetric with respect to* $\Re(\lambda) = \frac{1}{2}$ *whose borders are bounded independently of* h.

The preconditioned system can be solved using any Krylov subspace method. The convergence behavior of such methods depends implicitly on finding a polynomial that is small on the spectrum of the coefficient matrix. (Again see [42] or the paper by Van der Vorst in this volume.) The fact that the eigenvalues for the preconditioned system derived from (7.5.2) lie on both sides of the imaginary axis is a potential disadvantage of this preconditioner. An alternative that avoids this problem is the block triangular preconditioning operator

$$\begin{pmatrix} F & B^t \\ 0 & -\frac{1}{\nu}Q \end{pmatrix}. \tag{7.5.4}$$

For this choice, the associated generalised eigenvalue problem is

$$\begin{pmatrix} F & B^t \\ B & O \end{pmatrix} \begin{pmatrix} u \\ p \end{pmatrix} = \lambda \begin{pmatrix} F & B^t \\ 0 & -\frac{1}{\nu}Q \end{pmatrix} \begin{pmatrix} u \\ p \end{pmatrix}. \tag{7.5.5}$$

As above, one solution is $\lambda = 1$, now of multiplicity n_u. If $\lambda \neq 1$, then premultiplying the first block row of (7.5.5) by BF^{-1} and using the relation $Bu = -\lambda\,(\frac{1}{\nu}Q)p$ leads to the equation (7.5.3) for the other eigenvalues. Thus, we have the following result.

Theorem 7.5.2 *The eigenvalues of the discrete Oseen operator preconditioned by* (7.5.4) *consist of* $\lambda = 1$ *together with the generalised eigenvalues of* S *in* (7.5.3). *Therefore, the eigenvalues are bounded independently of* h *and they all have positive real part.*

The analysis in [16] shows that for a particular starting guess the i'th GMRES polynomial derived from the triangular preconditioning (7.5.4) is identical to the $(2i-1)st$ GMRES polynomial for the diagonal preconditioning (7.5.2). Experimental results for both GMRES and the quasi-minimal residual method (QMR) [20] indicate that this analysis is predictive for arbitrary initial guesses, i.e., the triangular method requires roughly half the iterations to converge [7]. Moreover, the inverse of the block triangular preconditioner can be expressed in factored form as

$$\begin{pmatrix} F & B^t \\ 0 & -\frac{1}{\nu}Q \end{pmatrix}^{-1} = \begin{pmatrix} F^{-1} & 0 \\ 0 & I \end{pmatrix} \begin{pmatrix} I & B^t \\ 0 & -I \end{pmatrix} \begin{pmatrix} I & 0 \\ 0 & \nu Q^{-1} \end{pmatrix},$$

so that the only overhead associated with using (7.5.4) instead of (7.5.2) is a matrix multiplication by B^t. Therefore, this preconditioner is typically more effective for the Oseen problem.

Since the eigenvalues for either of these preconditioners are independent of the mesh size, the asymptotic convergence rate of GMRES is also independent h [42]. Table 7.5 shows the iterations required by GMRES and QMR to solve the driven cavity problem on $\Omega = (-1,1) \times (-1,1)$ using the $Q_1\!-\!iso\!-\!Q_2$ discretization with an $n \times n$ non-uniform grid of elements for velocities. The initial guess was identically zero and the stopping criterion was

$$\frac{\left\| \begin{pmatrix} f \\ 0 \end{pmatrix} - \begin{pmatrix} F & B^t \\ B & 0 \end{pmatrix} \begin{pmatrix} u_k \\ p_k \end{pmatrix} \right\|_2}{\left\| \begin{pmatrix} f \\ 0 \end{pmatrix} \right\|_2} \leq 10^{-6}.$$

These results, which come from [7], indicate that the iteration counts are independent of the mesh size. (This is less evident for the smallest value $\nu = 1/100$ considered here; we believe that this is because finer meshes are needed for the asymptotic behavior to be displayed in this case.) See [7] for additional experimental results.

Iterations of GMRES

Grid	$\nu = 1$	$\nu = 1/10$	$\nu = 1/50$
16×16	18	25	45
32×32	19	31	69
64×64	17	32	93
128×128	14	31	110

Iterations of QMR

Grid	$\nu = 1$	$\nu = 1/10$	$\nu = 1/50$	$\nu = 1/100$
16×16	22	28	51	73
32×32	22	36	78	126
64×64	22	39	112	189
128×128	16	36	127	253

TABLE 7.5. Iterations for $Q_1{-}iso{-}Q_2$ finite elements applied to the Oseen equation with block triangular preconditioning.

7.5.2 Preconditioning II: Stokes solves

An alternative approach considered in [23] builds on the ideas of section 4 in a different way, by using a symmetric operator as a preconditioner for the Oseen equations. Here we consider one example from [23], the symmetric part of (7.5.1). This is a discrete Stokes operator

$$\begin{pmatrix} \nu A & B^t \\ B & 0 \end{pmatrix}. \tag{7.5.6}$$

Thus, using this with a Krylov subspace method entails solving the discrete Stokes equations at each step. See [23] for other examples of symmetric preconditioners as well as a discussion of their use for stationary iterative methods.

An analysis of the Stokes preconditioning is as follows. Once again, we have a generalised eigenvalue problem,

$$\begin{pmatrix} F & B^t \\ B & 0 \end{pmatrix} \begin{pmatrix} u \\ p \end{pmatrix} = \lambda \begin{pmatrix} \nu A & B^t \\ B & 0 \end{pmatrix} \begin{pmatrix} u \\ p \end{pmatrix}.$$

One solution is $\lambda = 1$, which has eigenvectors of the form $(u,p)^t$ where $Nu = 0$ and p is arbitrary. Any remaining eigenvalues satisfy

$$Fu + B^t p = \lambda \left((\nu A)u + B^t p \right)$$

where u is such that $Bu = 0$. If $(u,p)^t$ is any eigenvector, then taking the inner product with u leads to the expression for the corresponding

Iterations of GMRES

Grid	$\nu = 1$	$\nu = 1/10$	$\nu = 1/100$
8×8	5	11	26
16×16	4	12	39
32×32	4	12	45
64×64	4	12	45

Iterations of QMR

Grid	$\nu = 1$	$\nu = 1/10$	$\nu = 1/50$	$\nu = 1/100$
8×8	7	12	27	45
16×16	5	14	40	67
32×32	5	14	47	83
64×64	6	13	47	89

TABLE 7.6. Iterations for locally stabilised $Q_1 - P_0$ finite elements (with $\beta = 1/4$) applied to the Oseen equation with Stokes preconditioning.

eigenvalue

$$\lambda = 1 + \frac{(u, Nu)}{(u, (\nu A)u)} \cdot$$

(Note that if u exists it will be complex.) It follows that

$$|\Im(\lambda)| \leq \frac{1}{\nu}\rho(A^{-1}N).$$

Thus, we have established the following result.

Theorem 7.5.3 *The eigenvalues of the discrete Oseen operator* (7.5.1) *preconditioned by* (7.5.6) *consist of* $\lambda = 1$ *of multiplicity at least* n_p *together with at most* $n_u - n_p$ *eigenvalues of the form* $1 \pm i\eta/\nu$ *where* $|\eta| \leq \rho(A^{-1}N)$.

These eigenvalues lie on a vertical line segment in the complex plane with real part equal to 1. As noted in section 5.1, $\rho(A^{-1}N)$ is independent of the mesh size, so that the asymptotic convergence rate of GMRES will also be independent of h [23, 42].

Table 7.6 shows the results of numerical experiments with the Stokes preconditioner [23] applied to the driven cavity problem. Here the discretization is locally stabilised $Q_1 - P0$ with $\beta = 1/4$. (See section 6.) The stopping criterion and initial guess are as in section 5.2.

7.5.3 Discussion

We conclude this section with a brief discussion comparing the two classes of ideas presented here. Each of the approaches requires the solution of a key subproblem, the discrete convection-diffusion equation for the methods of section 5.1 and the discrete Stokes equations for the method of section 5.2.[3] We have not made a systematic comparison of these approaches and will refrain from making a recommendation here. For a practical computation we would expect the solution of either of the subproblems to be replaced by an approximate solution obtained using an iterative method. These computations could be done using the techniques of sections 3 or 4. This issue adds to the difficulty in making a comparison of the two approaches.

Finally, we point out that although both methodologies discussed here produce asymptotic convergence rates that are independent of the mesh size, they are dependent on the viscosity ν. This is seen in the lower bound of ν^2 for the real parts of the eigenvalues in Theorem 5.1 (which is shown to be tight in [7]) and the upper bound of $1/\nu$ in Theorem 5.3. In both cases the iteration counts appear to grow linearly in $1/\nu$, and therefore we expect these ideas to be most suitable problems with relatively high viscosity, i.e., low Reynolds numbers.

7.6 Test Problems and Software

In this section, the test problems used to illustrate the methodology in sections 3–5 are described. These problems can be constructed (and the solutions plotted) using MATLAB software which is available by anonymous `ftp` in the `tar` files

```
ftp://ftp.ma.man.ac.uk/pub/narep/convdiff.tar
ftp://ftp.ma.man.ac.uk/pub/narep/oseen.tar
```

The three test problems that are built-in are described below.

7.6.1 The Convection-Diffusion Problem

The directory `/convdiff/` contains two driver routines; `square_grid` and `stretch_grid`. These generate solutions to (7.2.1) using square or rectangular bilinear Q_1 elements. The "wind" $\mathbf{w}$ is defined within the function `transprt.m`, and for the test problem (see section 3.3) it is set to a constant vector $(-\sqrt{2}/2, \sqrt{2}/2)$. The boundary conditions are defined in the function `skewx.m`. In the test problem, the solution satisfies $u = 1$ on part

[3]As described, the techniques of section 5.1 also require the action of the inverse of the mass matrix. However, as we observed in section 4, this can be replaced with a less expensive computation using, say, the diagonal of the mass matrix, without affecting asymptotic convergence properties. Indeed, this choice was used for the results of section 5.1.

of the bottom boundary and on the right-hand wall, and $u = 0$ on the remainder.

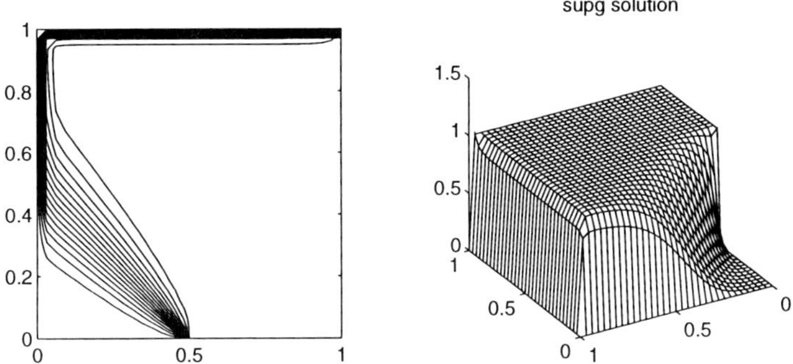

FIGURE 7.13. Convection skew to the mesh

Reducing the viscosity parameter ν increases the relative strength of the wind, and if ν is "small" there is an internal layer generated by the discontinuity on the inflow boundary, and a boundary layer at the left hand wall and along the top. The case $\nu = 1/100$ is illustrated in figure 7.13. This shows a uniform 32×32 grid solution corresponding to the streamline diffusion formulation (7.2.8), and was generated via `square_grid`. Note that for this combination of ν and h the standard Galerkin solution is oscillatory, unless stretched grids are used to resolve the boundary layer (via the routine `stretch_grid`).

7.6.2 The Stokes Problem

The directory `/oseen/` contains two driver routines; `square_mesh` and `stretch_mesh`. These generate finite element matrices associated with the Oseen operator using square (or rectangular) Q_1–P_0 elements. These matrices are "saved" on the datafile `system_nobc.mat`.

Having set up the system matrices, the Stokes flow test problem (see sections 4.4 and 4.5) can be solved using the driver `stokes`. The "leaky driven cavity" boundary conditions are defined in the function `ldcavf.m`; the vertical velocity is set to zero everywhere, whereas the horizontal velocity is set to unity on the lid, and is zero on the other boundaries. One of the interesting features of the problem is that the pressure is singular at the top corners, i.e. where the imposed velocity is discontinuous. Without convection the flow is (anti–)symmetric about the line $x = 0$, where the pressure must be identically zero. This feature can be exploited when generating the flow solution (see section 4). A typical flow is illustrated in figure 7.14. This shows a uniform 32×32 grid solution of the stabilised

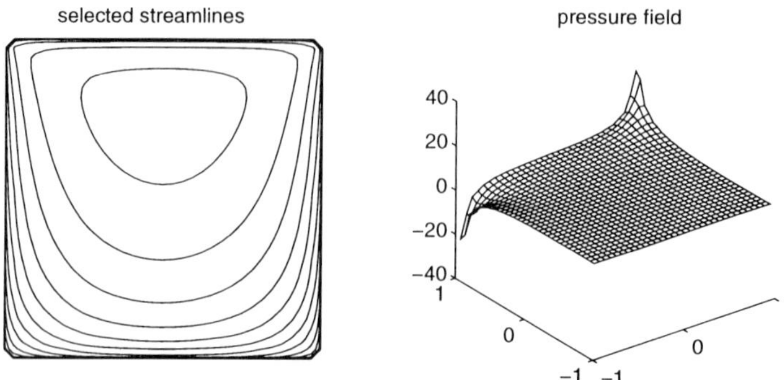

FIGURE 7.14. Stokes driven cavity flow

system (7.2.35) with the "optimal" stabilisation parameter $\beta = 0.058$. Using Q_1–P_0 the pressure solution becomes increasingly oscillatory as $\beta \to 0$, although a realistic velocity solution is obtained for this test problem without stabilisation (this is not true in general). If stretched grids are used (via the routine `stretch_mesh`) then secondary recirculations (so called "Moffatt eddies") can be observed in the bottom two corners.

7.6.3 *The Oseen Problem*

Having set up the system matrices as above (using `square_mesh` or `stretch_mesh`), the Oseen flow test problem (see sections 5.1 and 5.2) can be solved using the driver `osn`. Unlike the Stokes case where there is no convection, in the Oseen problem there is a "wind" which is defined within the function `wind.m`. For the test problem the wind is the "divergence-free vortex" $\mathbf{w} = (2y(1-x^2), -2x(1-y^2))$. As in the Stokes case, the "leaky cavity" boundary conditions are defined in the function `ldcavf.m`.

Reducing the viscosity parameter ν increases the relative strength of the wind, and if ν is "small" the centre of primary recirculation (which is on the line $x = 0$ in the Stokes case) is moved significantly to the right. The case $\nu = 1/50$ is illustrated in figure 7.15. This shows a stretched grid 32×32 grid solution of the stabilised system (7.5.1) with a stabilisation parameter $\beta = 1/4$. The secondary recirculations can be clearly observed here.

7.7 References

[1] O. Axelsson, *Iterative Solution Methods*, Cambridge University Press, New York, 1994.

[2] O. Axelsson and V. A. Barker, *Finite Element solution of boundary value problems: Theory and Computation*, Academic Press, New

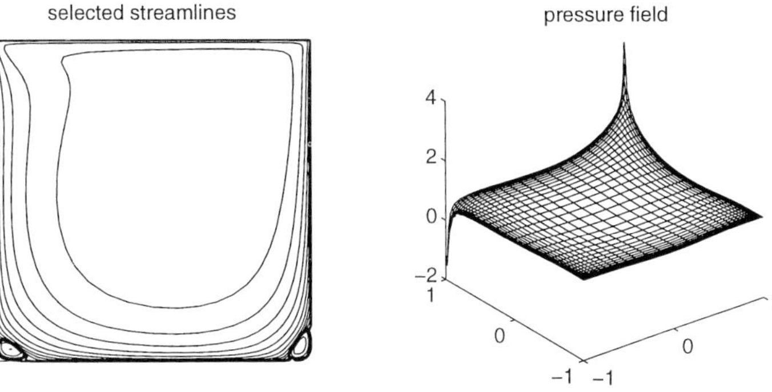

FIGURE 7.15. Oseen driven cavity flow

York, 1984.

[3] R. BEAUWENS, *Factorization iterative methods, M-operators and H-operators*, Numer. Math., 31 (1979), pp. 335–357.

[4] J. BRAMBLE AND J. PASCIAK, *Iterative techniques for time dependent Stokes problems*, in Solution Techniques for Large-Scale CFD Problems, W. Habashi, ed., John Wiley, 1995, pp. 201–216.

[5] E. DEAN AND R. GLOWINSKI, *On some finite element methods for the numerical solution of incompressible viscous flow*, in Incompressible Computational Fluid Dynamics, M. Gunzburger and R. Nicolaides, eds., 1993.

[6] J. DOUGLAS AND T. RUSSELL, *Numerical methods for convection dominated diffusion problems based on combining the method of characteristics with finite elements or finite differences*, SIAM J. Numer. Anal., 19 (1982), pp. 871–885.

[7] H. ELMAN AND D. SILVESTER, *Fast nonsymmetric iterations and preconditioning for Navier-Stokes equations*, SIAM J. Sci. Comput., 17 (1996), pp. 33–46.

[8] H. C. ELMAN, *Multigrid and Krylov subspace methods for the discrete Stokes equations*, Int. J. Numer. Meth. Fluids, 227 (1996), pp. 55–770.

[9] H. C. ELMAN AND M. P. CHERNESKY, *Ordering effects on relaxation methods applied to the discrete one-dimensional convection-diffusion equation*, SIAM J. Numer. Anal., 30 (1993), pp. 1268–1290.

[10] ——, *Ordering effects on relaxation methods applied to the discrete convection-diffusion equation*, in Recent Advances in Iterative Methods, G. H. Golub, A. Greenbaum, and M. Luskin, eds., Springer-Verlag, New York, 1994, pp. 45–57

[11] H. C. Elman and G. H. Golub, *Iterative methods for cyclically reduced non-self-adjoint linear systems*, Math. Comp., 54 (1990), pp. 671–700.

[12] ——, *Iterative methods for cyclically reduced non-self-adjoint linear systems, II*, Math. Comp., 56 (1991), pp. 215–242.

[13] ——, *Line iterative methods for cyclically reduced convection-diffusion problems*, SIAM J. Sci. Stat. Comput., 13 (1992), pp. 339–363.

[14] ——, *On the convergence of line iterative methods for cyclically reduced nonsymmetrizable linear systems*, Numer. Math., 67 (1994), pp. 177–190.

[15] H. C. Elman and M. H. Schultz, *Preconditioning by fast direct methods for nonselfadjoint nonseparable elliptic problems*, SIAM J. Numer. Anal, 23 (1986), pp. 44–57.

[16] B. Fischer, A. Ramage, D. Silvester, and A. Wathen, *Minimum residual methods for augmented systems*, Tech. Report No. 15, Mathematics Dept., Strathclyde University, June 1995. submitted to SIAM J. Matrix Anal. Appl.

[17] B. Fischer, A. Ramage, D. J. Silvester, and A. J. Wathen, *Optimal streamline upwinding for advection diffusion problems*, tech. report, Manchester Centre for Computational Mathematics, June 1996. in preparation.

[18] M. Fortin and R. Pierre, *Stability analysis of discrete generalised Stokes problems*, Numer. Meth. Partial Diff. Eq., 8 (1992), pp. 303–323.

[19] R. Freund, *On polynomial preconditioning for indefinite hermitian matrices*, Tech. Report 89.32, Research Institute for Advanced Computer Science, NASA Ames Research Center, August 1989.

[20] R. Freund and N. M. Nachtigal, *QMR: a quasi-minimal residual method for non-Hermitian linear systems*, Numer. Math., 60 (1991), pp. 315–339.

[21] V. Girault and P. Raviart, *Finite Element Methods for Navier-Stokes Equations*, Springer-Verlag, Berlin, 1986.

[22] G. H. Golub and C. F. Van Loan, *Matrix Computations*, John Hopkins, Baltimore, 1st ed., 1983.

[23] G. H. Golub and A. J. Wathen, *An Iteration for Indefinite Systems and its Application to the Navier-Stokes Equations*, Tech. Report AM-95-09, Mathematics Department, University of Bristol, 1995. To appear in SIAM J. Sci. Stat. Comput.

[24] I. S. GRADSHTEYN AND I. M. RYZHIK, *Tables of Integrals, Series and Products*, Academic Press, London, 4th ed., 1965. translated by A. Jeffrey.

[25] P. M. GRESHO AND R. L. LEE, *Don't suppress the wiggles – they're telling you something*, Computers and Fluids, 9 (1981), pp. 223–253.

[26] I. GUSTAFSSON, *A class of first order factorisation methods*, BIT, 18 (1978), pp. 142–156.

[27] C. JOHNSON, *Numerical Solution of Partial Differential Equations by the Finite Element Method*, Cambridge University Press, New York, 1987.

[28] ——, *The streamline diffusion finite element method for compressible and incompressible fluid flow*, in Numerical Analysis 1989, D. F. Griffiths and G. Watson, eds., Pitman Research Notes in Mathematics Series, 1989.

[29] O. A. KARAKASHIAN, *On a Galerkin-Lagrange multiplier method for the stationary Navier-Stokes equations*, SIAM. J. Numer. Anal., 19 (1982), pp. 909–923.

[30] N. KECHKAR AND D. SILVESTER, *Analysis of locally stabilised mixed finite element methods for the Stokes problem*, Math. Comp., 58 (1992), pp. 1–10.

[31] P. KLOUČEK AND F. RYS, *Stability of the fractional step θ-scheme for the nonstationary Navier-Stokes equations*, SIAM J. Numer. Anal., 31 (1994), pp. 1312–1335.

[32] J. A. MEIJERINK AND H. A. VAN DER VORST, *An iterative solution method for linear systems of which the coefficient matrix is a symmetric m-matrix*, Math. Comp., 31 (1977), pp. 148–162.

[33] G. MEINARDUS, *Approximation of functions: theory and numerical methods*, Springer, New York, 1967.

[34] J. ORTEGA, *Numerical Analysis: A Second Course*, Academic Press, New York, 1972.

[35] C. C. PAIGE AND M. A. SAUNDERS, *Solution of sparse indefinite systems of linear equations*, SIAM J. Numer. Anal., 12 (1975), pp. 617–629.

[36] S. V. PARTER, *On estimating the "rates of convergence" of iterative methods for elliptic difference operators*, Trans. Amer. Math. Soc., 114 (1965), pp. 320–354.

[37] ——, *Iterative methods for elliptic problems and the discovery of "q"*, SIAM Review, 28 (1986), pp. 153–175.

[38] S. V. Parter and M. Steuerwalt, *Block iterative methods for elliptic and parabolic difference equations*, SIAM J. Numer. Anal., 19 (1982), pp. 1173–1195.

[39] D. Peaceman and A. Rachford, *The numerical solution of parabolic and elliptic differential equations*, SIAM J., 3 (1955), pp. 28–41.

[40] A. Quarteroni and A. Valli, *Numerical Approximation of Partial Differential Equations*, Springer-Verlag, Berlin, 1994.

[41] Y. Saad, *Iterative Methods for Sparse Linear Systems*, PWS Publishing, Boston, 1996.

[42] Y. Saad and M. Schultz, *GMRES: a generalised minimal residual algorithm for solving nonsymmetric linear systems*, SIAM J. Sci. Stat. Comput., 7 (1986), pp. 856–869.

[43] D. Silvester, *Optimal low order finite element methods for incompressible flow*, Comp. Meths. Appl. Mech. Engrg., 111 (1994), pp. 357–368.

[44] D. J. Silvester and A. J. Wathen, *Fast and robust solvers for time-discretised incompressible Navier-Stokes equations*, in Numerical Analysis: proceedings of the 1995 Dundee biennial conference, D. F. Griffiths and G. A. Watson, eds., Longman, 1996. Pitman Research Notes in Mathematics Series 344.

[45] J. Simo and F. Armero, *Unconditional stability and long-term behavior of transient algorithms for the incompressible Navier-Stokes and Euler equations*, Comp. Meth. Appl. Mech. Eng., 111 (1994), pp. 111–154.

[46] A. Smith and D. J. Silvester, *Implicit algorithms and their linearisation for the transient Navier-Stokes equations*, Tech. Report 290, Manchester Centre for Computational Mathematics, June 1996.

[47] J. Stoer and R. Bulirsch, *Introduction to Numerical Analysis*, Springer-Verlag, New York, second ed., 1993.

[48] R. S. Varga, *Matrix Iterative Analysis*, Prentice-Hall, Englewood Cliffs, New Jersey, 1962.

[49] A. J. Wathen, *Realistic eigenvalue bounds for the Galerkin mass matrix*, IMA J. Numer. Anal., 7 (1987), pp. 449–457.

[50] A. J. WATHEN, B. FISCHER, AND D. J. SILVESTER, *The convergence rate of the minimum residual method for the Stokes problem*, Numer. Math., 71 (1995), pp. 121–134.

[51] Z. WOŹNICKI, *Two-Sweep Iterative Methods for Solving Large Linear Systems and their Application to the Numerical Solution of Multi-Group Multi-Dimensional Neutron Diffusion Equation*, PhD thesis, Institute of Nuclear Research, Swierk, Poland, 1973. Report N^{0}1447/CYFRONET/PM/A.

[52] D. M. YOUNG, *Iterative Solution of Large Linear Systems*, Academic Press, New York, 1970.

8

Decomposing a Signal into a Sum of Exponentials

Franklin T. Luk*
David Vandevoorde†

We examine a fundamental problem in control and signal processing: how to decompose an observed signal into a small set of decaying complex exponentials. In these notes, we discuss the problem in detail and survey the known work in the literature. We present a new matrix pencil approach, and show that the classical approaches of Prony and Kung are but special cases of our new class of iterative schemes. In addition, we present our observations and ideas, including a novel generalization of the Hankel SVD approach for signal decomposition and new insights into the characteristic factorization of a Hankel matrix.

8.1 Introduction

Many computational linear algebra algorithms were originally developed in more confined contexts such as signal processing or data fitting. The abstraction of these algorithms in terms of generic mathematical manipulations not only increases the usability of such methods, but often leads to enhanced insights in the original problem domain. In turn, new techniques may be produced. This type of multidisciplinary iteration also goes with generalization: related classes of problems are brought into a single framework of computational procedures.

A classic example of such an evolution is the method of *normal equations*, usually credited to Gauss, which arises when we attempt to apply a least squares fit of a linear model to given data. Originally, this application was mainly restricted to the realm of statistics; an approach called *principal components* [Hot36] was developed to handle cases where the data matrices

*Department of Computer Science, Rensselaer Polytechnic Institute, Troy, NY 12180, U.S.A. This research was supported in part by the Office of Naval Research under contract N00014-93-1-0268. The authors thank Professors Daniel Boley and Gene Golub for helpful discussions.

†Department of Computer Science, Rensselaer Polytechnic Institute, Troy, NY 12180, U.S.A.

are rank deficient. The concept grew beyond the statistical applications and a better tool, known as the *singular value decomposition* (SVD) [Bel73], was suggested as an alternative that would provide improved numerical accuracy. Golub and his colleagues (cf. [GK65] and [GR70]) are universally credited for championing the SVD cause. From there, the idea of singular values and corresponding vectors was further extended in terms of both the associated mathematical spaces (e.g., continuous dimensions [Rus87]) and the operators to be decomposed (e.g., the generalized SVD [VL75] and its variants [DMZ91]). These extensions also benefited the statistics community as illustrated by Larimore [Lar83], to quote but one example.

8.1.1 Problems of Interest

We wish to explore a similar evolutionary path for the following problem:

- Decompose a signal into a sum of a small set of decaying complex exponentials.

Although it has specific usage in modeling [EGLV80], [Kam79], this problem arises in a wide range of disciplines. Examples include system identification [KAR83], quantitative time domain analysis [BdBvO87], [SCSB76], speech processing [KT82] and coding theory [BBGL92].

A related problem that we will also examine is:

- Determine a structured, rank-revealing factorization of an arbitrary Hankel matrix.

The second problem is cast in a fairly generic form, but has nevertheless a wide range of applications in control [Glo84] and diverse aspects of signal and image processing, e.g., determination of *directions of arrival* [LQ93] and selection of appropriate regularization in ill-conditioned restoration problems [LV94].

Our two problems are related in that the two classical solution methods for the first problem synthesize Hankel matrices of various ranks. The first procedure can be traced back 200 years to Prony [Pro95] but the idea remains relevant today. The more recent second method is due to Kung *et al.* [KAR83]. However, Kung's proof and later proofs like Kahn's [Kah93] are tightly bound to particular applications. We investigate the more abstract mechanisms that make Kung's method tick. Our study has resulted in a remarkably simpler and purely algebraic proof of a new class of iterative algorithms that includes Kung's method.

8.1.2 Summary

In the ensuing sections, we will go into details of the two problems, discuss some of the known work in the field, and describe several of our observations and ideas. In particular, we introduce a new matrix pencil framework, and show how our new proof of Kung's method can inspire faster algorithms for the exponential fitting problem.

These notes are organized as follows. In Section 2, we describe structured matrices, followed in Section 3 by a description of our key problem and some of its properties, as well as as its relation to the structured factorization of a Hankel matrix. After reviewing some classical methods in Section 4, we present a new solution framework and derive specific algorithms in Sections 5 and 6. Section 7 introduces a novel denoising algorithm that has proven to be very effective in our applications. Conclusions follow in Section 8.

8.2 Structured Matrices

Relevant matrices and their associated properties are presented in this section. We make extensive use of linear vector spaces with real and complex valued vectors and matrices. Matrices are named with uppercase letters, and vectors with lowercase letters. We adopt MATLAB-like notations with respect to row and column indexing. For example, we use the colon ':' to indicate a range; hence, $A_{3,:}$ and $A_{:,2}$ refer, respectively, to the third row and second column of A, whereas $x_{4:10}$ denotes the fourth through tenth elements of vector x. We will find it useful to perform modulo arithmetic with matrix indices; for an index j, we define a modulus operator $\oplus$ by

$$\oplus j \equiv (j \bmod \mathrm{L}) + 1.$$

8.2.1 *Hankel-like Structures*

We often encounter matrix structures with constant elements along the diagonals or along the antidiagonals (sometimes called perdiagonals). Four cases can be distinguished:

1. Toeplitz matrices: Constant along each diagonal,

$$T^{\mathrm{K}\times\mathrm{L}}(s) \equiv \begin{pmatrix} s_{\mathrm{L}} & s_{\mathrm{L}-1} & s_{\mathrm{L}-2} & \cdots & s_1 \\ s_{\mathrm{L}+1} & s_{\mathrm{L}} & s_{\mathrm{L}-1} & \cdots & s_2 \\ s_{\mathrm{L}+2} & s_{\mathrm{L}+1} & s_{\mathrm{L}} & \cdots & s_3 \\ \vdots & \vdots & \vdots & \ddots & \vdots \\ s_{\mathrm{L}+\mathrm{K}-1} & s_{\mathrm{L}+\mathrm{K}-2} & s_{\mathrm{L}+\mathrm{K}-3} & \cdots & s_{\mathrm{K}} \end{pmatrix}.$$

2. Circulant matrices: Toeplitz matrices with wrap-around rows,

$$\widetilde{T}^{\mathrm{K}\times\mathrm{L}}(s) \equiv \begin{pmatrix} s_{\mathrm{L}} & s_{\mathrm{L}-1} & s_{\mathrm{L}-2} & \cdots & s_1 \\ s_1 & s_{\mathrm{L}} & s_{\mathrm{L}-1} & \cdots & s_2 \\ s_2 & s_1 & s_{\mathrm{L}} & \cdots & s_3 \\ \vdots & \vdots & \vdots & \ddots & \vdots \\ s_{\oplus(\mathrm{L}+\mathrm{K}-1)} & s_{\oplus(\mathrm{L}+\mathrm{K}-2)} & s_{\oplus(\mathrm{L}+\mathrm{K}-3)} & \cdots & s_{\oplus(\mathrm{K})} \end{pmatrix}.$$

3. Hankel matrices: Constant along each antidiagonal,

$$H^{K\times L}(s) \equiv \begin{pmatrix} s_1 & s_2 & s_3 & \cdots & s_L \\ s_2 & s_3 & s_4 & \cdots & s_{L+1} \\ s_3 & s_4 & s_5 & \cdots & s_{L+2} \\ \vdots & \vdots & \vdots & \ddots & \vdots \\ s_K & s_{K+1} & s_{K+2} & \cdots & s_{K+L-1} \end{pmatrix}.$$

4. Anticirculant matrices: Hankel matrices with wrap-around rows,

$$\widetilde{H}^{K\times L}(s) \equiv \begin{pmatrix} s_1 & s_2 & s_3 & \cdots & s_L \\ s_2 & s_3 & s_4 & \cdots & s_1 \\ s_3 & s_4 & s_5 & \cdots & s_2 \\ \vdots & \vdots & \vdots & \ddots & \vdots \\ s_{\oplus(K)} & s_{\oplus(K+1)} & s_{\oplus(K+2)} & \cdots & s_{\oplus(K+L-1)} \end{pmatrix}.$$

Matrices are representations of transformations in specific bases, and it is usually possible to renumber the basis vectors of the row space so that a matrix with one of the first two structures takes on one of the last two forms. This may seem inconsequential, but a quick glance at the latter two structures reveals that they are symmetric with all the associated numerical advantages; e.g., the guarantee of real eigenvalues and orthogonal eigenvectors when the matrix itself is real [GVL89].

The above four forms are pervasive in the context of one-dimensional ("time domain") modeling or signal-processing. For two-dimensional applications, e.g., image restoration, the natural generalization of the Toeplitz form would be the block Toeplitz with Toeplitz block (BTTB) structure. Such a matrix is obtained by substituting every s_k in $T^{K\times L}(s)$ by a Toeplitz structured block. These matrices are sometimes called *level-2 Toeplitz* matrices [CN96].

A recurring class of Toeplitz equations has a special right-side vector, and is known as the Yule-Walker [GVL89] problem:

$$\begin{pmatrix} s_L & s_{L-1} & \cdots & s_1 \\ s_{L+1} & s_L & \cdots & s_2 \\ s_{L+2} & s_{L+1} & \cdots & s_3 \\ \vdots & \vdots & \ddots & \vdots \\ s_{L+K-1} & s_{L+K-2} & \cdots & s_K \end{pmatrix} \begin{pmatrix} x_1 \\ x_2 \\ \vdots \\ x_L \end{pmatrix} = - \begin{pmatrix} s_{L+1} \\ s_{L+2} \\ s_{L+3} \\ \vdots \\ s_{L+K} \end{pmatrix}. \tag{2.1.1}$$

By reordering the unknowns, we obtain the Hankel form of this problem:

$$\begin{pmatrix} s_1 & s_2 & \cdots & s_L \\ s_2 & s_3 & \cdots & s_{L+1} \\ s_3 & s_4 & \cdots & s_{L+2} \\ \vdots & \vdots & \ddots & \vdots \\ s_K & s_{K+1} & \cdots & s_{L+K-1} \end{pmatrix} \begin{pmatrix} x_L \\ x_{L-1} \\ \vdots \\ x_1 \end{pmatrix} = - \begin{pmatrix} s_{L+1} \\ s_{L+2} \\ s_{L+3} \\ \vdots \\ s_{L+K} \end{pmatrix}. \tag{2.1.2}$$

Of anecdotal interest is the fact that Prony presented this special matrix equation in 1795 [Pro95]. His paper predates the 1927 work of Yule [Yul27] and the 1931 work of Walker [Wal31] by more than a century. To give Prony due credit, we should properly refer to (2.1.1) and (2.1.2) as the Prony-Yule-Walker equations.

8.2.2 *Vandermonde Matrices*

Another commonly encountered structure is the class of Vandermonde matrices which are distinguished by columns that form geometric sequences:

$$V^{\mathrm{K}\times\mathrm{L}}(z) \equiv \begin{pmatrix} 1 & 1 & \cdots & 1 \\ z_1 & z_2 & \cdots & z_{\mathrm{L}} \\ z_1^2 & z_2^2 & \cdots & z_{\mathrm{L}}^2 \\ \vdots & \vdots & \ddots & \vdots \\ z_1^{\mathrm{K}-1} & z_2^{\mathrm{K}-1} & \cdots & z_{\mathrm{L}}^{\mathrm{K}-1} \end{pmatrix}. \tag{2.2.1}$$

Let

$$p \equiv (p_1, p_2, \ldots, p_{\mathrm{K}})^T.$$

Then $p^T V^{\mathrm{K}\times\mathrm{L}}(z)$ is a row vector whose l-th entry equals a polynomial determined by the coefficients p_k evaluated at the knot z_l, i.e.,

$$[p^T V^{\mathrm{K}\times\mathrm{L}}(z)]_l = \sum_{k=1}^{\mathrm{K}} p_k\, z_l^{k-1}.$$

Similarly, define

$$q \equiv (q_1, q_2, \ldots, q_{\mathrm{L}})^T,$$

so that the k-th entry of the column vector $V^{\mathrm{K}\times\mathrm{L}}(z)q$ computes the $(k-1)$-st *moment* of q with respect to the set of knots $\{z_l\}$:

$$[V^{\mathrm{K}\times\mathrm{L}}(z)q]_k = \sum_{l=1}^{\mathrm{L}} q_l\, z_l^{k-1}.$$

Later on, we will exhibit a close relationship between the Vandermonde and Hankel matrices and use it to simplify our formal reasoning.

A useful Vandermonde-like structure is the class of *confluent* Vandermonde matrices. Denote by $V^{(\mathrm{M}_i)}(z_i)$ a $\mathrm{K} \times \mathrm{M}_i$ matrix with its (k,l)-entry given by

$$\left[V^{(\mathrm{M}_i)}(z_i)\right]_{k,l} = \begin{cases} 0, & \text{for } k < l \le \mathrm{M}_i\,; \\ \binom{k-1}{l-1} z_i^{k-l}, & \text{for } k \ge l. \end{cases}$$

Confluent Vandermonde are matrices of the form:

$$\left(V^{(\mathrm{M}_1)}(z_1),\ V^{(\mathrm{M}_2)}(z_2),\ \ldots,\ V^{(\mathrm{M}_d)}(z_d)\right).$$

They arise when the knots $\{z_l\}$ are not distinct, causing a Vandermonde matrix of the form (2.2.1) to be singular. In other words, the block $V^{(\mathrm{M}_i)}(z_i)$ is generated by a knot z_i with multiplicity M_i, and d denotes the number of *distinct* knots.

8.2.3 Fourier Matrix

The $\mathrm{L} \times \mathrm{L}$ Fourier matrix F is a unitary Vandermonde matrix with

$$F_{k,l} = e^{i(2\pi/\mathrm{L})(k-1)(l-1)},$$

where $i^2 = -1$. If $s \equiv (s_1, s_2, s_3, \ldots, s_{\mathrm{L}})^T$ is a finite sequence of samples, then the *Discrete Fourier Transform* (DFT) of s is given by Fs. For simplicity, we will use s to denote both the sample sequence and the corresponding column vector.

Let us demonstrate an important connection between the DFT and circulant matrices. Consider an $\mathrm{L} \times \mathrm{L}$ circulant matrix

$$C = \widetilde{T}^{\mathrm{L}\times\mathrm{L}}(s) \equiv \begin{pmatrix} s_{\mathrm{L}} & s_{\mathrm{L}-1} & s_{\mathrm{L}-2} & \cdots & s_1 \\ s_1 & s_{\mathrm{L}} & s_{\mathrm{L}-1} & \cdots & s_2 \\ s_2 & s_1 & s_{\mathrm{L}} & \cdots & s_3 \\ \vdots & \vdots & \vdots & \ddots & \vdots \\ s_{\mathrm{L}-1} & s_{\mathrm{L}-2} & s_{\mathrm{L}-3} & \cdots & s_{\mathrm{L}} \end{pmatrix},$$

constructed from s. By the circulant characteristic, viz., $C_{k+1,l} = C_{1,\oplus(l-k)}$, we find for the $(k+1)$-st element of the vector $CF_{:,l}$:

$$\begin{aligned} [CF_{:,l}]_{k+1} &= \sum_{m=1}^{\mathrm{L}} C_{k+1,m}\, e^{i(2\pi/\mathrm{L})(m-1)(l-1)} \\ &= \left(\sum_{m=1}^{\mathrm{L}} C_{1,\oplus(m-k)}\, e^{i(2\pi/\mathrm{L})(\oplus(m-k-1))(l-1)}\right) e^{i(2\pi/\mathrm{L})k(l-1)} \\ &= \left(\sum_{m=1}^{\mathrm{L}} C_{1,m}\, e^{i(2\pi/\mathrm{L})(m-1)(l-1)}\right) F_{k+1,l}\ . \end{aligned}$$

Define

$$\lambda_l = \sum_{m=1}^{\mathrm{L}} C_{1,m}\, e^{i(2\pi/\mathrm{L})(m-1)(l-1)},$$

a quantity which equals the DFT of the first row of C. Then

$$CF_{:,l} = \lambda_l F_{:,l}\ .$$

So the l-th column of the Fourier matrix is a right eigenvector of any circulant matrix and corresponds to the eigenvalue λ_l. Since the Fast Fourier Transform (FFT) algorithm computes a DFT in $O(\text{L} \log \text{L})$ operations, the eigenvalue decomposition of a circulant matrix can be determined with the same computational complexity:

$$C = F \Lambda_C F^H.$$

The solution of a circulant system of linear equations is thus composed of three FFT's and the solution of a diagonal system, at a total of $O(\text{L} \log \text{L})$ complexity.

8.2.4 Exploiting Structure

We want to take advantage of the problem structure to reduce computational and storage costs. Ideally, this should maintain or even improve accuracy. We have already seen how such improvement is possible for circulant matrices. But circulant systems relate to the property of periodicity, and so can only approximate nonperiodic phenomena. The corresponding limitation when using the FFT algorithm is widely known in applied signal processing. A common, if somewhat *ad hoc*, method to reduce the ill effects of treating a segment of an arbitrary signal as a period of a cyclic function is to taper the ends of the segment, thereby reducing the likely discontinuity resulting from the periodic assumption.

Toeplitz and Hankel matrices offer improved accuracy in the modeling of acyclic phenomena [LV94]. However, no known method matches the FFT approach for circulant problems in terms of both speed and accuracy. Since the 1940's, Levinson [Lev46] and others have developed $O(\text{L}^2)$ methods to solve square Toeplitz equations. During the 1980's, several asymptotically fast $O(\text{L}(\log \text{L})^c)$ Hankel and Toeplitz solvers appeared; see, e.g., [AG88] and [BGY80]. Unfortunately, they suffer from high complexity (a large constant is hidden under the O symbol) and limited accuracy (cf. [Bun85]). An attractive category of methods relies on fast iterative schemes, typically based on conjugate gradients, and combined with effective circulant preconditioners [CN96]. These algorithms lead to relatively simple and accurate methods, not only for square systems but also for overdetermined problems in a least squares sense; this class of techniques has a computational complexity of $O(\text{L}(\log \text{L})^2)$.

Reducing problems to the Vandermonde form is often frowned upon because the resultant Vandermonde matrix is usually ill-conditioned. We can make two important remarks on this matter. First, Higham [Hig87] has shown that even with a very high condition number, a Vandermonde system can still be stably solved. Second, it has already been mentioned that the Fourier matrix possesses both an explicit Vandermonde structure and an implicit unitary property which guarantees a minimal condition number.

We have observed that the transition to a high condition number is fairly gradual and that when the components of z lie not too far from the unit circle and not too close to each other, a very reasonable condition number can be expected. Various fast direct methods exist for Vandermonde-like matrices as well.

Unfortunately, very few methods for structured matrices are known for the other classic problems such as numerical rank estimation, eigenvalue and singular value decompositions (and their generalizations), and the approximate solution of rectangular linear systems. In [LV96], we propose a new algorithm to take advantage of the matrix structures that arise in the exponential decomposition problem.

8.3 Signal Decomposition

As mentioned in the Introduction, our basic problem is the decomposition of a signal into a small, and preferably minimal, set of complex exponentials. Usually, a limited sequence

$$s = (s_1, s_2, \ldots, s_{\mathrm{N}})^T$$

of N noisy samples is available and the physics of the problem imply that the noiseless exponentials are *damped* sinusoids. Our formal signal model is thus

$$s_k = \sum_{r=1}^{\mathrm{R}} a_r z_r^{k-1}, \tag{3.1}$$

where the coefficients $\{a_r\}$ and the knots $\{z_r\}$ are 2R unknowns to be determined. Although we usually work with this form, applications often require that we calculate values of the characteristic frequencies f_r and the damping coefficients d_r:

$$s_k = \sum_{r=1}^{\mathrm{R}} a_r e^{(-d_r + i2\pi f_r)(k-1)t_s}.$$

The sampling period t_s is known. This decomposition reduces to the DFT if we impose the constraint that $d_r = 0$ and $f_r = (r-1)/(\mathrm{N}t_s)$. Hence we certainly can fit L samples exactly when $\mathrm{R} = \mathrm{N}$. However, the number of available parameters is doubled because the $\{z_r\}$ are no longer fixed. So it does not seem unreasonable to expect to fit N samples with no more than $\mathrm{R} = \lceil \mathrm{N}/2 \rceil$ terms. Note also that if N is odd, the decomposition is not unique even when R is minimal; this non-uniqueness can be removed by fixing z_1.

8.3.1 Rank Properties

It is instructive to cast the exponential decomposition problem in terms of Vandermonde matrices. We observe from (3.1) that s_k is the $(k-1)$-st moment of the coefficients $\{a_r\}$ with respect to the knots $\{z_r\}$. Indeed,

$$s = V^{\mathrm{N}\times\mathrm{R}}(z)a. \tag{3.1.1}$$

We are faced with the problem of recovering 2R unknowns from the nonlinear equations (3.1), suggesting a need for at least 2R samples s_k. This assumes however that the value of R is known *a priori*, which is rarely the case in practice. Define an $\mathrm{R}\times\mathrm{R}$ diagonal matrix D_a^{R} by

$$D_a^{\mathrm{R}} \equiv \mathrm{diag}(a_1, a_2, \ldots, a_{\mathrm{R}}),$$

and consider the product $V^{\mathrm{K}\times\mathrm{R}}(z)D_a^{\mathrm{R}}(V^{\mathrm{L}\times\mathrm{R}}(z))^T$. Some elementary algebra reveals that the (k,l) element of the resultant matrix equals s_{k+l-1}. Moreover, since $k+l-1$ is constant along each antidiagonal, we get a Hankel matrix decomposition:

$$H \equiv \begin{pmatrix} s_1 & s_2 & \cdots & s_{\mathrm{L}} \\ s_2 & s_3 & \cdots & s_{\mathrm{L}+1} \\ s_3 & s_4 & \cdots & s_{\mathrm{L}+2} \\ \vdots & \vdots & \ddots & \vdots \\ s_{\mathrm{K}} & s_{\mathrm{K}+1} & \cdots & s_{\mathrm{K}+\mathrm{L}-1} \end{pmatrix} = V^{\mathrm{K}\times\mathrm{R}}(z)\, D_a^{\mathrm{R}}\, (V^{\mathrm{L}\times\mathrm{R}}(z))^T. \tag{3.1.2}$$

The importance of this simple property is exhibited by noticing that the rank of H is equal to the number R of nonzero elements in a, so long as R is no larger than $\max(\mathrm{K}, \mathrm{L})$. Henceforth, we will also refer to R as the *rank* of the given signal. We can determine the value of R using any numerical rank estimation scheme at our disposal. We can also use (3.1.2) as a simple tool to construct Hankel matrices with a given rank.

It is natural to ask if every Hankel matrix has a decomposition of the form (3.1.2). Although the answer is negative, it is not difficult to prove the following more general result [LV96].

Theorem 1 *Let W^c be an $\mathrm{N}\times\mathrm{R}$ ($\mathrm{N} = \mathrm{K}+\mathrm{L}$) confluent Vandermonde matrix :*

$$W^c = \left(V^{(\mathrm{M}_1)}(z_1), V^{(\mathrm{M}_2)}(z_2), \ldots, V^{(\mathrm{M}_d)}(z_d)\right),$$

where the M_j are positive integers such that $\mathrm{M}_1 + \mathrm{M}_2 + \cdots + \mathrm{M}_d = \mathrm{R}$, and the z_j are distinct complex numbers. For $j = 1, \ldots, d$, define $A^{[j]}$ as "left" triangular $\mathrm{M}_j \times \mathrm{M}_j$ Hankel matrices:

$$A^{[j]} = H^{\mathrm{M}_j\times\mathrm{M}_j}(a^{[j]}),$$

with

$$a^{[j]} = \left(a_1^{[j]}, a_2^{[j]}, \ldots, a_{\mathrm{M}_j}^{[j]}, 0, \ldots, 0\right)^T$$

and $a^{[j]}_{\mathrm{M}_j} \neq 0$. *Then the matrix* H, *defined by*

$$H \equiv W^c_{1:K,:}\, \mathrm{diag}(A^{[1]}, A^{[2]}, \ldots, A^{[d]})\, (W^c_{1:L,:})^T, \tag{3.1.3}$$

is Hankel structured and has its rank equal to $\min(\mathrm{K}, \mathrm{L}, \mathrm{R})$.

Obviously, the class of Hankel matrices that can be factored in the form of (3.1.3) is larger than the class of Hankel matrices decomposable in the form of (3.1.2), because (3.1.2) is just a special case of (3.1.3) with all M_j equal to 1. Although the methods for determining the z_j values will be derived in terms of the less general decomposition (3.1.2), they can be applied to compute (3.1.3). The value M_j then corresponds to the multiplicity of z_j in these methods. Just like (3.1.2), the decomposition (3.1.3) is not unique.

8.3.2 Illustrating Non-Uniqueness

Consider a 4×4 Hankel matrix:

$$H = \begin{pmatrix} 0 & 0 & 0 & 1 \\ 0 & 0 & 1 & 4 \\ 0 & 1 & 4 & 10 \\ 1 & 4 & 10 & 20 \end{pmatrix}.$$

This matrix can be decomposed in the form of (3.1.2) in many different ways; for example, we may either pick

$$V^{4\times 4}(z) = \begin{pmatrix} 1 & 1 & 1 & 1 \\ -1+2i & 1-2i & 3+2i & 3-2i \\ -3-4i & -3+4i & 5+12i & 5-12i \\ 11-2i & 11+2i & -9+46i & -9-46i \end{pmatrix}$$

and

$$D^{\mathrm{R}}_a = ((1-i)/128)\, \mathrm{diag}(1, i, -i, -1),$$

or choose

$$V^{4\times 4}(z) = \begin{pmatrix} 1 & 1 & 1 & 1 \\ 0 & 2 & 1+i & 1-i \\ 0 & 4 & 2i & -2i \\ 0 & 8 & -2+2i & -2-2i \end{pmatrix}$$

and

$$D^{\mathrm{R}}_a = (1/4)\, \mathrm{diag}(1, i, -i, -1).$$

We can also find a decomposition in the form of (3.1.3), with $d = 1$, $z_1 = 1$ and $\mathrm{M}_1 = 4$ in Theorem 1:

$$H = \begin{pmatrix} 1 & 0 & 0 & 0 \\ 1 & 1 & 0 & 0 \\ 1 & 2 & 1 & 0 \\ 1 & 3 & 3 & 1 \end{pmatrix} \begin{pmatrix} 0 & 0 & 0 & 1 \\ 0 & 0 & 1 & 0 \\ 0 & 1 & 0 & 0 \\ 1 & 0 & 0 & 0 \end{pmatrix} \begin{pmatrix} 1 & 0 & 0 & 0 \\ 1 & 1 & 0 & 0 \\ 1 & 2 & 1 & 0 \\ 1 & 3 & 3 & 1 \end{pmatrix}^T.$$

Now, add one row to H to get a 5×4 matrix:

$$\hat{H} = \begin{pmatrix} 0 & 0 & 0 & 1 \\ 0 & 0 & 1 & 4 \\ 0 & 1 & 4 & 10 \\ 1 & 4 & 10 & 20 \\ 4 & 10 & 20 & 35 \end{pmatrix}.$$

Although we can decompose $\hat{H}$ in the form of (3.1.3):

$$\hat{H} = \begin{pmatrix} 1 & 0 & 0 & 0 \\ 1 & 1 & 0 & 0 \\ 1 & 2 & 1 & 0 \\ 1 & 3 & 3 & 1 \\ 1 & 4 & 6 & 4 \end{pmatrix} \begin{pmatrix} 0 & 0 & 0 & 1 \\ 0 & 0 & 1 & 0 \\ 0 & 1 & 0 & 0 \\ 1 & 0 & 0 & 0 \end{pmatrix} \begin{pmatrix} 1 & 0 & 0 & 0 \\ 1 & 1 & 0 & 0 \\ 1 & 2 & 1 & 0 \\ 1 & 3 & 3 & 1 \end{pmatrix}^T,$$

we are unable to do it in the form of (3.1.2) unless we allow R to exceed 4.

8.3.3 Preliminary Noise Removal

The exponential decomposition problem turns out to be particularly sensitive to errors in the samples $\{s_k\}$. Clearly, given that every exponential component is damped, only a limited number of samples will be useful. To maximize the number of usable samples, we may initiate the overall computation with a noise removal procedure. There is a wealth of literature and algorithms that deal with the issue of noise removal. Here we show only how the SVD provides a powerful way to obtain such a result and we describe Cadzow's method to restore the Hankel structure to the solution. Assume an additive error model:

$$H = H^{[s]} + H^{[n]}, \tag{3.3.1}$$

where $H^{[s]}$ and $H^{[n]}$ denote, respectively, the signal and noise components of H (or more generally, the significant and nonsignificant parts). The simplest, and very usable, noise models imply that the noise subspace is orthogonal to the signal subspace. Hence an orthonormal basis exists such that the noise subspace is spanned by a subset of the basis vectors while the pure signal space is spanned by the remaining basis vectors. Assuming that we can determine this orthonormal basis, we still need a criterion to separate the components of the noise basis from those of the signal basis. It is a common practice to consider the coefficients of the signal decomposition in this basis and deem the components associated with "small" coefficients as noise contribution.

The SVD of the data matrix H:

$$H - U\Sigma V^H, \tag{3.3.2}$$

provides a convenient and numerically stable way to determine the noise subspace. To decide which coefficients are "small," we look for a "gap" in the entries on the diagonal of Σ:

$$\sigma_1 \geq \cdots \geq \sigma_{\mathrm{R}} >> \sigma_{\mathrm{R}+1} \geq \cdots \geq \sigma_{\min(\mathrm{K},\mathrm{L})} \geq 0.$$

We refer to R as the *numerical rank* of H. Then

$$\begin{cases} H^{[s]} & = U_{:,1:\mathrm{R}}\, \Sigma_{1:\mathrm{R},1:\mathrm{R}}\, V^H_{:,1:\mathrm{R}}, \\ H^{[n]} & = U_{:,\mathrm{R}+1:\mathrm{L}}\, \Sigma_{\mathrm{R}+1:\mathrm{L},\mathrm{R}+1:\mathrm{L}}\, V^H_{:,\mathrm{R}+1:\mathrm{L}}. \end{cases}$$

Any explicit structure of H is almost certainly lost in $H^{[s]}$. Cadzow [Cad88] proposed to iterate on a two-step procedure, with signal extraction via the SVD followed by Hankel structure restoration via averaging along the antidiagonals. He shows that under relatively mild assumptions this procedure will converge, and he reports very good noise removal performance even when the signal to noise ratio is 1. Cadzow's idea is quite similar to the projection theory of Youla and Webb [YW82], a well-known result in image restoration.

8.3.4 *Counterexample to Cadzow's Method*

Although Cadzow proves that his algorithm converges under very mild assumptions, the $H^{[s]}$ that is obtained from his method is not necessarily a good one. Consider the following example:

$$H = \begin{pmatrix} 1+\epsilon & 0 & 0 \\ 0 & 0 & 1 \\ 0 & 1 & 0 \end{pmatrix},$$

where ϵ is a very small positive number. Suppose that we wish to approximate H with a rank two Hankel matrix. An initial SVD of H finds

$$H^{(1)}_{\mathrm{SVD}} = \begin{pmatrix} 1+\epsilon & 0 & 0 \\ 0 & 0 & 0 \\ 0 & 1 & 0 \end{pmatrix},$$

and the subsequent averaging on the antidiagonal gives

$$H^{(1)}_{\mathrm{avg}} = \begin{pmatrix} 1+\epsilon & 0 & 0 \\ 0 & 0 & 0.5 \\ 0 & 0.5 & 0 \end{pmatrix}.$$

To avoid "convergence to zero," a re-scaling may be used at every iteration to maintain the Frobenius norm of the Hankel matrix:

$$H^{(1)}_{\mathrm{scaled}} \approx \begin{pmatrix} \sqrt{2} & 0 & 0 \\ 0 & 0 & \sqrt{0.5} \\ 0 & \sqrt{0.5} & 0 \end{pmatrix}.$$

The iterates will converge to a rank one Hankel matrix, given by

$$H^{(\infty)} \approx \begin{pmatrix} \sqrt{3} & 0 & 0 \\ 0 & 0 & 0 \\ 0 & 0 & 0 \end{pmatrix}.$$

One can check that the following rank two matrix:

$$\hat{H} = \begin{pmatrix} 0 & 0 & 0 \\ 0 & 0 & \sqrt{1.5} \\ 0 & \sqrt{1.5} & 0 \end{pmatrix}$$

may provide a better approximation to H: it possesses the desired rank and is closer in the Frobenius norm.

Clearly, the quality of any denoising procedure depends heavily on the applicable noise model. Consider for example a linear system where high-powered noise is present but is limited to the high frequencies of the Fourier spectrum, while a fairly faint signal is located in a low frequency band. A low-pass filter may provide good denoising in this case, while SVD-based procedures (including Cadzow's) will most likely confuse $H^{[s]}$ and $H^{[n]}$.

8.4 Classical Algorithms

In this section, we present the root-finding method of Prony [Pro95] and subsequent improvements, as well as an SVD approach of Kung [KAR83].

8.4.1 Prony's Method

In 1795, Prony [Pro95] tackled the signal decomposition problem, ignoring the effects of noise, by considering the unknown polynomial $q_{\mathrm{R}}(z)$ whose zeros are the knots $\{z_r\}$ we are looking for.

The polynomial $q_{\mathrm{R}}(z)$ can be written out in terms of its unknown coefficients:

$$q_{\mathrm{R}}(z) = z^{\mathrm{R}} + \gamma_{\mathrm{R}-1} z^{\mathrm{R}-1} + \gamma_{\mathrm{R}-2} z^{\mathrm{R}-2} + \cdots + \gamma_1 z + \gamma_0, \tag{4.1.1}$$

with the assumption that:

$$\forall r \in \{1, 2, \ldots, \mathrm{R}\} : q_{\mathrm{R}}(z_r) = 0. \tag{4.1.2}$$

Prony used (3.1) to generate combinations of s_k that reduce to zero by (4.1.2). Indeed, for any $m \in \{1, 2, \ldots, \mathrm{N} - \mathrm{R}\}$:

$$s_{\mathrm{R}+m} + \sum_{r=0}^{\mathrm{R}-1} \gamma_r s_{r+m} = a_1\, z_1^m\, q_{\mathrm{R}}(z_1) + a_2\, z_2^m\, q_{\mathrm{R}}(z_2) + \cdots + a_{\mathrm{R}}\, z_{\mathrm{R}}^m\, q_{\mathrm{R}}(z_{\mathrm{R}}) = 0,$$

which is expressible as a Yule-Walker equation:

$$\begin{pmatrix} s_1 & s_2 & \cdots & s_{\mathrm{R}} \\ s_2 & s_3 & \cdots & s_{\mathrm{R}+1} \\ s_3 & s_4 & \cdots & s_{\mathrm{R}+2} \\ \vdots & \vdots & \ddots & \vdots \\ s_{\mathrm{N}-\mathrm{R}} & s_{\mathrm{N}-\mathrm{R}+1} & \cdots & s_{\mathrm{N}-1} \end{pmatrix} \begin{pmatrix} \gamma_0 \\ \gamma_1 \\ \vdots \\ \gamma_{\mathrm{R}-1} \end{pmatrix} = - \begin{pmatrix} s_{\mathrm{R}+1} \\ s_{\mathrm{R}+2} \\ s_{\mathrm{R}+3} \\ \vdots \\ s_{\mathrm{N}} \end{pmatrix}. \tag{4.1.3}$$

Strictly speaking, Prony considered only the even-determined case. That is, he assumed that R is given and picked $\mathrm{N} = 2\mathrm{R}$ so that (4.1.3) is square. We can summarize the computation as follows:

Prony's Method

1. Solve a square Yule-Walker equation (4.1.3), i.e., $\mathrm{N} = 2\mathrm{R}$.
2. Determine the roots $\{z_r\}$ of the polynomial $q_{\mathrm{R}}(z)$ of (4.1.1).
3. Find the coefficients $\{a_r\}$ by solving a square Vandermonde system:

$$s_{1:\mathrm{R}} = V^{\mathrm{R}\times\mathrm{R}}(z)a.$$

8.4.2 Subsequent Improvements

Prony developed his method by largely ignoring the issue of noise, or other errors in the given data, to obtain a square Yule-Walker system. When considering the presence of noise, it is tempting to simply approximate steps 1 and 3 above in a least squares sense. However, this still leaves us with the problem of determining the rank of the given data.

Lanczos [Lan56] presented cases where this approach leads to very poor estimates of the unknown $\{z_r\}$. Both finding the roots of an explicit polynomial and solving an overdetermined Vandermonde system of equations in a least squares sense are prone to unacceptable sensitivity. Kumaresan and Tufts [KT82] proposed to regularize the least squares problem with a truncated SVD. Cadzow [Cad88] additionally showed how projection methods can be used to denoise the available samples in this case.

Osborne and Smyth [OS91] argued that this particular least squares formulation is intrinsically deficient, but that an alternative least squares scheme can be worked out that exhibits better numerical properties. Their method leads to a nonlinear eigenvalue problem that can be solved using a Gauss-Newton or Levenberg-Marquardt scheme, but at a higher computational cost. They additionally showed how to regard Prony's method as one instance of a more general family of solutions to nonlinear interpolation problems. Ruhe [Ruh80] treated a special case of the exponential fitting problem by resorting to nonlinear optimization techniques. However, his method covers only the case where all exponentials are real and positive.

8.4.3 Kung's Method

Perhaps the most remarkable breakthrough since Prony's classical essay is the SVD-based method presented by Kung *et al.* in 1983 [KAR83]. One important property of this approach is that no polynomial coefficients are explicitly computed. The algorithm has been refined by later researchers, e.g., [VH93], but their refinements are mostly limited to individual steps of the algorithm. Kung's method was discovered in the context of state-space based signal modeling; subsequent proofs persisted in using the vast arsenal of state-space theory. For the sake of documentation, we mention here that in state-space based linear system theory one expresses an *output vector* $o(t)$ of a discrete-time linear system in terms of the *state vector* $x(t)$ and the *input vector* $i(t)$:

$$\begin{cases} o(t) & = Ax(t) + Bi(t), \\ x(t+\Delta t) & = Cx(t) + Di(t). \end{cases}$$

The state vector $x(t)$ embodies the history of the system at time t, which explains why only the input $i(t)$ at time t is explicitly present in the model equations. The operators A, B, C and D characterize the system.

Kung's method, which is also referred to as the Hankel SVD or HSVD algorithm, can be summarized as follows. We begin by computing the SVD of the Hankel matrix H, as in (3.3.2). Although this can be done accurately, it also sacrifices the Hankel structure. That is, the Hankel structure cannot be exploited in computing or storing U, Σ and V. Kung proceeds by truncating the decomposition to the R greatest singular values:

$$H \approx U_{:,1:\mathrm{R}} \Sigma_{1:\mathrm{R},1:\mathrm{R}} V_{:,1:\mathrm{R}}^{H}. \tag{4.3.1}$$

The right side of (4.3.1) would not be Hankel structured if any noise were present in the matrix H. The number R of exponential components corresponds to the numerical rank of H and can be determined as described in Section 3.

A characteristic matrix X is then computed from either

$$U_{2:\mathrm{K},1:\mathrm{R}} = U_{1:\mathrm{K}-1,1:\mathrm{R}} X \tag{4.3.2}$$

or

$$U_{2:\mathrm{K},1:\mathrm{R}} \Sigma_{1:\mathrm{R},1:\mathrm{R}}^{1/2} = U_{1:\mathrm{K}-1,1:\mathrm{R}} \Sigma_{1:\mathrm{R},1:\mathrm{R}}^{1/2} X. \tag{4.3.3}$$

These systems are usually overdetermined and so can be solved using least squares [KAR83] or total least squares [VH93] methods.

If no error, noise, round-offs, etc., is present, we can prove that the eigenvalues of X equal the $\{z_r\}$ values in (3.1). The coefficients $\{a_r\}$ can be determined by applying any linear fitting method, e.g., least squares, to (3.1.1). We summarize Kung's method as follows.

Kung's Method

1. Compute an SVD of H: $H = U\Sigma V^H$.
2. Solve equation (4.3.2) or (4.3.3) for X.
3. Compute an eigenvalue decomposition of X: $X = YD_\zeta^{\mathrm{R}}Y^{-1}$.
4. Set $D_z^{\mathrm{R}} = D_\zeta^{\mathrm{R}}$.
5. Find $\{a_r\}$ by solving the Vandermonde equations (3.1.1), using either ordinary or total least squares.

An important consequence of the HSVD method is that no polynomial coefficients are explicitly computed. Subsequent refinements of the HSVD method are mostly restricted to Step 2, e.g., [VH93]. Although Kung claims in his paper [19] that the numerical stability of the SVD contributes considerably to the good behavior of the HSVD method, we have observed that the critical step in Kung's method lies not in the SVD but in the determination of the eigenvalues of X. Our stability assumption, that the eigenvalues have magnitude no greater than one, may provide a better basis to discuss the stability of Kung's procedure.

8.5 Matrix Pencil Framework

We present here a new matrix pencil formulation for computing $\{z_r\}$ in our model (3.1). We also develop a new proof of Kung's algorithm with the following three desirable properties:

1. Simple: Existing proofs rely on rather specialized previous work;
2. Purely Algebraic: Our proof does not need the tools specific to state-space based signal processing;
3. General: We will present a more general approach to the problem.

For the purpose of generalization, we introduce two arbitrary but nonsingular transformations F and G, where F is $(\mathrm{K}-1)\times(\mathrm{K}-1)$ and G is $\mathrm{L}\times\mathrm{L}$. Define two new $(\mathrm{K}-1)\times\mathrm{L}$ matrices $C^{[1]}$ and $C^{[2]}$ by

$$\begin{cases} C^{[1]} & \equiv FH_{1:\mathrm{K}-1,:}G, \\ C^{[2]} & \equiv FH_{2:\mathrm{K},:}G. \end{cases}$$

Consider the matrix pencil problem:

$$C^{[2]}y = \zeta C^{[1]}y.$$

Suppose we have found an $\mathrm{L}\times\mathrm{R}$ matrix Y of linearly independent eigenvectors and an $\mathrm{R}\times\mathrm{R}$ diagonal matrix D_ζ^{R} of eigenvalues $\{\zeta_l\}$. So,

$$C^{[2]}Y = C^{[1]}YD_\zeta^{\mathrm{R}}. \tag{5.1}$$

We will prove that $\{\zeta_l\} = \{z_r\}$.

8.5.1 Key Assumption

We assume that the $\mathrm{K} \times \mathrm{L}$ matrix H has the decomposition (3.1.2). Let

$$W \equiv V^{\mathrm{N}\times\mathrm{R}}(z).$$

Then (3.1.2) becomes

$$H = W_{1:\mathrm{K},1:\mathrm{R}} D_a^{\mathrm{R}} W_{1:\mathrm{L},1:\mathrm{R}}^T \ . \tag{5.1.1}$$

The decomposition is key to the development in this section. Define

$$D_z^{\mathrm{R}} \equiv \mathrm{diag}(z_1, z_2, \ldots, z_{\mathrm{R}}).$$

Since

$$W_{2:\mathrm{K},:} = W_{1:\mathrm{K}-1,:} D_z^{\mathrm{R}} \ ,$$

we can split (5.1.1) as follows:

$$H_{1:\mathrm{K}-1,:} = W_{1:\mathrm{K}-1,:} D_a^{\mathrm{R}} W_{1:\mathrm{L},:}^T$$

and

$$H_{2:\mathrm{K},:} = W_{1:\mathrm{K}-1,:} D_z^{\mathrm{R}} D_a^{\mathrm{R}} W_{1:\mathrm{L},:}^T = W_{1:\mathrm{K}-1,:} D_a^{\mathrm{R}} D_z^{\mathrm{R}} W_{1:\mathrm{L},:}^T \ .$$

Hence

$$C^{[1]} = (F W_{1:\mathrm{K}-1,:} D_a^{\mathrm{R}})(W_{1:\mathrm{L},:}^T G) \tag{5.1.2}$$

and

$$C^{[2]} = (F W_{1:\mathrm{K}-1,:} D_a^{\mathrm{R}}) D_z^{\mathrm{R}} (W_{1:\mathrm{L},:}^T G) \ . \tag{5.1.3}$$

8.5.2 Existence and Uniqueness

First, we prove that we can always find a solution to (5.1). We use the fact that $W_{1:\mathrm{L},:}^T = (W_{1:\mathrm{R},1:\mathrm{R}}^T | \ldots)$ to get

$$W_{1:\mathrm{L},:}^T \begin{pmatrix} W_{1:\mathrm{R},1:\mathrm{R}}^{-T} \\ 0 \end{pmatrix} = I^{\mathrm{R}\times\mathrm{R}}.$$

By choosing

$$Y = G^{-1} \begin{pmatrix} W_{1:\mathrm{R},1:\mathrm{R}}^{-T} \\ 0 \end{pmatrix}, \tag{5.2.1}$$

we obtain

$$W_{1:\mathrm{L},:}^T G Y = I^{\mathrm{R}\times\mathrm{R}}.$$

So, from (5.1.2) and (5.1.3), we get

$$C^{[1]} Y = F W_{1:\mathrm{K}-1,:} D_a^{\mathrm{R}}$$

and

$$C^{[2]} Y = (F W_{1:\mathrm{K}-1,:} D_a^{\mathrm{R}}) D_z^{\mathrm{R}} = C^{[1]} Y D_z^{\mathrm{R}},$$

establishing a solution to (5.1) with $D_\zeta^{\rm R} = D_z^{\rm R}$ and Y given by (5.2.1).

Second, we wish to determine the properties of the pair of matrices $D_\zeta^{\rm R}$ and Y that solves (5.1). We will show how they are related to $D_z^{\rm R}$, $W_{1:{\rm L},:}^T$ and G. Use (5.1.2) and (5.1.3) to substitute for $C^{[1]}$ and $C^{[2]}$ in (5.1):

$$(FW_{1:{\rm K}-1,:}D_a^{\rm R})D_z^{\rm R}(W_{1:{\rm L},:}^T GY) = (FW_{1:{\rm K}-1,:}D_a^{\rm R})(W_{1:{\rm L},:}^T GY)D_\zeta^{\rm R}. \quad (5.2.2)$$

Define an ${\rm R} \times {\rm R}$ matrix B by

$$B \equiv (W_{1:{\rm L},:}^T G)Y.$$

This matrix B is nonsingular because G is nonsingular and both $W_{1:{\rm L},:}$ and Y have full column rank. Since $W_{1:{\rm K}-1,:}$ also has full column rank, and F and $D_a^{\rm R}$ are nonsingular, we may simplify (5.2.2) to get

$$D_z^{\rm R} B = BD_\zeta^{\rm R}. \quad (5.2.3)$$

The (k,l) entry in (5.2.3) satisfies

$$z_k B_{k,l} = \zeta_l B_{k,l}. \quad (5.2.4)$$

But not all elements in a row or a column of B can be zero. So (5.2.4) can be satisfied only if for each z_k there is an equal ζ_l and vice versa. In other words, there must exist a permutation Π such that

$$D_z^{\rm R} = \Pi^T D_\zeta^{\rm R} \Pi.$$

Furthermore, from (5.2.3), we get

$$D_z^{\rm R} B\Pi = B\Pi(\Pi^T D_\zeta^{\rm R} \Pi),$$

and so

$$D_z^{\rm R}(B\Pi) = (B\Pi)D_z^{\rm R}.$$

Hence the matrix $B\Pi$ must be diagonal. Define this diagonal matrix by

$$D_\beta^{\rm R} \equiv B\Pi.$$

Then Y satisfies

$$(W_{1:{\rm L},:}^T G)Y\Pi = D_\beta^{\rm R}. \quad (5.2.5)$$

Note that

$$(W_{1:{\rm L},:}^T G)\left(Y\Pi(D_\beta^{\rm R})^{-1}\right) = I^{{\rm R}\times{\rm R}},$$

and so the diagonal entries of $D_\beta^{\rm R}$ are normalization scalars for the columns of $Y\Pi$. We present a new algorithm for computing $\{z_r\}$, with $\Pi = I$.

Matrix Pencil Framework

1. Choose matrices F and G.
2. Solve the matrix pencil (5.1): $C^{[2]}Y = C^{[1]}YD_\zeta^{\rm R}$.
3. Set $D_z^{\rm R} = D_\zeta^{\rm R}$.
4. Solve (3.1.1) in some approximate sense.

8.5.3 Notes

The pencil framework leaves many aspects of the solution unspecified. The choice of F and G, as well as the method for finding the eigenvalues, will affect the efficiency and accuracy of the overall procedure. In addition, it is not clear how to select F and G so as to get a well-conditioned pencil.

The last step in our generic algorithm is a delicate one, for it is difficult to justify the computation of the coefficients $\{a_r\}$ by solving an overdetermined Vandermonde system using ordinary or total least squares. Some improvement is achieved by weighting these procedures to compensate for the varying impact of noise on the different samples. In general, more sophisticated methods are desired. However, we will not discuss this further. Instead, we will concentrate on the determination of the knots $\{z_r\}$ in the rest of these notes.

8.6 Eigenvalue Procedures

In this section, we consider an important case of the matrix pencil framework where

$$\mathrm{L} = \mathrm{R}.$$

We still have $\mathrm{K} > \mathrm{R}$. The matrix Y is $\mathrm{R} \times \mathrm{R}$ and nonsingular and (5.1) becomes

$$C^{[2]} = C^{[1]}(Y D_{\zeta}^{\mathrm{R}} Y^{-1}).$$

Define an $\mathrm{R} \times \mathrm{R}$ matrix X by

$$X \equiv Y D_{\zeta}^{\mathrm{R}} Y^{-1}. \tag{6.1}$$

Note that (6.1) defines an eigenvalue decomposition of X, and also that X satisfies

$$C^{[1]} X = C^{[2]}. \tag{6.2}$$

The matrix equation (6.2) is usually overdetermined. In the presence of noise, X may be solved by the methods of least squares [KAR83] or total least squares [VH93].

We may also construct $W_{1:\mathrm{R},1:\mathrm{R}}$ without using $\{z_r\}$, because (5.2.5) reduces to

$$W_{1:\mathrm{R},1:\mathrm{R}} = G^{-T} Y^{-T} D_{\beta}^{\mathrm{R}}.$$

The diagonal matrix D_{β}^{R} follows trivially from the property that the first row of W consists of all ones. Also,

$$W_{1:\mathrm{R},1:\mathrm{R}}^{-1} = (D_{\beta}^{\mathrm{R}})^{-1} Y^T G^T.$$

Eigenvalue Method

1. Choose matrices F and G.
2. Find X by solving (6.2): $C^{[1]}X = C^{[2]}$.
3. Compute an eigenvalue decomposition of X: $X = YD_{\zeta}^{\mathrm{R}}Y^{-1}$.
4. Set $D_z^{\mathrm{R}} = D_{\zeta}^{\mathrm{R}}$.
5. Find $\{a_r\}$ by solving the Vandermonde equation (3.1.1) in some approximate sense.

In the next subsections, first we demonstrate how specific choices of F and G will result in the Prony's and Kung's methods, and then we show how new methods with improved numerical properties can be constructed.

8.6.1 Reduction to Prony's Method

Suppose further that $\mathrm{K} = \mathrm{R} + 1$; so both matrices $C^{[1]}$ and $C^{[2]}$ are $\mathrm{R} \times \mathrm{R}$. Choose

$$F = I^{\mathrm{R}\times\mathrm{R}} \quad \text{and} \quad G = H_{1:\mathrm{R},1:\mathrm{R}}^{-1}.$$

Hence,

$$C^{[1]} = I^{\mathrm{R}\times\mathrm{R}}\, H_{1:\mathrm{R},1:\mathrm{R}}\, H_{1:\mathrm{R},1:\mathrm{R}}^{-1} = I^{\mathrm{R}\times\mathrm{R}},$$

and (6.2) simplifies to

$$\begin{aligned} X &= C^{[2]} = I^{\mathrm{R}\times\mathrm{R}}\, H_{2:\mathrm{R}+1,1:\mathrm{R}}\, H_{1:\mathrm{R},1:\mathrm{R}}^{-1} \\ &= \begin{pmatrix} 0 & 1 & 0 & \cdots & 0 & 0 \\ 0 & 0 & 1 & \cdots & 0 & 0 \\ 0 & 0 & 0 & \cdots & 0 & 0 \\ \vdots & \vdots & \vdots & \ddots & \vdots & \vdots \\ 0 & 0 & 0 & \cdots & 0 & 1 \\ -\gamma_0 & -\gamma_1 & -\gamma_2 & \cdots & -\gamma_{\mathrm{R}-2} & -\gamma_{\mathrm{R}-1} \end{pmatrix}, \end{aligned}$$

where the $\{\gamma_i\}$ satisfy the Prony-Yule-Walker equation (4.1.3). Hence $C^{[2]}$ is a companion matrix to the Prony polynomial $q_{\mathrm{R}}(z)$ of (4.1.1), and our eigenvalue method is identical to Prony's method.

8.6.2 Reduction to Kung's Method

In Kung's method, we compute an SVD of a $\mathrm{K} \times \mathrm{R}$ Hankel matrix H:

$$H = U\Sigma V^H.$$

Now, pick

$$F = I^{(\mathrm{K}-1)\times(\mathrm{K}-1)} \quad \text{and} \quad G = V\Sigma^{-1}.$$

We find that

$$C^{[1]} = U_{1:\mathrm{K}-1,:} \qquad \text{and} \qquad C^{[2]} = U_{2:\mathrm{K},:} \ .$$

So (6.2) reduces to (4.3.2):

$$U_{1:\mathrm{K}-1,:}X = U_{2:\mathrm{K},:} \ .$$

Similarly, if we choose

$$F = I^{(\mathrm{K}-1)\times(\mathrm{K}-1)} \quad \text{and} \quad G = V\Sigma^{-1/2},$$

then

$$C^{[1]} = U_{1:\mathrm{K}-1,:}\Sigma^{1/2} \qquad \text{and} \qquad C^{[2]} = U_{2:\mathrm{K},:}\Sigma^{1/2},$$

and we get (4.3.3), the balanced realization of Kung's method:

$$U_{1:\mathrm{K}-1,:}\Sigma^{1/2}X = U_{2:\mathrm{K},:}\Sigma^{1/2}.$$

8.6.3 HQRD Method

We propose to replace the SVD with a less expensive decomposition. One possible choice is the QR decomposition of a $\mathrm{K} \times \mathrm{R}$ Hankel matrix H:

$$H = QR, \tag{6.3.1}$$

where Q is $\mathrm{K} \times \mathrm{K}$ unitary, and R is $\mathrm{K} \times \mathrm{R}$ right triangular. Choose

$$F = I^{(\mathrm{K}-1)\times(\mathrm{K}-1)} \quad \text{and} \quad G = R^{-1}.$$

We find that

$$C^{[1]} = Q_{1:\mathrm{K}-1,:} \qquad \text{and} \qquad C^{[2]} = Q_{2:\mathrm{K},:} \ ,$$

so that (6.2) reduces to

$$Q_{1:\mathrm{K}-1,:}X = Q_{2:\mathrm{K},:} \ . \tag{6.3.2}$$

HQRD Method

1. Compute QR decomposition of H, cf. (6.3.1): $H = QR$.
2. Choose $F = I^{(\mathrm{K}-1)\times(\mathrm{K}-1)}$ and $G = R^{-1}$.
3. Solve equation (6.3.2) for X: $Q_{1:\mathrm{K}-1,:}X = Q_{2:\mathrm{K},:}$.
4. Compute an eigenvalue decomposition of X: $X = YD_{\zeta}^{\mathrm{R}}Y^{-1}$.
5. Set $D_z^{\mathrm{R}} = D_{\zeta}^{\mathrm{R}}$.
6. Find $\{a_r\}$ by approximating the Vandermonde equations (3.1.1).

8.6.4 Numerical Behavior

Note that perturbation bounds for the determination of eigenvalues, e.g., the Bauer-Fike theorem [GVL89], are often strongly dependent on the corresponding eigenvectors. Hence, we expect that in our generalization, the F and G transformations could be selected to reduce the effect of numerical perturbations on the computed eigenvalues.

In any case, we should not forget the remark of Lanczos [Lan56] on the exponential fitting problem: "It would be idle to hope that some other modified mathematical procedure could give better results, since the difficulty lies not with the manner of evaluation but with the extraordinary sensitivity of the exponents and amplitudes to very small changes of the data, which no amount of least-square or other form of statistics could remedy." Indeed, we observe that unless the $\{z_r\}$ are reasonably close to and inside of the unit circle, both Prony's and Kung's method fail abysmally.

Example 1 *This example is taken from Van Huffel's paper [VH93]. A simulated signal of numerical rank five, typical of the signals obtained in nuclear magnetic resonance applications, is chosen. Figure 6.1 shows its spectrum: the sampling frequency is 10kHz and the signal-to-noise ratio is approximately 2.4 (not dB). Perhaps the most interesting challenge presented by this signal is the accurate resolution of the two peaks near 500Hz. The computations are carried out on the first 130 samples. The effectiveness of applying Kung's method after one iteration of Cadzow's denoising strategy is illustrated in Figures 6.2 and 6.3.*

8.7 A New Denoising Strategy

We have already pointed out that Cadzow's denoising procedure and the HSVD algorithm sacrifice the problem structure and are computationally expensive. We have found that in many cases the assumption of stable roots, i.e.,

$$|z_r| \leq 1,$$

can be exploited to improve the accuracy and reliability of any specialization of our generic pencil approach.

8.7.1 Zero Rejection

To this end, we choose to overestimate the rank R of the available samples. Just as in Prony's original proposal, we choose R so that (4.1.3) is square. A significant difference is as follows.

- We fix N, the number of samples, and choose $\text{R} = \lfloor \text{N}/2 \rfloor$.
- Prony fixes R, the rank of the matrix, and chooses $\text{N} = 2\text{R}$.

With noise, the Hankel matrix we get is always nonsingular. As a side benefit, we can determine the coefficients of (4.1.3) using any of the special Yule-Walker solvers. Figure 7.1 shows the cloud of zeros of the Prony polynomial obtained from the 65×65 data matrix of Example 1.

Under the usual assumption that the noise space is approximately orthogonal to the signal subspace, it is not unreasonable to expect that a few components will still fit much of the embedded signal, whereas most other zeros serve as outlets to fit the noise. Our numerical experiments seem to confirm this conjecture. Given our stability criterion, it is logical to discard those zeros of the polynomial that are greater than one in magnitude and hence cannot correspond to a stable component of the signal. Figure 7.2 shows the result.

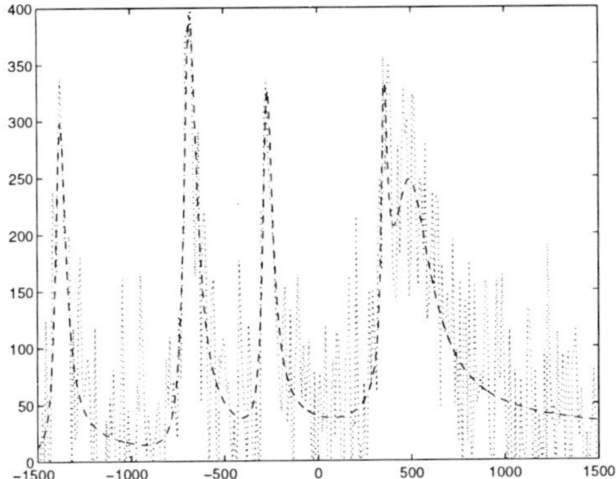

Figure 6.1 Real part of DFT of signal of Example 1: dashed line depicts spectrum of noiseless signal.

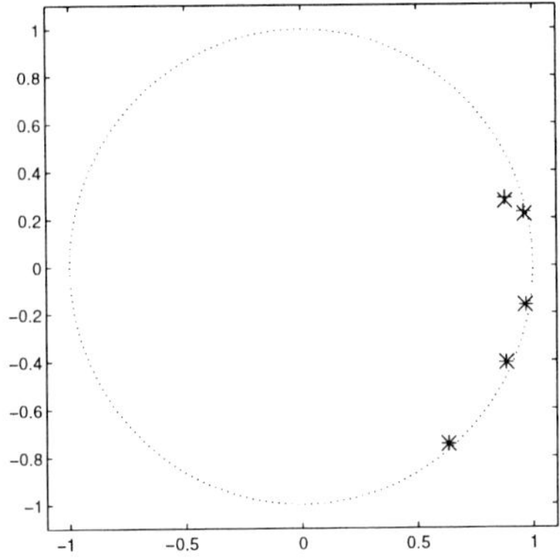

Figure 6.2 Kung's method with Cadzow's denoising: computed (+) and exact ($\times$) z values are in good agreement.

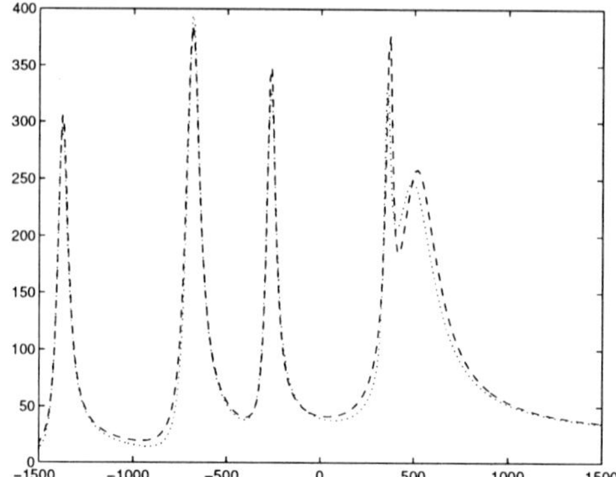

Figure 6.3 Kung's method with Cadzow's denoising: dashed line displays computed DFT spectrum while dotted curve depicts noiseless spectrum.

8.7.2 Zero Selection

At this point, a significant portion of the noise has been rejected from the given data, but we are still left with many noise-contributed zeros. This is an appropriate time to recall that zeros on the unit circle in the complex plane result in sharp peaks in the Fourier spectrum; zeros slightly inside that circle can be characterized in the same sense even though the sharpness decreases with increased damping.

We have developed a straightforward "maximum-finder" to decide what is a peak and what is not. Our selection process works by mapping zeros on DFT peaks and not vice versa; i.e., we do not look for a peak and then search for a zero that represents it best. The procedure takes advantage of the fact that the set of DFT frequencies is a sequence of equidistant values. The zeros, on the other hand, are not equidistant and are unordered.

Peak Selection Method

For each zero z	$O(\mathrm{R})$ iterations
Compute angle $arg(z)$ and map it in frequency domain	$O(1)$ time
Select closest DFT value and a few neighboring points	$O(1)$ time
Compute a "peak quality" with these samples	$O(1)$ time
Select z if its "peak quality" is high enough	$O(1)$ time

So a total of $O(\mathrm{R})$ work is required. The process is both asymptotically fast and relatively simple; it works well even with a peak quality based on a single sample. The converse process of mapping peak frequencies onto the z values is significantly more costly. Figure 7.3 portrays the zeros identified as belonging to the signal rather than to the noise; these are the zeros corresponding to high peaks in the DFT of Figure 7.2.

8.7.3 Notes

Our experience is that Kung's method combined with one step of Cadzow's cleanup algorithm is comparable in robustness to our zero selection strategy applied to Prony's method. Note that although we have made the illustration with Prony's method, our new strategy can in fact be applied to any of the eigenvalue procedures described in Section 5.

Unlike Cadzow's procedure, the approach presented in this section is an *a posteriori* estimation strategy. Indeed, it is possible to initiate our overall computation with a few Cadzow iterations, and to end it with the selection procedure outlined above.

It is our intention to explore alternative zero selection techniques. In particular, we believe that the stochastic character of noise should enable us to use correlation techniques for different signal subsequences; for example, if an ergodic noise model is assumed.

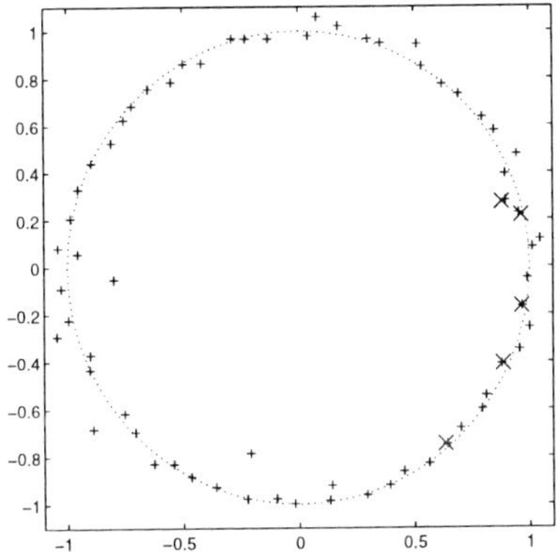

Figure 7.1 Overestimating signal rank: '+' denotes computed roots (65 total) and '×' denotes exact roots (5 total).

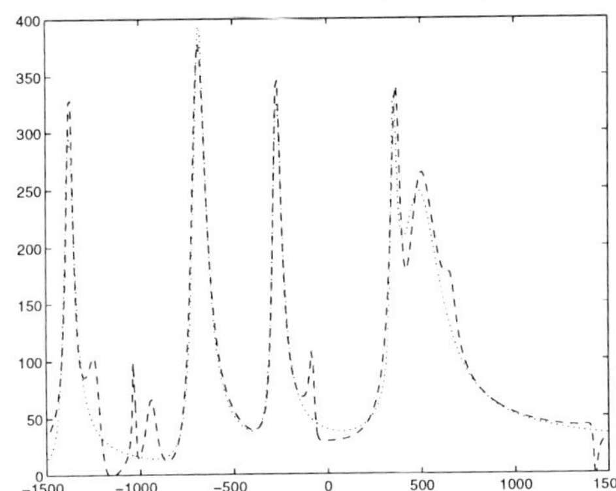

Figure 7.2 Removing unstable roots and recomputing new samples: dashed line shows new spectrum and dotted curve denotes noiseless spectrum.

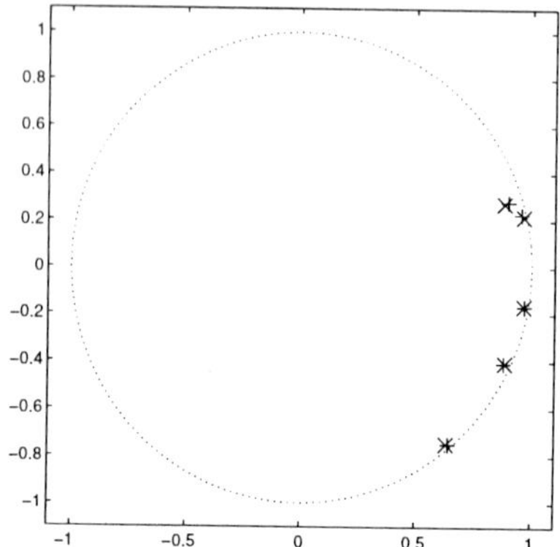

Figure 7.3 Applying peak selection strategy: Selected (+) and exact (×) z values overlap.

8.8 Conclusions

These notes survey various aspects of computing an exponential decomposition. Although our emphasis has been on linear algebraic iterative methods, we remind the reader of the extensive literature on nonlinear approaches. Once again, we stress the important fact that this problem is very sensitive to errors in the data.

We have presented a close relationship between the exponential decomposition and the Hankel structure. The connection allows us to formulate a framework for a large class of solution procedures. In addition, we demonstrate how discarding the unstable components of an exponential decomposition will lead to a considerable reduction of the effects of noise in the frequency domain. This in turn will provide us with a new tool for finding a structured rank-revealing decomposition of a Hankel matrix [LV96].

8.9 References

[AG88] G.S. Ammar and W.B. Gragg. Superfast solution of real positive definite Toeplitz systems. *SIAM J. Matrix Anal. Appl.*, 9:61–76, 1988.

[BBGL92] D.L. Boley, R.P. Brent, G.H. Golub, and F.T. Luk. Algorithmic fault tolerance using the Lanczos method. *SIAM J. Matrix Anal.*, 13:312–332, 1992.

[BdBvO87] H. Barkhuijsen, R. de Beer, and D. van Ormondt. Improved algorithm for noniterative time-domain model fitting to exponentially damped magnetic resonance signals. *J. Magn. Reson.*, 73:553–557, 1987.

[Bel73] E. Beltrami. Sulle funzioni bilineari. *Giornale di Mathematiche*, 11:98–106, 1873. (English translation by D.L. Boley. On Bilinear Functions. In M. Moonen and B. De Moor, editors, *SVD and Signal Processing III*, pages 9-18, Amsterdam, The Netherlands, 1995. Elsevier).

[BGY80] R.P. Brent, F. Gustavson, and D. Yun. Fast solution of Toeplitz systems of equations and computation of Padé approximants. *J. Algo.*, 1:259–295, 1980.

[Bun85] J.R. Bunch. Stability of methods for solving Toeplitz systems of equations. *SIAM J. Sci. Stat. Comp.*, 6:349–364, 1985.

[Cad88] J.A. Cadzow. Signal enhancement – a composite property mapping algorithm. *IEEE Trans. Acoust., Speech, Signal Processing*, 36:49–62, 1988.

[CN96] R.H. Chan and M.K. Ng. Conjugate gradient methods for Toeplitz systems. *SIAM Review*, 38, 1996. to appear.

[DMZ91] B. De Moor and H. Zha. A tree of generalizations of the ordinary singular value decompositions. *Lin. Alg. and its Applic.*, 147:469–500, 1991.

[EGLV80] J.W. Evans, W.B. Gragg, and R.J. Le Veque. On least squares exponential sum approximation with positive coefficients. *Math. Comp.*, 34:203–212, 1980.

[GK65] G.H. Golub and W. Kahan. Calculating the singular values and pseudo-inverse of a matrix. *SIAM J. Num. Anal. Ser. B*, 2:205–224, 1965.

[Glo84] K. Glover. All optimal Hankel-norm approximations of linear multivariable systems and their l^∞-error bounds. *Int. J. Control*, 39:1115–1193, 1984.

[GR70] G.H. Golub and C. Reinsch. Singular value decomposition and least squares solutions. *Numer. Math.*, 14:403–420, 1970.

[GVL89] G.H. Golub and C.F. Van Loan. *Matrix Computations (second edition)*. Johns Hopkins University Press, Baltimore, 1989.

[Hig87] N.J. Higham. Error analysis of the Björck-Pereyra algorithms for solving Vandermonde systems. *Numer. Math.*, 50:613–632, 1987.

[Hot36] H. Hotelling. Relations between two sets of variates. *Biometrika*, 28:321–377, 1936.

[Kah93] M. Kahn. A state space method for estimating frequencies and dampings. Technical report, Supercomputer Facility, Australian National University, Feb. 1993.

[Kam79] D.W. Kammler. Least squares approximation of completely monotonic functions by sums of exponentials. *SIAM J. Numer. Anal.*, 16:801–818, 1979.

[KAR83] S.Y. Kung, K.S. Arun, and B.D. Rao. State-space and singular value decomposition-based approximation methods for the harmonic retrieval problem. *J. Opt. Soc. Am.*, 73:1799–1811, 1983.

[KT82] R. Kumaresan and D.W. Tufts. Estimating the parameters of exponentially damped sinusoids and pole-zero modeling in noise. *IEEE Trans. Acoust. Speech and Sig. Proc.*, 30:833–840, 1982.

[Lan56] C. Lanczos. *Applied Analysis.* Prentice-Hall, Englewood Cliffs, 1956.

[Lar83] W.E. Larimore. System identification, reduced-order filtering and modeling via canonical variate analysis. In H.S. Rao and P. Dorato, editors, *Proc. 1983 Amer. Control Conf.*, volume 2, pages 445–451, Piscataway, 1983. IEEE.

[Lev46] N. Levinson. The Wiener rms (root mean square) error criterion in filter design and prediction. *J. Math. and Phys.*, 25:261–278, 1946.

[LQ93] F.T. Luk and S. Qiao. A new matrix decomposition for signal decomposition. In G.H. Golub M.S. Moonen and B.L.R. De Moor, editors, *Linear Algebra for Large Scale and Real-Time Applications*, pages 241–247, Dordrecht, The Netherlands, 1993. Kluwer Academic Publishers.

[LV94] F.T. Luk and D. Vandevoorde. Reducing boundary distortion in image restoration. In F.T. Luk, editor, *Advanced Signal Processing Algorithms, Architectures, and Implementations V*, volume 2296, pages 554–565, Bellingham, 1994. SPIE.

[LV96] F.T. Luk and D. Vandevoorde. Computing a structured factorization of a Hankel matrix. Technical report, Department of Computer Science, Rensselaer Polytechnic Institute, June 1996.

[OS91] M.R. Osborne and G.K. Smyth. A modified Prony algorithm for fitting functions defined by difference equations. *SIAM J. Sci. Stat. Comput.*, 12:362–382, 1991.

[Pro95] R. Prony. Essai expérimental et analytique sur les lois de la dilatabilité et sur celles de la force expansive de la vapeur de l'eau et de la vapeur de l'alkool, à différentes températures. *J. de l'Ecole Polytechnique*, 1:24–76, 1795.

[Ruh80] A. Ruhe. Fitting empirical data by positive sums of exponentials. *SIAM J. Sci. Stat. Comput.*, 1:481–498, 1980.

[Rus87] C.K. Rushforth. Signal restoration, functional analysis, and Fredholm integral equations of the first kind. In H. Stark, editor, *Image Restoration: Theory and Application*, pages 1–27, Orlando, 1987. Academic Press.

[SCSB76] M.R. Smith, S. Cohn-Sfetcu, and H.A. Buckmaster. Decomposition of multicomponent exponential decays by spectral analytic techniques. *Technometrics*, 18:467–482, 1976.

[VH93] S. Van Huffel. Enhanced resolution based on minimum variance estimation and exponential data modeling. *Signal Processing*, 33:333–355, 1993.

[VL75] C.F. Van Loan. A general matrix eigenvalue algorithm. *SIAM J. Num. Anal.*, 12:819–834, 1975.

[Wal31] G. Walker. On periodicity in series of related terms. *Proc. Royal Soc. London Ser. A*, 131A:518–532, 1931.

[Yul27] G.U. Yule. On a method of investigating periodicities in disturbed series, with special reference to Wolfer's sunspot numbers. *Philos. Trans. Roy. Soc. London Ser. A*, 226A:267–298, 1927.

[YW82] D.C. Youla and H. Webb. Image restoration by the method of convex projections: Part I – Theory. *IEEE Trans. Med. Imaging*, 1:81–94, 1982.

9

Iterative Methods for Total Variation Image Restoration

Tony F. Chan*
Pep Mulet†

9.1 Introduction

Image processing refers to the analysis and extraction of information from images. Some example of the tasks are: restoration, compression and segmentation [21]. Applications can be found in many areas, including medical diagnosis, satellite surveying and surveillance.

Since the number of pixels for even a two dimensional image with modest resolution often exceeds several hundred thousands, image processing involves many computationally intensive tasks. The requirements for real time response and multi-frame/band analysis add further to the need for fast and efficient computational algorithms.

Traditionally, the standard methods involve computation in the frequency domain [21], facilitated by efficient FFT (and more recently wavelet) algorithms. In the last few years, there has been a new movement towards a more PDE-based approach; see for example the recent review article by Alvarcz and Morel [1] and the books by Morel and Solimini [28] and ter Haar Romeny [34]. The new models are motivated by a more systematic approach to restoring images with sharp edges, as well as for image segmentation. The image is diffused (denoised) according to a nonlinear anisotropic diffusion PDE, designed to diffuse less near edges [29]. The PDEs are often designed to possess certain desirable geometrical properties such as affine invariance and causality. Total variation based image restoration methods belong to this new class of models.

From a computational standpoint, the PDE formulations leads to compu-

---

*Dept. of Mathematics, UCLA, Los Angeles, CA 90095-1555, USA. Email: chan@@math.ucla.edu. URL: http://www.math.ucla.edu/~chan.

This work has been supported in part by the ONR under contract N00014-96-1-0277 and the NSF under contract DMS-9626755.

†Departamemt de Matematica Aplicada, Univ. de Valencia, Dr. Moliner, 50, 46100 Burjassot, Spain. Email: mulet@@godella.matapl.uv.es.

This author was also supported by Spanish DGICYT grants EX94 28990695 and PB94-0987.

tational problems which are often of a different nature from the traditional frequency domain and algebraic approach and calls for new computational techniques which can better exploit the fundamental nature of the governing nonlinear PDEs. As yet, the nonlinear diffusion models are considered somewhat expensive compared to traditional methods and naturally a part of the research in this area is directed towards making them more efficient while retaining and improving their desirable geometric properties.

In this chapter, we shall consider the use of iterative methods for total variation models for image restoration. Iterations are needed for solving the nonlinear problem, as well as for the linear algebra problems which arise at each step. Analogous to the situation for solving large discretized PDEs in several dimensions, the size of the problems are large enough that direct solution methods are too costly and iterative methods can be much more efficient.

The goal of this chapter is two fold. First, we would like to give a concise and self-contained overview of the total variation image restoration models and some existing numerical methods for solving them. Second, we would like to review briefly some of our own research in this area. Specifically, we will discuss two topics: the use of primal-dual Newton methods for solving a minimization problem involving the highly nonlinear total variation functional, and preconditioning techniques for differential-convolution type operators.

9.2 Image restoration

The recording of an image usually involves a degradation process: a *blurring*, due to atmosphere turbulence, camera misfocus or relative movement, followed by a random *noise*, due to errors of the physical sensors or to quantization. The actual degradation model will depend on many factors, but a commonly used model is that of a linear blurring operator and additive Gaussian white noise, which is the one we will consider in this paper. Others may involve nonlinear blurring operators, multiplicative noise, noise with more complicated distributions and with possible correlation with the image.

The aim of image restoration is the estimation of the ideal true image from the recorded one. The *direct problem* of computing the imaging system response (blurred image) from a given image is often assumed to be known and *well-posed.* The usual model for it is the convolution by a given kernel or *point spread function (PSF)*, which, in most of the cases, implies that the *inverse problem* of computing the true image from the observations is an *ill-posed problem.* A general principle for dealing with the instability of the inverse problem is that of *regularization*, which mainly consists in restricting the set of admissible solutions and including some *a priori information*

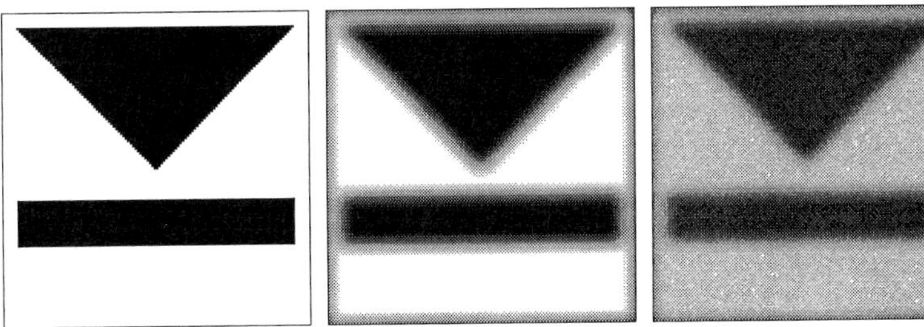

FIGURE 9.1. From left to right: original image, blurred image, observed image

(non negativity, smoothness, existence of edges, etc.) in the formulation of the problem. Both the accurate modeling of the imaging system and the choice of regularization will be essential for a satisfactory image restoration process.

Throughout this paper, the word *image* will refer to a real function u, defined on $\Omega = [0,1]^2$, with $u(x,y)$ being the intensity or grey level at pixel (x,y).

9.2.1 Degradation model

Let u denote the ideal image to be estimated and z the observed image. We model the imaging system as

$$z = Ku + n,$$

where K is a linear operator (the blur operator) and n is a random field (noise), of which we may know some of its statistics (see Fig 9.1).

The operator K is usually a convolution by a known point spread function h, i.e.,

$$Ku(x,y) = \int_\Omega h(x-\overline{x}, y-\overline{y})u(\overline{x},\overline{y})\, d\overline{x}\, d\overline{y},$$

corresponding to a *spatially invariant* blurring process. Another interesting case, which can be seen as a special case of the latter, is when the operator K is the identity, which corresponds to *denoising*.

The noise n is assumed to be a *Gaussian white noise*, i.e., the values $u(x,y)$ are uncorrelated random variables with a normal distribution with mean 0 and variance σ^2, for all $(x,y) \in \Omega$.

9.2.2 Regularization

The inverse problem $Ku = z$ is ill-posed nature in most cases due to the compactness of the convolution operators with square integrable kernels. This means that the problem has no solution and/or this solution does not depend continuously on the data z. Since this data will contain measurements errors, direct solutions of the inverse problem, in case of existence, will be useless. We thus need to modify this inverse problem in order to turn it into a well-posed problem. This can be achieved by *regularizing* it. We can cite as regularization procedures the following (more references can be found in the survey paper [24]):

- **Direct Regularization Methods.**
  - Tikhonov Regularization.
  - Truncated matrix factorizations, e.g. SVD and RRQR.
  - Mollifier Methods.
- **Iterative Regularization Methods.**
  - Lanczos and Conjugate Gradient regularization.

Some of these methods require parameter specifications. We cite some parameter choice methods:

- Discrepancy Principle.
- Generalized Cross-Validation.
- L-curve.
- Methods based on Error Estimation.

9.2.3 Tikhonov Regularization

We will focus on the paradigm of Tikhonov-like regularization in this paper. We consider two variants:

1. **Noise level constrained formulation**: it consists in the solution of the constrained optimization problem

$$\min_u R(u)$$

$$\text{subject to} \quad \|Ku - z\|^2 = \sigma^2,$$

   where R is a functional, the *regularization functional*, that measures the *irregularity* of u. The quadratic constraint acts as the restriction of the solution set, since $\|Ku - z\|^2 = \|n\|^2 \approx \sigma^2$, and the functional R incorporates the *a priori* information of the image. This approach is equivalent to the discrepancy principle.

2. **Unconstrained formulation**: if there is no good estimate of the variance of the noise, then we may consider the unconstrained optimization problem

$$\min_u f(u) \equiv \frac{1}{2}\|Ku - z\|^2 + \alpha R(u),$$

where α controls the tradeoff between a good fit to the data and an irregular solution. It can be shown that ideally α should be chosen to be the reciprocal of the Lagrange multiplier for the previous problem.

The regularization functionals used more often are quadratic functionals of the type $R(u) = ||Qu||^2$ where $Q = I$ (the identity operator) or $Q = \nabla$ (the H^1 semi-norm). This choice essentially gives a linear least squares problem, but has the drawback of penalizing discontinuities in u. Therefore, this is not a good choice if we are interested in edge restoration.

9.2.4 Algorithms for Tikhonov regularization with quadratic regularization functionals

If we consider a quadratic regularization functional $R(u) = ||Qu||_2^2$, with a linear functional Q, then we can write

$$f(u) = \alpha||Qu||_2^2 + ||Ku - z||_2^2 = \left\|K\sqrt{\alpha}Qu - z0\right\|_2^2 = ||Au - f||_2^2.$$

Thus the problem reduces to an ordinary least squares problem. In order to improve efficiency, one must try to exploit the Toeplitz (or block-Toeplitz) structure of A^TA. Since A can be very large (if the picture contains $n \times n$ pixels, then K is $n^2 \times n^2$) we need efficient algorithms.

Direct methods

Fast direct methods for structured least squares problems is an actively researched area. The classical Levinson and Schur algorithms [20] can reduce the complexity from $\mathcal{O}(\backslash^{\ni})$ to $\mathcal{O}(\backslash^{\in})$ or $\mathcal{O}(\backslash \log^{\in} \backslash)$. The stability of these fast and super-fast algorithms are still not fully understood [4]. More recently, fast algorithms based on Gaussian-elimination with partial pivoting for Cauchy matrices and matrices with displacement structures have been developed [25, 19].

Iterative methods: Circulant Preconditioners

An alternative to direct methods is preconditioned iterative methods. This approach was first proposed by Strang [33]. The idea is the following. The conjugate gradient algorithm on the normal equations $A^TAu = A^Tz$ has the advantage of only needing the products Av and A^Tv. If K, Q are Toeplitz matrices then efficient algorithms can be employed to compute Av, A^Tv, for instance, by embedding these matrices into circulant ones and then using FFT's. However, the convergence rate will be governed by the condition number of A and hence preconditioning will be essential for the feasibility of this approach.

Some preconditioners based on circulant matrices have been proposed:

- [33, 5]: It consists in using a circulant matrix C to approximate a Toeplitz matrix T by first copying the main diagonals of T to C and then completing C to form a circulant matrix. This is illustrated in the fashion outlined in the following picture.

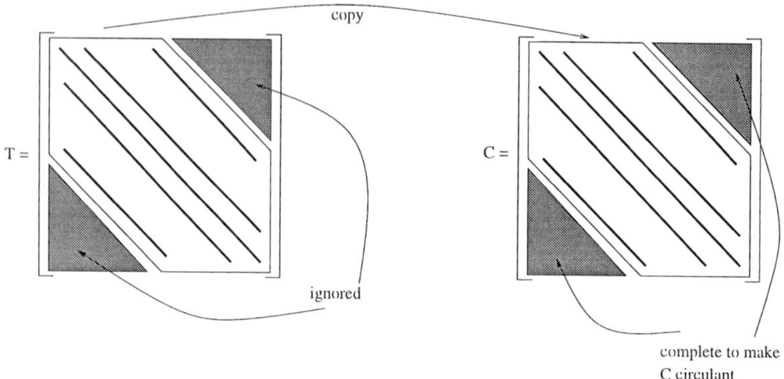

- [14]: This is a variational approach. For a given matrix T (not necessarily Toeplitz), it consists in finding the circulant matrix C which minimizes the Frobenius norm of the difference $C - T$, i.e., $C = argmin\, C circulant \|T - C\|_F$. The solution amounts to averaging the entries in the corresponding diagonals of T, as illustrated in the picture below.

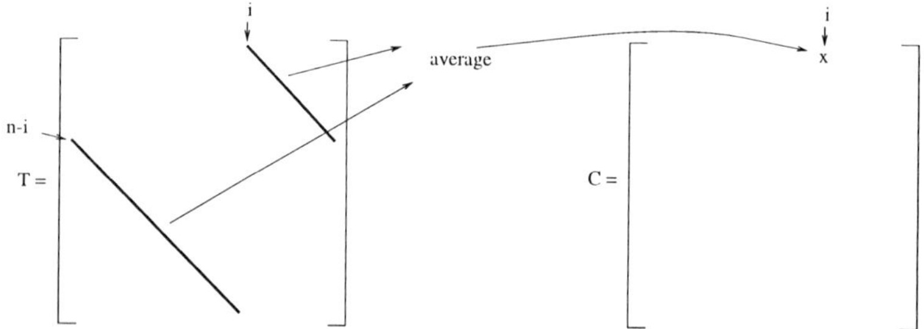

  This preconditioner has the following advantages:

  - It preserves SPD matrices.
  - It is applicable to general A, not just Toeplitz T.

- Many other variants have been proposed (cf. the recent survey [13]). There is also a vast literature on convergence theory for fast transform preconditioners:

  - For Toeplitz matrices in the Wiener Class (i.e. whose entries in a row is absolutely summable), the spectrum of $M^{-1}A$ clusters around 1 [9], [26], [3].

- For elliptic operators, the condition number of the preconditioned matrix, $\mathcal{K}(\mathcal{M}^{-\infty}\mathcal{A})$, can be bounded by $\mathcal{O}(\infty)$ independent of h, if the type of boundary conditions of the preconditioner matches those of the elliptic operator, and $\mathcal{O}(\backslash)$ otherwise [6], [17], [10].

For applications of some of these preconditioning techniques to Toeplitz least squares problems, see [8, 7].

More recently, an idea that has received a lot of interest is to use early truncation of the conjugate gradient method applied to the normal equation for $Ku = z$ as a regularization procedure [23].

9.3 Total Variation Restoration

9.3.1 Total Variation Regularization

The main disadvantage of using quadratic regularization functionals such as the H^1 semi-norm of u is the inability to recover sharp discontinuities (edges). Mathematically, this is clear because discontinuous functions do not have bounded H^1 semi-norms. To remedy this, Rudin, Osher and Fatemi proposed in [31] the use of the *Total Variation*

$$TV(u) = \int_{\Omega} |\nabla u| dx \;=\; \int_{\Omega} \sqrt{u_x^2 + u_y^2}\, dx\, dy$$

as a regularization functional. The intuition for the use of this functional is that it measures the *jumps* of u, even if it is discontinuous. It is also connected to minimal surface problems.

The following result illustrates the advantage of the use of the Total Variation versus the quadratic functional given by the H^1-semi-norm for a simplified 1D problem (cf. Fig 9.2)

Lemma. Let $S = \{f | f(0) = a, f(1) = b\}$.

1. For the total variation norm, we have that $\min_{f\in S} TV(f) = b - a$ is achieved by any monotone, not necessarily continuous, $f \in S$.

2. For the H^1 semi-norm, we have that $\min\limits_{f\in S} \int_0^1 (f'(x))^2\, dx$ has the unique solution $f(x) = a(1 - x) + bx$.

Thus, the total variation norm allows more functions, including discontinuous ones, to approximate a given noisy function, whereas the H^1 semi-norm "prefers" the linear function over others.

To illustrate the suitability of Total Variation denoising we depict a comparison of several signal denoising algorithms in Fig. 9.3 .

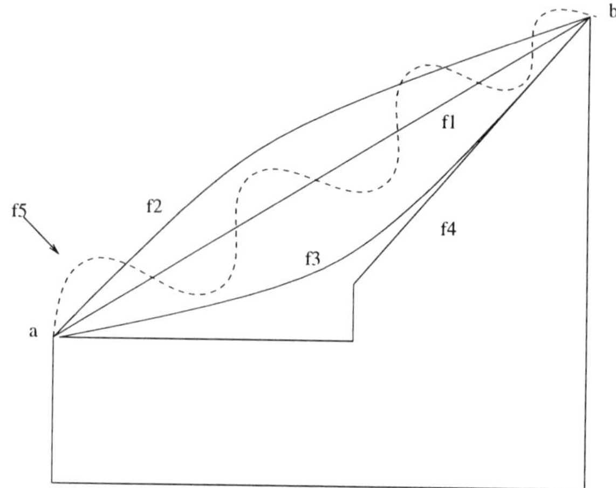

FIGURE 9.2. $TV(f_5) > TV(f_1) = TV(f_2) = TV(f_3) = TV(f_4)$

For simplicity, we will consider the unconstrained formulation of the Tikhonov regularization. The problem is then:

$$\min_u f(u) \equiv \alpha TV(u) + \frac{1}{2}||Ku - z||_2^2, \tag{9.1}$$

which is a strictly convex problem. To minimize $f(u)$, we will need its gradient. By using integration by parts with Neumann boundary conditions on (9.1), it can be shown that the gradient of the Total Variation functional is

$$\nabla TV(u) = -\nabla \cdot \left(\frac{\nabla u}{|\nabla u|} \right).$$

Thus, the first order optimality condition for the problem (9.1) is given by:

$$\nabla f(u) = g(u) = -\alpha \nabla \cdot \left(\frac{\nabla u}{|\nabla u|} \right) + (K^* K u - K^* z) = 0. \tag{9.2}$$

This is the problem that we need to solve to find the restored image u.

9.3.2 Computational challenges

The solution of (9.2) poses several computational difficulties:

- The problem size can be very large, due to the large number of pixels, even for normal resolutions. This is even worse if 3D or video images are considered. The use of iterative methods for linear systems is necessary for most of the cases.
- The TV functional is non-differentiable at locations where $|\nabla u| = 0$, which often happens in real images. This makes necessary the use

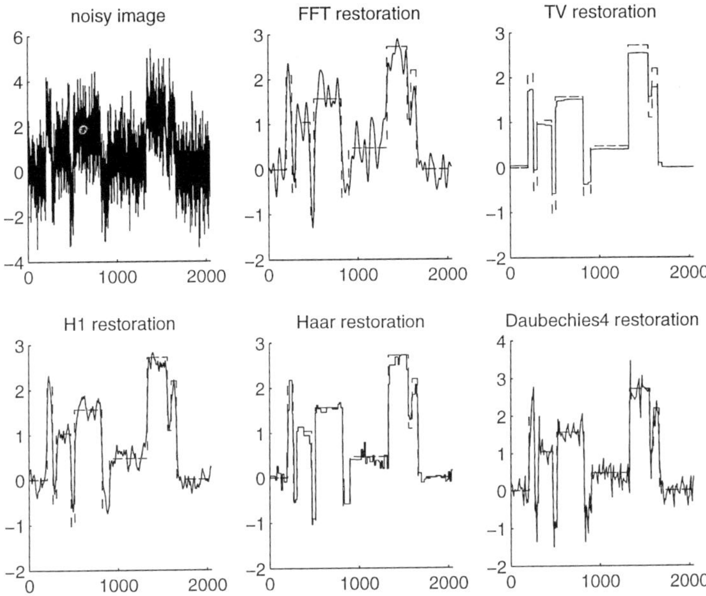

FIGURE 9.3. Original image plotted with dashed line. The Fourier, Haar and Daubechies plots correspond to truncated spectral decompositions and the H^1 restoration to Tikhonov regularization using the H^1-semi-norm as functional.

of numerical regularization, consisting in replacing the term $|\nabla u|$ by $\sqrt{|\nabla u|^2 + \beta}$ for a *small enough* positive artificial parameter β.

- The operator $\nabla \cdot \left(\frac{\nabla}{|\nabla u|}\right)$ is highly nonlinear.
- The ill-conditioning of the operators $\nabla \cdot \left(\frac{\nabla}{|\nabla u|}\right)$ (when linearized, it is a second order elliptic operator) and K^*K (it is a convolution operator) can cause numerical difficulties.
- When linearized, the elliptic operator $\nabla \cdot \left(\frac{\nabla}{|\nabla u|}\right)$ has highly varying coefficients, due to the presence of the term $\frac{1}{|\nabla u|}$.
- If $K \neq I$, then one needs to efficiently precondition the sum of an elliptic operator and a blurring operator. While good preconditioning techniques are available for the individual operators, preconditioning the sum is much more difficult. We will consider this issue in Section 9.5.

9.3.3 TV Restoration Algorithms

There are several iterative schemes which have been proposed in the literature for solving the optimality equation (9.2).

- The *Time Marching* scheme proposed in Rudin, Osher and Fatemi [31, 30] consists in finding the steady state of the parabolic PDE

$$u_t = -g(u), \; u(0) = z.$$

  It is equivalent to the steepest descent method. It can use the *projected gradient* method to handle noise constraints. The drawback of this method is that explicit methods can be slow due to stability constraints. Implicit methods can also be applied but then one has to deal with the nonlinearity and the solution of the resulting linear systems.

- In Vogel [35], the *Fixed Point Lagged Diffusivity Iteration* is introduced. This method consists in linearizing the nonlinear differential term in (9.2) by lagging the diffusion coefficient $1/|\nabla u|$ behind one iteration. Thus, u^{n+1} is obtained as the solution to the linear integro-differential equation

$$(-\alpha\nabla \cdot \left(\frac{\nabla}{|\nabla u^n|}\right) + \; K^*K)u^{n+1} = K^*z.$$

  This algorithm can be interpreted within the framework of *generalized Weiszfeld's methods*, as introduced in [37]. As proved in that paper, this method is monotonically convergent, in the sense that the objective function evaluated at the iterates forms a monotonically descreasing sequence, and that the convergence rate is linear. In practice, this method is very robust.

- In Vogel and Oman [36] and Chan, Chan and Zhou [12], *Newton-type methods* are considered. Unfortunately, the domain of convergence of Newton's method applied to solve equation (9.2) is extremely small, specially when the regularizing parameter β is small. A *continuation method* is needed to make this method work efficiently. However, controlling the continuation process is not easy and many heuristic must be used to make it an overall efficient procedure.

9.4 A Primal-dual Method for Total Variation Minimization

The difficulty of the linearization of (9.2) is caused by the singularity of the square root that appears at the denominator. In this section, we briefly describe a special linearization technique that we have developed in [15] which has proven to be very efficient and robust compared to the methods mentioned earlier.

9.4.1 A linearization based on a dual variable

The key idea is to introduce a smooth auxiliary (dual) variable:

$$w \equiv \frac{\nabla u}{|\nabla u|}.$$

Note that $|w| = 1$. Then we rewrite (9.2) as a system of equations in the (w, u) variables:

$$\left.\begin{array}{rcl} f(w,u) = & w|\nabla u| - \nabla u & = 0 \\ g(w,u) = & -\alpha\nabla \cdot w + K^*Ku - K^*z & = 0 \end{array}\right\}$$

and linearze it instead of the u-system:

$$-\alpha\nabla \cdot \left(\frac{\nabla u}{|\nabla u|}\right) + K^*Ku - K^*z = 0.$$

This approach is similar to the primal-dual method of [16, 2] for minimizing a sum of Euclidean norms.

We cite in the following some heuristic explanations for the advantages of this technique:

- The vector $\frac{\nabla u}{|\nabla u|}$ is the normal vector to the level sets of u (cf. Fig 9.4). Thus, w is smooth and well-defined if the level sets are smooth whereas if we linearize it we will lose this property. Moreover, $\nabla \cdot w = \kappa$ is the curvature of the level sets of u and the time marching method can be regarded as a curvature-driven flow. Hence, w is a natural variable to use in a geometry-based image restoration algorithm.

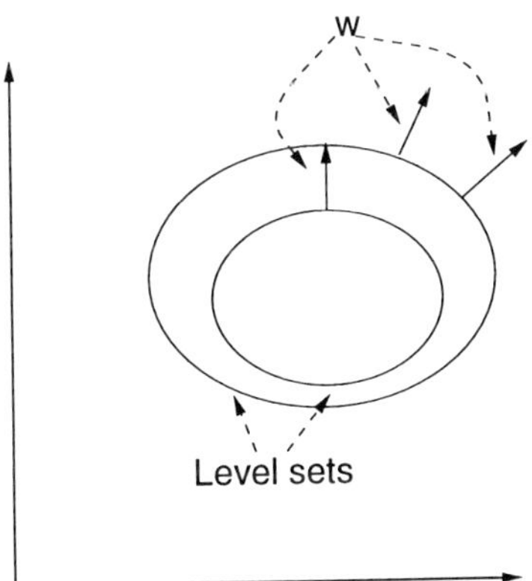

FIGURE 9.4. w variable is the normal to the level sets of u.

- The better global convergence behavior of Newton's method for (w, u)-system is due to the fact that this system is more *globally linear* than the u-system. To back up this assertion, we consider the following scalar example: we use Newton's method to find the solution of $f(x) = w - \frac{x}{\sqrt{x^2+\beta}} = 0 \equiv$ $g(x) = w\sqrt{x^2+\beta} - x = 0$ for $w = 0.9999, \beta = 1e-4, x \equiv \nabla u$ (cf. Fig 9.5). We find that Newton's method applied to $g(x) = 0$ converges for

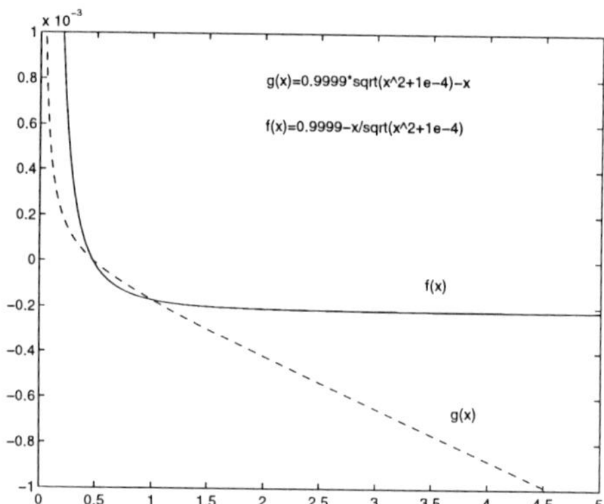

FIGURE 9.5. Plot of $f(x)$ and $g(x)$. The vertical axis has been scaled by 10^{-3}.

any initial guess > 0 and applied to $f(x)$ diverges for any initial guess > 2. Note also that the graph of $g(x)$ resembles a linear function in almost all the horizontal axis, whereas the graph of $f(x)$ resembles a rational function.

The linearization of the u-system yields

$$\left[-\alpha\nabla \cdot \left(\frac{1}{|\nabla u|}(I - \frac{\nabla u \nabla u^T}{|\nabla u|^2})\nabla\right) + K^*K)\right] \delta u = -g(u)$$

and that of the (w, u)-system yields:

$$\begin{bmatrix} |\nabla u| & -(I - \frac{w\nabla u^T}{|\nabla u|})\nabla \\ -\alpha\nabla\cdot & K^*K \end{bmatrix} \begin{bmatrix} \delta w \\ \delta u \end{bmatrix} = - \begin{bmatrix} f(w,u) \\ g(w,u) \end{bmatrix},$$

which after eliminating the δw variable yields:

$$\begin{aligned} \left[-\alpha\nabla \cdot \left(\tfrac{1}{|\nabla u|}(I - \tfrac{w\nabla u^T}{|\nabla u|})\nabla\right) + K^*K\right] \delta u &= -g(u), \\ \delta w = \tfrac{1}{|\nabla u|}(I - \tfrac{w\nabla u^T}{|\nabla u|})\nabla\delta u - w + \tfrac{\nabla u}{|\nabla u|}. \end{aligned}$$

Note that the cost of solving the (w, u)-system is comparable to that for the u-system. The main cost in solving the system in δu in both cases.

9.4.2 *Dual definition for TV norm*

The definition of the Total Variation as

$$TV(u) = \int_\Omega |\nabla u| dx = \int_\Omega \sqrt{u_x^2 + u_y^2}\, dx\, dy$$

is not well-defined if u is non-smooth. An equivalent dual definition that is well-defined for non-smooth u is the following (cf. [18]):

$$TV(u) = \max\{\int_\Omega u \nabla \cdot w\, dx\, dy : w = (w_1, w_2),\ w_i \in C^\infty(\Omega),\ |w|_\infty \leq 1\}.$$

Hence, another formulation for the TV restoration problem is:

$$\min_u \max_{|w| \leq 1} \alpha \int_\Omega u \nabla \cdot w\, dx\, dy + \frac{1}{2}||Ku - z||^2.$$

It can be shown that this leads to an equivalent primal-dual formulation.

9.4.3 *Numerical Results for the Primal-Dual Method*

We will now compare numerically the convergence behavior of the primal-dual method to that of the primal-Newton method under several circumstances. The continuation procedure used for both algorithms is based on a trial and error strategy, together with backtracking heuristics to minimize the work caused by possible errors. The line search on the u variable is based on sufficient decrease for the L^2-norm of the gradient of the objective function. The line search on the w-variable is a simple and inexpensive algorithm to ensure that $|w(x, y)| < 1$ throughout the iteration, by taking a fraction of $\inf_{(x,y) \in \Omega}(\sup\{s : |w(x, y) + s\delta w(x, y)| < 1\})$ (see [15] for details).

We show in Table 9.1 the work required by both algorithms to find the solution of the Total Variation restoration problem for the noisy image (K = *identity*) displayed in Fig 9.6, obtained from the true image, also displayed in Fig 9.6, by adding a random noise whose distribution is normal of zero mean and variance $\sigma^2 \approx 1200$, resulting in a SNR$\approx$ 1. The target parameter β we have used is 0.01×256^2. The parameter α has been set to 1.18, which has been computed from the Lagrange multiplier obtained from the solution of the constrained problem. The stopping criterion for the intermediate Newton's iterations, i.e., during the continuation, before solving the problem for the target β, has been to achieve a relative decrease of 10^{-2} for the L^2-norm of the gradient of the objective function. The relative tolerance for the last subproblem has been 10^{-4}.

We use the conjugate gradient method preconditioned by the incomplete Cholesky decomposition as linear solver. We have used an adaptive relative tolerance on the Euclidean norm of the residual as recommended in [27, Eq. 6.18]

$$\eta_n = \{0 .1 if n = 0, \min(0.1, 0.9||g_n||^2/||g_{n-1}||^2).$$

FIGURE 9.6. Original, noisy and denoised image, SNR≈ 1.

In order to be able to apply the conjugate gradient method to the (w, u)-system, we have replaced its Jacobian by its symmetric part, which is an SPD matrix. Furthermore, this Jacobian converges to a symmetric matrix and therefore the convergence is still locally quadratic by results in [27].

In Table 9.1, under the columns headed by NWT, we have the total number of Newton iterations required to solve the problem and under the columns headed by CG, we have the total number of (inner) CG iterations, i.e., matrix vector multiplications, including all iterations incurred by the backtracking procedure.

	primal-dual Newton		primal Newton	
	NWT	CG	NWT	CG
continuation, no line search	25	186	70	265
continuation, line search on u	25	187	53	210
continuation, line search on u&w	13	51	53	210
no continuation, line search on u&w	12	58	Not converged	

TABLE 9.1. NWT≡# nonlinear iterations,CG≡# CG iterations

The conclusions that can be drawn from this experiment are:

- The most crucial factor for the primal-dual method is controlling the dual variable w via the step length algorithm mentioned above. In fact, our experience is that this algorithm with the dual step length control is globally convergent for the parameters α and β in a reasonable range. A line search for the primal variables almost always

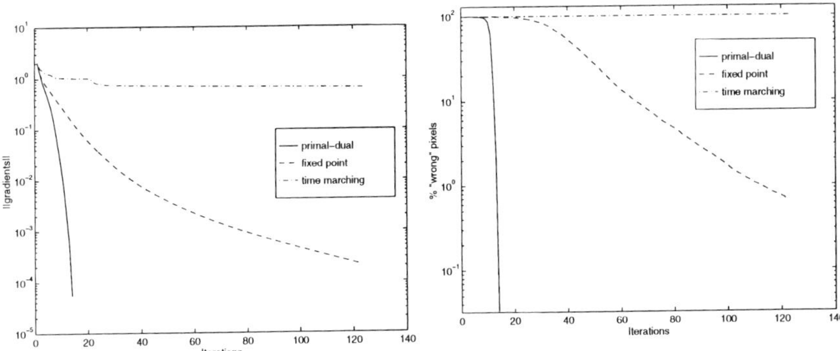

FIGURE 9.7. Convergence comparison of primal-dual, fixed point and time-marching: at the left we plot the Euclidean norm of the gradient of the Total Variation restoration functional versus iterations and at the right we plot the percentage of pixels that differ by more than 1 pixel value(the *pixel tolerance*, since the dynamic range of the image is 256) from a pre-computed high accuracy solution.

yields unit step lengths.

- The primal-dual method with the dual step length control algorithm does not need continuation to converge, although using it might be slightly beneficial in terms of work.
- The primal-dual method with the dual step length control has a much better convergence behavior than the primal method.

In our second experiment, we compare the primal-dual Newton, fixed point and time marching methods for a picture of a satellite contaminated with artificial noise [1]. For the primal-dual Newton method, we use the step length algorithm for the dual variables, no continuation and the same parameters as in the previous experiment. The same parameters are used for the fixed point method, except that we have used a fixed linear relative residual decrease $\eta_n = 0.1$ (it is in this case *optimal* according to our experience). We have used a line search based on sufficient decrease for the time marching method. The stopping criterion for the time marching method is based on the iteration count since we have not been able to achieve the prescribed accuracy in a reasonable amount of time. In Fig. 9.7 we plot the convergence history of this experiment and in Fig. 9.8 we display some of the intermediate results for the primal-dual and fixed point methods.

The conclusions we draw from this experiment are:

[1]given to us by Prof. Bob Plemmons of Wake Forest University

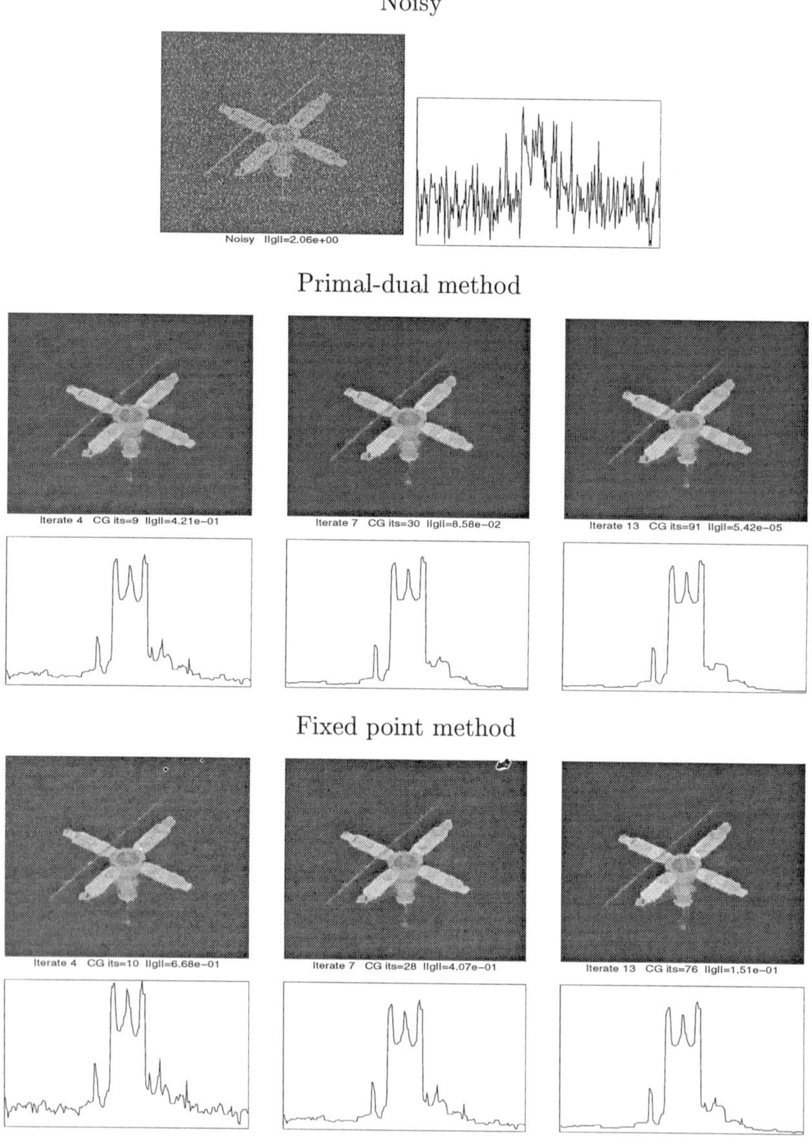

FIGURE 9.8. The size of the original image is 256×256 and the signal to noise ratio is approximately 1. The plots are horizontal cross sections of the corresponding images at pixel 130.

- The primal-dual algorithm is quadratically convergent, as can be seen from the left plot in Fig. 9.7, whereas the others are at best linearly convergent.
- The primal-dual algorithm behaves similarly to the fixed point method in the early stages, but in a few iterations can attain high accuracy.
- Both the primal-dual and fixed point are able to restore the large scale features of the image after only a few iterations, but the fixed point method takes much longer to completely restore the finer scale features, as it can be seen from the right plot in Fig. 9.7 and the horizontal cross sections plots in Fig. 9.8.
- The cost per iteration, as well as the memory requirement, of the primal-dual method is between 30 and 50 per cent more than for the fixed point iteration. The cost of an iteration of the time marching method is roughly the same as that of an inner CG iteration for the primal-dual method, but both the primal-dual method and the fixed point method require far less (CG) iterations than the time marching method.

9.5 Cosine Transform Preconditioners For Total Variation Image Denoising and Deblurring

In this section, we will treat the issues regarding preconditioning the conjugate gradient algorithm used for the solution of the linear systems arising in Total Variation image restoration.

9.5.1 Preconditioners for Total Variation image restoration

Let us denote by $A = \alpha L + K^*K$, where L corresponds to the differential operator of either the fixed point iteration or Newton's method.

When $K = I$, the matrix A will correspond to an elliptic differential operator and there are many possible preconditioners for it, e.g. SSOR, ILU, MG, DD [22, 32]. We have satisfactorily used ILU preconditioning in our numerical experiments. In [35, 36] a multigrid preconditioner is used.

When $K \neq I$, there exist many preconditioners for each of the summands of A separately, e.g. the elliptic preconditioners mentioned above and circulant preconditioners [33] for the convolution operator K^*K. However, a preconditioner for the sum of both terms is not easy to find.

In Vogel and Oman [36], the following operator splitting/ADI type preconditioner is proposed:

$$M = (K^*K + \gamma_1 I)^{1/2}(\alpha L_\beta(u^n) + \gamma_2 I)(K^*K + \gamma_1 I)^{1/2}.$$

This approach works well when α is very large or very small, but the intermediate cases are often the ones which are of interest.

In the rest of this section, we shall briefly describe a preconditioning technique that we have developed in [11] which can be applied to both terms.

9.5.2 *Optimal Discrete Cosine Transform Preconditioners*

Recall that the matrix of the one-dimensional n-discrete cosine transform is

$$C_n = \sqrt{\frac{2-\delta_{i1}}{n}} \cos\left(\frac{(i-1)(2j-1)\pi}{2n}\right), \quad 1 \le i, j \le n,$$

which is an orthogonal matrix. Furthermore, if v is an n-vector, then the product $C_n v$ can be performed in $O(n \log n)$ operations.

We denote the set of matrices which can be diagonalized by the discrete cosine transform by

$$\mathcal{B}_{n \times n} = \{C_n^t \Lambda_n C_n \mid \Lambda_n \text{ is a real diagonal matrix}\}.$$

Given A_n, we define the optimal cosine transform preconditioner for A_n, $c(A_n)$, as the minimizer B_n of the Frobenius norm of the difference $B_n - A_n$, $||B_n - A_n||_F$, over all matrices $B_n \in \mathcal{B}_{n \times n}$. The matrix $c(A_n)$ is always symmetric and it can be shown that $c(A_n)$ is positive definite whenever A_n is so. More details can be found in [10].

The construction cost of $c(A_n)$, for different types of matrix A_n, is shown in Table 9.2.

A_n	cost of constructing $c(A_n)$
general	$O(n^2)$
Toeplitz (from K)	$O(n \log n)$
Diagonally sparse (from L)	$O(n \log n)$

TABLE 9.2. Construction cost for the optimal cosine transform preconditioner.

The matrix of the two-dimensional $n \times n$-discrete cosine transform is $C_n \otimes C_n$ and the optimal two-dimensional cosine transform preconditioner for a matrix $A_{n \times n}$ is similarly defined as the matrix $c_2(A_{n \times n}) = B_{n \times n}$ that can be diagonalized by the two-dimensional cosine transform and which minimizes the Frobenius norm $||B_{n \times n} - A_{n \times n}||_F$, over all those matrices that can be diagonalized by the two-dimensional cosine transform. The construction cost of $c_2(A_{n \times n})$, for different types of matrix $A_{n \times n}$, is shown in Table 9.3.

A_n	cost of constructing $c(A_n)$
general	$O(n^4)$
2d Toeplitz (from K)	$O(n^2 \log n)$
2d diagonally sparse (from L)	$O(n^2 \log n)$

TABLE 9.3. Construction cost for the two-dimensional optimal cosine transform preconditioner.

9.5.3 Cosine Transform for TV denoising and deblurring

Since the matrix A is a general dense matrix, from Table 9.3 we deduce that the cost of constructing the preconditioner $c_2(A)$ would be n^4, which can easily be prohibitively large. Therefore some simplifications have to be done in order to reduce this cost. Since

$$c_2(A) = c_2(K^*K + \alpha L) = c_2(K^*K) + c_2(\alpha L) \approx c_2(K)^* c_2(K) + \alpha c_2(L),$$

we propose

$$M = c_2(K)^* c_2(K) + \alpha c_2(L)$$

as a preconditioner for A.

If $c_2(K) = C^t \Lambda_1 C$ and $c_2(L) = C^t \Lambda_2 C$, then

$$M = (C_n \otimes C_n)^t \ (\Lambda_1^* \Lambda_1 + \alpha \Lambda_2) \ (C_n \otimes C_n),$$

which is easily invertible in $O(n^2 \log n)$ operations using the FFT. We can also incorporate the technique of diagonal scaling for taking into account large variations of the coefficients of the differential operator. Note that the cost per conjugate gradient step is $O(n^2 \log n)$.

9.5.4 Numerical Results for Cosine Transform Preconditioners

In this section we compare the work used to perform 5 fixed point iterations using the conjugate gradient method with different preconditioners. This work is measured in inner CG iterations.

The test and original images are displayed in Fig 9.9 and the parameters that have been used are: $\beta = 0.01$, $\alpha = 10^{-6}$ and conjugate gradient relative tolerance $= 10^{-2}$. The results are shown in Table 9.4. It can be seen that the diagonally scaled cosine transform preconditioner can reduce the number of iterations by a factor of more than 4 compared with using no preconditioning at all. Other tests in [11] show that the performance is also insensitive to the value of α.

Preconditioner	#. CG iterations
Cosine + diag. scal.	204
Cosine only	348
diag. scal. only	462
no preconditioning	922

TABLE 9.4. Comparison of the work for the TV image restoration with different preconditioners

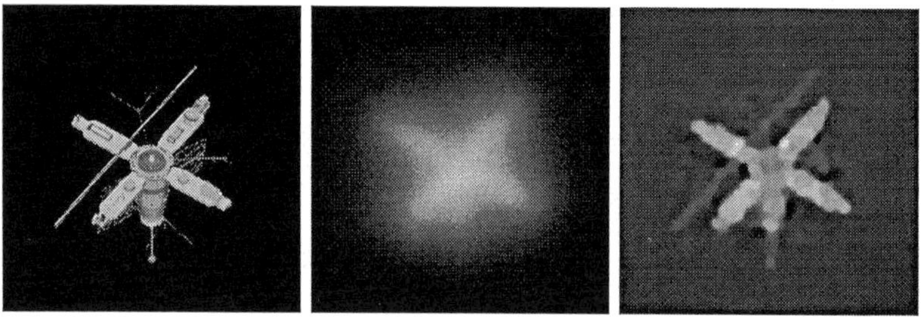

FIGURE 9.9. Original image, blurred and deblurred satellite image.

9.6 Conclusion

These nonlinear PDE based image restoration methods are very promising. Their primary advantage is a systematic way of preserving sharp edges. There is a nice mathematical framework and the techniques can be extended to more general image processing tasks such as segmentation and object recognition. There is also the possibility of combining them with the FFT/wavelet based methods as well. While these methods are still relatively expensive compared to traditional transform and frequency based methods, well established PDE and CFD numerical schemes can be used to make them more efficient. In particular, iterative methods can play a significant role and we have attempted to show some of the possibilities in this chapter.

9.7 References

[1] L. Alvarez and J. Morel, *Formalization and computational aspects of image analysis*, Acta Numerica, (1994), pp. 1–59.

[2] K. D. Andersen, *Minimizing a sum of norms (large scale solution of symmetric positive definite linear systems)*, PhD thesis, Odense University, 1995.

[3] E. Boman and I. Koltracht, *Fast transform based preconditioners for Toeplitz equations*, SIAM J. Matrix Anal., 16 (1995), pp. 628–645.

[4] J. BUNCH, *Stability of methods for solving Toeplitz systems of equations*, SIAM J. Sci. Stat. Comp., 6 (1985), pp. 349–364.

[5] R. CHAN, *Iterative Methods for Overflow Queueing Models I* , Stud. Appl. Math., 74 (1987), pp. 171–176.

[6] R. CHAN AND T. CHAN, *Circulant Preconditioners for Elliptic Problems*, J. Numer. Linear Algebra Appls., 1 (1992), pp. 77–101.

[7] R. CHAN, J. NAGY, AND R. PLEMMONS, *Block circulant preconditioners for 2-dimensional deconvolution problems*, in Proceedings to the Conference on Advanced Signal Processing Algorithms, Architectures and Implementation III, Vol. 1770, San Diego, CA, July, 1992, F. T. Luk, ed., SPIE, 1992.

[8] ——, *Circulant preconditioned toeplitz least squares iterations*, SIAM J. Matrix Anal. Appl., 15 (1994), pp. 80–97.

[9] R. CHAN AND G. STRANG, *Toeplitz Equations by Conjugate Gradients with Circulant Preconditioner*, SIAM J. Sci. Statist. Comput., 10 (1989), pp. 104–119.

[10] R. CHAN AND C. WONG, *Sine Transform Based Preconditioners for Elliptic Problems.* submitted.

[11] R. H. CHAN, T. F. CHAN, AND C.-K. WONG, *Cosine transform based precontioners for total variation minimization problems in image processing*, in Proc. 2nd IMACS Symposium on Iterative Methods, Blagoevgrad, Bulgaria, June 1995, S. Margenov and P. Vassilevski, eds., IMACS, 1995, pp. 311–329.

[12] R. H. CHAN, T. F. CHAN, AND H. M. ZHOU, *Continuation method for total variation denoising problems*, in Advanced Signal processing algorithms, F. T. Luk, ed., vol. 2563, SPIE - The Int'l Society for Optical Eng., 1995, pp. 314–325.

[13] R. H. CHAN AND M. NG, *Conjugate gradient methods for toeplitz systems*, Tech. Rep. 94-8(35), Dept. of Math., The Chinese University of Hong Kong, 1994. To appear in SIAM Review.

[14] T. CHAN, *An Optimal Circulant Preconditioner for Toeplitz Systems*, SIAM J. Sci. Statist. Comput., 9 (1988), pp. 766–771.

[15] T. F. CHAN, G. H. GOLUB, AND P. MULET, *A nonlinear primal-dual method for Total Variation-based image restoration*, in ICAOS'96, 12th Int'l Conf. on Analysis and Optimization of systems: Images, wavelets and PDE's, Paris, June 26 28, 1996, M. Berger, R. Deriche, I. Herlin, J. Jaffre, and J. Morel, eds., no. 219 in Lecture Notes in Control and Information Sciences, 1996, pp. 241–252.

[16] A. R. CONN AND M. L. OVERTON, *A primal-dual interior point method for minimizing a sum of euclidean norms.* preprint, 1995.

[17] G. FIORENTINO AND S. SERRA, *Tau preconditioners for high order elliptic problems*, in Proc. 2nd IMACS Symposium on Iterative Methods, Blagoevgrad, Bulgaria, June 1995, IMACS, 1995, pp. 342–353.

[18] E. GIUSTI, *Minimal Surfaces and Functions of Bounded Variations*, Birkhäuser, 1984.

[19] I. GOHBERG, T. KAILATH, AND V. OLSHEVSKY, *Fast Gaussian elimination with partial pivoting for matrices with displacement structure*, Math. Comp., (1995). To appear.

[20] G. GOLUB AND C. V. LOAN, *Matrix Computations, 2nd Edition*, The Johns Hopkins Univ. Press, 1989.

[21] R. GONZALEZ AND R. WOODS, *Digital image processing*, Addison Wesley, 1993.

[22] W. HACKBUSCH, *Iterative solution of large linear systems of equations*, Springer Verlag, NY, 1994.

[23] M. HANKE, *Conjugate gradient type methods for ill-posed problems*, Notes in Math., No. 327, Longman, Harlow, 1995.

[24] M. HANKE AND P. C. HANSEN, *Regularization methods for large scale problems*, Surv. Math. Ind., 3 (1993), pp. 253–315.

[25] G. HEINIG, *Inversion of generalized cauchy matrices and other classes of structured matrices*, in Linear algebra in signal processing, vol. 69 of IMA volumes in mathmatics and its applications, 1994, pp. 95–114.

[26] T. HUCKLE, *Circulant and Skew-circulant Matrices for Solving Toeplitz Matrix Problems*, SIAM J. Matrix Anal. Appl., 13 (1992), pp. 767–777.

[27] C. T. KELLEY, *Iterative Methods for Linear and Nonlinear Equations*, vol. 16 of Frontiers in Applied Mathematics, SIAM, 1995.

[28] J. M. MOREL AND S. SOLIMINI, *Variational methods for image segmentation*, Birkhauser, Boston, 1995.

[29] P. PERONA AND J. MALIK, *Scale space and edge detection using anisotropic diffusion*, IEEE Trans. Pattern Anal. Mach. Intelligence, 12 (1990), pp. 629–639.

[30] L. RUDIN AND S. OSHER, *Total variation based image restoration with free local constraints*, in Proc. Int. Conf. on Image Processing, Austin, TX, 1994, 1994, pp. 31–35.

[31] L. RUDIN, S. OSHER, AND E. FATEMI, *Nonlinear total variation based noise removal algorithms*, Physica D, 60 (1992), pp. 259–268.

[32] Y. SAAD, *Iterative methods for sparse linear systems*, PWS Publ. Co., Boston, 1996.

[33] G. STRANG, *A Proposal for Toeplitz Matrix Calculations*, Numer. Math., 51 (1987), pp. 143–180.

[34] B. M. TER HAAR ROMENY, *Geometry-Driven Diffusion in Computer Vision*, Kluwer Academic Publishers, 1994.

[35] C. R. VOGEL, *A multigrid method for total variation-based image denoising*, in Computation and Control IV, K. Bowers and J. Lund, eds., vol. 20 of Progess in Systems and Control Theory, Birkhauser, 1995.

[36] C. R. VOGEL AND M. E. OMAN, *Iterative methods for total variation denoising*, SIAM J. Sci. Statist. Comput., 17 (1996), pp. 227–238.

[37] H. VOSS AND U. ECKHARDT, *Linear convergence of generalized Weiszfeld's method*, Computing, 25 (1980).

Index